TAKE TEN

 KU-713-355

CONTEMPORARY BRITISH
FILM DIRECTORS

TAKE TEN

CONTEMPORARY BRITISH
FILM DIRECTORS

Jonathan Hacker and David Price

YLOS

Oxford   New York
OXFORD UNIVERSITY PRESS

Oxford University Press, Walton Street, Oxford OX2 6DP
Oxford New York Toronto
Delhi Bombay Calcutta Madras Karachi
Petaling Jaya Singapore Hong Kong Tokyo
Nairobi Dar es Salaam Cape Town
Melbourne Auckland
and associated companies in
Berlin Ibadan

Oxford is a trade mark of Oxford University Press

Interviewees' replies © the several interviewees 1991
Other text and editing of interviews © Jonathan Hacker and David Price 1991

First published 1991
First issued as an Oxford University Press paperback 1992

All rights reserved. No part of this publication may be reproduced,
stored in a retrieval system, or transmitted, in any form or by any means,
electronic, mechanical, photocopying, recording, or otherwise, without
the prior permission of Oxford University Press

This book is sold subject to the condition that it shall not, by way
of trade or otherwise, be lent, re-sold, hired out or otherwise circulated
without the publisher's prior consent in any form of binding or cover
other than that in which it is published and without a similar condition
including this condition being imposed on the subsequent purchaser

British Library Cataloguing in Publication Data
Hacker, Jonathan.
Take ten: contemporary British film directors.
1. British cinema films. Directing
I. Title II. Price, David
791.430223
ISBN 0–19–811217–3
ISBN 0–19–285251–5 (Pbk)

Library of Congress Cataloging in Publication Data
Hacker, Jonathan.
Take ten: contemporary British film directors/Jonathan Hacker
and David Price.
Includes bibliographical references and index.
1. Motion picture producers and directors—Great Britain.
I. Price, David. II. Title.
PN1998.2.H33 1991
791.43'0223'092241—dc20
ISBN 0–19–811217–3
ISBN 0–19–285251–5 (Pbk)

Typeset by Wyvern Typesetting Ltd, Bristol
Printed in Great Britain by
The Alden Press, Oxford

LEEDS POLYTECHNIC
1700488267
A√
36080   26.10.92
13.5.93
791.4302330922

To our parents

PREFACE

The director is rarely, if ever, the sole creative force in such a collaborative medium as film. But with his passion and vision, he generally has the pivotal role, and is the subject of this book. When examining the variety of British directors included here, the reader will see just how misleading the term 'British film' can be. The book places ten contemporary British directors side by side. But whilst the reader is able to trace certain common themes, comparisons between the directors are actually characterized by a startling degree of diversity of style and opinion.

The directors chosen for the book represent a broad cross-section of British film-makers. The selection is not meant to be definitive. Our criteria were that they all had a sizeable body of work behind them, and that they were all still making films. It was tempting to include several great retired British directors such as David Lean, Michael Powell,[1] Roy Boulting, and Charles Crichton. This was especially so in the case of David Lean, for although he has only done one film in the last twenty years, he is currently engaged in the adaptation of Conrad's *Nostromo*. Similarly Crichton has recently come out of retirement to make *A Fish Called Wanda*. However, to have discussed their distinguished bodies of work would have taken us as far back as the 1940s to the generation before television and its profound influences.

The book consists of a detailed introduction and ten chapters on the ten directors. The introduction attempts to give a context for the problems that the directors have faced and the decisions they have made in their careers. It analyses some of the cultural and economic forces at work on the British film community and looks back into British film history to find the roots of many of today's problems. Chapters are divided into four parts. The essay section outlines the career, aims, and stylistic traits of the director. The edited discussion presents his views—often widely divergent from those of his fellow-directors—on the role of the director and on British film, and frequently gives some unusual insights into the director's work. The filmography and bibliography are as complete as is possible given our limited resources, but we hope that they will be a convenient research source for anyone wishing to study

---

[1] Sadly, David Lean and Michael Powell have died since this book was written.

these directors further. For a more general bibliography on British cinema we recommend the one compiled by Susan Daws in *British Cinema History*, edited by James Curran and Vincent Porter (Weidenfeld and Nicolson, 1983), although it is now a little out of date.

Acknowledgements

The authors wish to acknowledge their gratitude to the directors included in this volume who were kind enough to give their time to be interviewed.

We are also grateful to the writers and publishers of the works quoted in the text and listed in the bibliographies.

The photographs used in this work are the copyright of the following companies or individuals: Lindsay Anderson, Anglo International Films, British Broadcasting Corporation, British Film Institute, Central Independent Television, Channel 4, Columbia Pictures Corporation, Curzon Film Distributors, the Trustees of the Imperial War Museum, London, Edie and Elau Landau, Bill Forsyth, Goldcrest Film and Television, Mike Laye, MCA/Universal Pictures, MGM/UA, Oasis Film Production, Orion Pictures Corporation, Paramount Pictures Corporation, Pathé Releasing, stills from the films *Whistle Down the Wind*, *Bugsy Malone*, and *Bad Timing* by courtesy of the Rank Organization plc, Rex Features, The Samuel Goldwyn Company, John Schlesinger, Warner Brothers, Weintraub Screen Entertainment, Zenith Productions. We have made every attempt to contact the copyright holders for all photographic material. Should there be any instance in which we have been unsuccessful, we invite the copyright holder to contact us direct. For help and advice in obtaining photographs for the book, thanks are due to Nigel Algar, Lindsay Anderson, Sue Anderson, George Biggs, Rachel Blackman, Don Boyd, Elizabeth Burn, Jane Carmichael, Fliss Coombs, Margaret Duerden, Louise Fawkler-Corbett, Norman Flicker, Elizabeth Frith, Diana Frost, George Helyer, Karen Henson, John Herron, David Hersch, Lyn Jamieson, Yvon Kartak, Lesley Key, Margaret Kirby, Edie and Elau Landau, Mike Laye, Arthur Leese, Glen Marks, Angela Murphy, Susan Nezami, John Pierce, David Rosenbaum, Eric Senat, Sheila Sutherland, Alison Webb, and Alison Wrather. We are also grateful to the staff of the British Film Institute Reference and Stills Libraries, and to Kevin Brownlow and David Gill at Thames, Louise Coulter, Diana Hawkins, Margaret Humphreys, Kate Lenahan, Stephanie Sheahan, and also to Peter Webb for letting one of the authors take time off work to research and write for this book.

For the constant personal and professional support and encourage-

x · ACKNOWLEDGEMENTS

ment we are ever grateful to Ross McKibbin; to Kim Thornton and Billy Kosco for their invaluable advice on the introduction; and of course to our families for always being behind us — Adam for helping with the typing, Fiona for letting us use her office, Julie for her patience while we took over her living room, and Jocelyn, Alison and Ian, and our parents, most especially Peter for his tireless guidance at every stage.

CONTENTS

LIST OF ILLUSTRATIONS

A BRITISH FILM CULTURE?
A BRITISH FILM INDUSTRY?

A BRITISH audience, which might naturally be reluctant to watch foreign films, has always openly welcomed American cinema. This unconscious disposition has been so long-standing that the taste of the British audience for film—their cinematic education—has been effectively formed by Hollywood's finely crafted, glamorous entertainment. The result is that, today, a British film, despite its greater relevance for a British audience, has little appeal. A state of affairs has been reached where parochial comedies like *Porky's* about American High School teenagers are more popular than a British equivalent like *Rita, Sue and Bob Too*. The implications of this situation for British directors are disturbing, since the tastes of the audience inevitably influence the type of films they make. How is it then that British directors and their audience have come to be so alienated? Before one looks at this, however, one must first consider what a British film is, and what value it might have—matters often left implicit in debates about British cinema.

Any attempt at a sharp definition of 'British style in film', in this sense, is likely to be more misleading than illuminating. The endeavour is comparable to trying to define national character—the sharper the definition the more likely it is to approximate to a caricature. It is vain to look for a smooth line or curve on a graph when all that exists is a cluster of points. One can do no more than isolate an array of resembling patterns. Anything to do with national characteristics is necessarily indeterminate. However, several criteria can still be usefully identified.

In financial terms one could define a British film as one in which most of the investment was British—*Platoon* or *Last Emperor* would be two examples. A further criterion is where the bulk of the budget was spent: perhaps on British resources—technicians, equipment, and studios. The *Superman* movies would qualify under this definition. In so far as one considers British film merely as an industry of some £250 million turnover (£100 million incidentally owned by American subsidiaries) and employing some 30,000 people (full and part time, in production, distribution, and exhibition), a successful film is one that brings employment and profits to Britain. Economically, however, the strength of British movies has been steadily declining since the war.

But film can be considered separately from these financial considerations. It is not merely an industry, it is also part of a culture. *The Wolves of Willoughby Chase*, for example, was shot in Czechoslovakia for budgetary reasons, despite the fact that it is set in Victorian England, and *Chariots of Fire* was funded from America and Egypt, yet both are considered British. When trying to identify the cultural criteria defining a British film two aspects should be separated: those with a British subject matter, and those with a discernibly British style.

Our concept of British films must be able to encompass foreign film-makers making films with strongly British issues or a British setting. Alternatively it should include films which reflect British values and social mores, even though they may have a foreign subject matter. The origins of a film's aesthetics, its dramatic structure, and visual or acting style provide another ingredient when trying to gauge the Britishness of a British film. This is, in some senses, the most subtle and powerful ingredient of all. But it is a particularly complex matter to define for this country. Partly this is because the co-operative nature of film-making hinders direct analysis of a film's aesthetic origins; the actors may be American, the music French, and so forth. It is still possible to discern certain aesthetic traditions in the film histories of countries like America, Russia, or Japan, but much harder with Britain.

These difficulties should not suggest that the existence of British films in all the above senses is not valuable. The economic advantages are obvious, and need not be elaborated. In as much as it is important to make more people aware of the political, social, and moral issues that surround them, the need for films which explore British subject matter and values or cultural traditions is evident. Given the place television and feature films hold in British society[1] the material shown should not be just escapist fantasy. At the very least, even if British films merely reflect and do not examine the moral and social values of the nation, they will foster a distinctive sense of identity and community—something which foreign films can never do. More ambitiously, they may also attempt to encourage thought and emotion about issues which have a national context. Of course, some foreign films, whether American, Continental, or Japanese, can cross many of the cultural barriers without reducing the depth or possible relevance of their meaning. But it is essential that Britain should not rely on these other sources alone. Foreign films will reflect, even if critically, their own culture's values and conditions; yet these values, if absorbed by the British audience, can be

---

[1] Every day around five different films are shown on broadcast television, around one million pre-recorded video tapes are rented, 20,000 more are bought, and there are some 230,000 cinema admissions. (Source: *Cultural Trends*, Policy Studies Institute, 1989.)

socially very inappropriate. British producer, David Puttnam, has written,

It seems unarguable that much of the power of fantasy that exists today, the expectation of instant gratification, the commitment to selfhood over all, the waning concern for reason, or disciplined achievement—is related to the cheap fantasy life so sedulously manufactured in Hollywood. Coming as it does from the United States, which for the most part created the systems to cope with these expectations, this type of cinema (and television) is in my view creating long term havoc in communities with different, sharper economic and social problems—countries like Britain in which these promoted expectations are increasingly unrealisable.[2]

We should now return to our original question. How is it that British cinema has been unable to develop a stronger sense of direction? British Screen's director, Simon Relph,[3] recently asked some of the major advertising agencies to come up with a concept for an image to promote British cinema; not one of them could offer anything constructive. Although there have been no strong cinematic movements with their own individual references and values, discernible threads can be observed in certain areas; in comedy there is Ealing under Michael Balcon, Carry On in the 1970s, and more recently Monty Python; in horror, Hammer films; and, arising largely from the theatre, certain traditions of social realism in the late fifties and early sixties. In particular, there are the influences of documentary film-makers like Grierson, Jennings, or Free Cinema on narrative cinema. But British film, unlike British theatre or even British documentary, has never had its own distinctive directors of the calibre of Eisenstein, Ford, or Bergman. Through their individuality, and even interpretation of their cultural milieu, these types of directors furthered the development of a national film tradition. Britain never had such distinctive paradigms.[4] Michael Powell and David Lean, arguably two of Britain's archetypal film directors, have had only limited influence in this country.

However, the explanation of the relative weakness of the indigenous British film culture can hardly be directed at an absence of cinematic talent as a whole. Although Britain has had virtually no cinematic geniuses of such calibre as to mould or give direction to British film, it has a wide range of internationally respected directors, writers, actors, and technicians. One might even say that it has more than its fair share, for many other countries' producers come to Britain to take advantage of this talent.

[2] George Perry, *The Great British Picture Show* (Pavilion, 1985).
[3] Left British Screen in December 1990; independent producer.
[4] Raymond Durgnat remarks that 'if a clearly marked personal style is one's criterion of interest, then few British films reward the concern given to such directors as Dreyer, Bunuel, Franju, and Renoir'. Raymond Durgnat, *A Mirror for England* (Faber and Faber, 1970).

The superficial explanation for why American film culture has had such easy access into Britain is the common English language. The resistance of an audience to the artificiality of dubbing or distraction of subtitles clearly reduces the chances of the films of non-English speaking nations from being accepted on an equal basis to Anglo-Saxon productions. This places America in a comparatively stronger position than most other nations to exert an influence on British film—even if the different accent might initially have been a barrier. But why has the influence of other English speaking nations such as Ireland, Australia, or Canada, which would have had a similar advantage, been so negligible? Moreover, this common language might just as well have meant that American audiences would have accepted typical British films for what they were, rather than regarding them frequently as parochial or quaint. The fundamental reason for the dominance of American cinema is, in fact, the strength of American culture itself.

The latter part of the twentieth century is distinguished by the replacement of centuries of Europe's 'cultural imperialism' by America's. American culture is an optimistic, immigrant one capable of communicating on a very basic level to a wide range of peoples and reflecting the openness of its society. Film was the perfect medium for its dissemination. Not surprisingly, the culture was readily embraced by early cinema-going audiences. Perhaps if fiction film had been socially accepted in Britain in these early stages as a new and highly expressive art form rather than being treated with a certain indifference or snobbishness, things would have been very different. As it was, the English language left Britain open to the influence of American culture. The combined effects of America's cultural strength and the potential vulnerability due to a common language have also been felt by Australia and Canada—possibly to an even greater degree than Britain.

Britain's critics might have done something to help construct a firmer cultural identity in British film; one more capable of withstanding the American competition. But most of them traditionally have taken their responsibility to be limited to being a consumers' guide to new film releases—a guide to what is, in their opinion, 'worth' seeing. For movies other than most horror/sex/exploitation films the reviews are crucial for the viability of the film at the box office. But the real problem is not the quality of individual reviews but that the media determine the way in which film is understood. One should not be surprised that British films are increasingly felt by much of the public to be more foreign than Hollywood films if the values which many of the critics use to judge the films are so totally consumption orientated—as if a film were just a product to be marked out of ten and nothing more influential. An element of this sort of reviewing is clearly useful. But critical appreciation by the standards of the dominant, popularist film culture of America is highly detrimental. In contrast, French appreciation of film is as an art form, not just as a way to pass a couple of hours. Save for individuals such as Derek Malcolm, or Alexander Walker, the critics in Britain, however committed they are to British film, have done little to encourage intelligent appreciation. They rarely help a wider audience to stand back and actually *understand* the films they watch or to address the issues tackled within the film itself. The passivity of their whole approach supports the status quo, merely acting as a filter to the marketing of the films themselves. Much of the blame is, of course, editorial because lack of space determines the extent of any discussion. But one does wish that they would contribute to a more serious understanding of film, in particular British film and the constraints within which it operates.

America has also had powerful economic advantages which have helped to mould its international dominance. With a domestic market equal to half the international cinema market (some 1,175 million cinema admissions per year in 1985), it has always been able to maintain a huge industrial base. Potential profits from domestic admissions alone enable Hollywood to make any investments it feels necessary, technical or artistic, on a film. Britain, on the other hand, has always had a much smaller public at hand. Because it has never found it easy to raise production finance, it has had to watch costs. The relative cheapness necessarily implied by this has made it hard to compete with American rivals. Although cost and quality are never synonymous, some potentially profitable quality films fail because of lack of sufficient investment. Often proper locations are not used, the shooting period is too rushed, the special effects are badly done, the film has to be shot on 16 mm., or the best technicians can not be afforded. The effects of the lack of capital are even more profoundly felt, ironically, after the film has been made and is being marketed and then released. This aspect of film-making, so crucial to a film's potential profitability, has been traditionally weak in Britain. But one can easily understand British parsimony here because the costs of sufficient release prints and advertising for all the cinemas can easily mount up to as much as the cost of the film itself.

The inability to raise finance has left British cinema—in the three areas of production, distribution, and exhibition—vulnerable to being bought up or taken over by American film companies who see the advantages of such investments in another English speaking country. The local benefits of capital investment from whatever source are not, however, to be sneered at. When the exchange rate is advantageous, Britain is seen as a cost effective and convenient place to make American films using the comparatively cheap but highly skilled British technicians and craftsmen and the British studios. This brings employment and ensures the maintenance of a substantial pool of skills, studios, and facilities. At more optimistic times when British subjects are fashionable in America, Hollywood has even welcomed the idea of trying to make indigenous British films both through its British subsidiaries and direct investment. In 1967 90 per cent of money in British films was American. Films of the 'Swinging Sixties' like *Alfie* for Paramount, *Blow Up* for MGM, and the Beatles and Bond Films for United Artists brought in considerable profits. This sort of investment is, however, double edged. In the first place, the profits are taken by American not British companies. Moreover, fluctuations in the strength of the pound, the American economy, and even the unpredictable changes in the fashion of British films for the American market result in this foreign capital being pulled out with little compunction, as happened in the early seventies.[5] Having come to rely on it, the industry was left stranded.

The difficulty of competing with American cinema culturally or economically on Britain's own terms (let alone by making films in Hollywood's own style in an attempt to capture their market—it can be like taking coals to Newcastle) is not a recent phenomenon. It has been a dominating factor to differing degrees for most of this century and it is, in fact, one of British film history's primary elements of continuity. But over the years this has been exacerbated; in particular because of factors such as the inadequacy of government support, the rise of television, the decline of British studios, and the development of a duopoly in cinema exhibition.

There has been a lack of clear thinking or commitment to British film by the government whether in import controls, direct financial aid, levies, or tax incentives. In most other European countries government support of the film industry would have been forthcoming in the hope of stimulating an indigenous film culture and industry, or at least of preventing or limiting this decline. But in Britain government intervention has been, at best, hesitant or uncoordinated. Nothing was done to make the television companies pay a fair price for films or to limit the number of American films shown on television. On the other hand action was taken to stem the flood of Hollywood films released in Britain's cinemas as far back as 1927, when distributors and exhibitors were required to handle a minimum quota of British films.[6] The result, however, was the emergence of the cheaply made 'quota quickie' and the

[5] The success of films like *Easy Rider, Woodstock, Midnight Cowboy, The Graduate*, and *Butch Cassidy and the Sundance Kid* represented a switch in fashion for the American production companies, many of whom were losing money anyway and were therefore unwilling to take perceived risks in British product.

[6] For renters and exhibitors respectively; 1927, $7\frac{1}{2}$% and 5%; 1938, 20% for both; 1948, 30% and 25%.

basing of American productions in Britain to qualify under the quota. The act was not abolished until 1982.

Direct financial support from the government was instituted on a small scale with the establishment of the National Film Finance Corporation (NFFC) in 1949. As a form of semi-commercial industrial credit organization, its function was to top up budgets of viable projects which could not otherwise raise the money and to help develop scripts. Recent scripts developed include *Loose Connections, Defence of the Realm*, and *Another Country*, and examples of recent films made possible are *Moonlighting, Britannia Hospital, Gregory's Girl*, and *Babylon*. But in 1984 it was privatized (under the joint control of Rank, Thorn-EMI, and Channel 4) creating the British Screen Finance Corporation, and National Film Development Fund. Limited funding, being topped up from private sources, of £2 million a year will stop in 1993.[7] Over its thirty years of hand-to-mouth existence it cost the Treasury only some £11 million in written-off loans while it helped fund some 360 different films. Another source of direct government investment in film production has been through the £1 million per year given to the BFI Production Board for experimental films, the Regional Arts Association, various local authority sources, and the Arts Council of Great Britain.[8] Although the small sums of money given have been particularly valuable for local community film-making, this sort of tiny financing has had no significant long-term impact for British film-makers or British film culture. More hopeful is the European Co-Production Fund, which was initiated in 1990 with a £5 million government grant to be spread over three years.

Another area of attempted government involvement was through the creation of the Eady levy on exhibition. Under an Act of 1957 a twelfth of all cinema receipts was put into the British Film Fund which was then distributed to films qualifying as British in proportion to their success at the box office (in 1983 £2.7 million out of the £4.8 million collected), as well as to the NFFC, the National Film School, the Children's Film Foundation, and the British Film Institute Production Fund. As with the quota the definition allowed films like *Superman* to qualify as British because they were shot in British studios with British technicians. Successful American style films such as this took most of this fund, and it did little to encourage films with a more British perception, which needed it most. The BFI's film finance has been based on a policy of

---

[7] It helps finance around a dozen projects annually. £350,000 goes to the national film development fund for script development, £150,000 to fund short films, and £1.5 million towards the production of British feature films.

[8] In 1972 the Arts Council established a fund for films which were concerned with questions of fine arts and aesthetics.

financing experimental films and can trace its origins back to the early 1950s and Michael Balcon's Experimental Film Fund. Recent successes include *The Draughtsman's Contract*, *Ascendancy*, *The Gold Diggers*, *Flight to Berlin*, and the Terence Davies trilogy. But, given the decline in audiences, the money which the levy provided rapidly dwindled, and it was abolished.[9] Money for the BFI and National Film School now comes from the Office of Arts and Libraries but little goes on production to help British films. The BFI's production funding now, not surprisingly, comes largely from Channel 4.

The final area of government involvement is in the area of reducing the tax burden for film investment—an area crucial in determining the City's relationship with the British film industry. Although highly significant for a few years, much of what was done by the Conservative government relating to the tax position of film, was done for industrial production in general rather than because of any commitment to cinema. The 1979 tax write-off system, by defining the entire cost of the production as capital investment, enabled financial companies to write off 100 per cent of film production costs in the first year.[10] This cushioned the position of the investor who, entering into a sale and lease back deal, could still make a profit on a loss-making film. Films such as *The French Lieutenant's Woman* and *Local Hero* received money from Chemco Merchant Bank largely because of this. But over 1984–6 the government decided to phase out capital allowances. Only the Business Expansion Scheme, introduced in 1981, remained as a small scale (maximum £40,000) alternative. Even with the lower tax levels of 1989, 30 per cent of film production expenditure returns to the Exchequer in the form of VAT, PAYE, NHI, and other taxes (not to mention the new withholding tax on foreign entertainers). As Alexander Walker put it, 'The tax man giveth and the tax man taketh away.'[11]

Government policy has never been very clear headed, but it has increasingly failed to differentiate between an industrial and a cultural policy. As far back as 1952 the PEP report on the film industry read, 'If the public considers it desirable for political, cultural, or economic reasons that British films should be produced, then it must be prepared for the Government not only to protect the industry indefinitely, but also to aid it financially for as far ahead as can be seen.' In fact, little has been done to stem the decline and foster truly indigenous film-making. At best it has encouraged American majors to base some of their productions

[9] In particular, under lobbying by the Cinematograph Exhibitors' Association.
[10] In 1980 it was tightened to apply only to investment in British films, or those with British content—to stop some of the problems seen before of money going to UK-based American films.
[11] Alexander Walker, *National Heroes* (Harrap, 1985).

here, and therefore bring capital into the country. Every other European country has had far more significant financial injections or tax incentives to maintain and stimulate healthy domestic film production. Even Australia introduced in 1989 the £35 million (A$70 million) Australian Film Finance Corporation to invest in qualifying Australian film, and the tax incentives for film investment are higher in America than Britain. Ironically, film in Britain has always had to compete with television, which in the case of both the BBC (with its licence fee) and the ITV companies (through their own franchises) *has* had firm government support, and with other subsidized arts like theatre, opera, or ballet. Mamoun Hassan while at the NFFC commented, 'recently a journalist who had found the Answer suggested that film should be treated no differently from the ball bearing industry. We should be so lucky. The ball bearing industry has enjoyed considerable investment from the private sector, and has benefited from government intervention.'[12] Simon Relph believes that 'It's almost impossible for a country of our size to maintain a genuinely indigenous character for their film industry unless they do have some subsidy. We need something—tax concessions or a levy—which in effect boosts the home market.'[13]

The steadily falling cinema audience has been of chief importance to the decline of British cinema. In 1945 there were 1,585 million visits a year while 1984 saw the all time low of 55 million; in other words a drop from thirty visits a year down to one a year per head of population. What was steadily destroying the cinema going habit, of course, was the introduction of television, colour television, and then video (Britain developed in the 1980s one of the highest video-owning rates anywhere in the world) as an alternative to the big screen. It would, however, be a mistake to see this as a waning in the desire by British audiences to see films. If anything, more films are watched than ever before. But the results for British film-making were threefold.

Traditionally television companies—both the BBC and ITV—have been reluctant to invest any money directly in British films. They have preferred to invest in their own drama productions, many of exceedingly high standard. But the drama was affected by the very specific nature of the television companies, the methods of commission, the budgets available, the subject matter which the drama departments favoured, and the restricted scale implied by the small screen of television. The feature films to be shown on television were purchased after production, but the vast majority of these films were American and not British or even European. In following this policy of satisfying what they believed were the audience's tastes for Hollywood products, the television companies

[12] *Guardian*, 21 Apr. 1983.
[13] John Walker, *The Once and Future Film* (Methuen, 1985).

were seemingly unaware that audiences tend to watch what they are shown, and in so doing are, in fact, having their tastes created. Moreover, the rates paid by the television stations for the broadcast rights are some of the lowest in the Western world. The BBC's average payment for a feature film in 1985 was £20,000, and Channel 4 set a ceiling of £15,000 as its top price for a feature.

The declining theatrical audiences have hit the distributors and exhibitors hard. Cinemas closed one after the other over the 1950s and 1960s. Money could no longer be found for capital investment on cinemas such as refurbishment or updated projection technology; this in turn contributed to the decline of audiences. The limited market implies a small budget for those who want to make British films. However, traditionally the distributors have felt it necessary under the squeeze to play safe and only distribute or pre-purchase before production the distribution rights to films with mass appeal. These are films which fit their criteria for 'commercial' — in other words, generally American. Any British films which interest them will therefore tend to involve stars and high production value, and concurrently a higher budget. Even with money from television and video, these budgetary demands actually make it impossible to recoup the higher production costs in Britain alone. This has had the result of forcing the producer of a British film to pre-sell the distribution rights to an American distributor as well. Many financial organizations, potentially interested in funding a British film, will not even consider financing a project until this has been done. But the demands in the editorial content by an American distributor which has its own market to consider can be even harsher than a British one. In an attempt to win this finance British producers are frightened off anything which might alienate these investors, and are attracted to subject matter which might result, for instance, in Anglo-American content (in the last few years including films such as *A Fish Called Wanda*, *Memphis Belle*, *High Spirits*, *Stormy Monday*, or *Dealers*) or likely to cast American actors who could not really be justified on anything other than financial grounds (Denzel Washington in *For Queen and Country*, Glenne Headly in *Paperhouse*, or Bridget Fonda in *Scandal* are possible recent examples). These sorts of compromises have become almost unavoidable, although some films have made these changes more successfully than others. As might be expected, this gives the home audience little chance to see a genuinely indigenous product on the cinema screens or to develop any ability to appreciate it. This in turn makes it even harder for a genuinely indigenous film to strike a chord with its public.

For many years various tensions — not least a natural conservatism strengthened by the low prices which television traditionally paid for feature films — inhibited the television stations from directing some of their capital into film. There was a persistent conviction that television drama was inappropriate for the big screen. Furthermore, investment decisions were retarded by an inability to decide upon a satisfactory period between when the film could be released at the cinema and when it could then be shown on television without affecting the audiences for either.[14] A bright light appeared in November 1982 when Jeremy Isaacs, as Chief Executive of Channel 4, declared that his new channel would help restore British film through investing in and commissioning independent producers for the 'Film on Four' slot.[15] The ITV companies, and then the BBC in 1988, followed suit in the never-ending search for 'new product'. By the late 1980s the number of feature films facilitated by television money alone was huge—indeed a British film without any television money at all has become the exception rather than the rule. Channel 4 has had theatrical successes like *My Beautiful Laundrette* and *Letter to Brezhnev*, and critical acclaim for scores of others. The BBC produced *War Requiem* and *Black Eyes*; Granada has made films like *The Fruit Machine* and *Tree of Hands*; Zenith (established as the film division of Central Television) made films such as *Insignificance*, *Prick Up Your Ears*, or *Wetherby*; and London Weekend Television has recently produced *A Handful of Dust*, and *The Tall Guy*. The development of cable and satellite television suggests that this trend will continue unabated. But the recent changes in the old ITV levy system (which encouraged these companies to invest in high quality productions in return for the monopoly commercial broadcast rights in a geographical area) have dampened much of the optimism.

These recent developments in television have done much to stem the decline of indigenous British film. The fundamental question, however, is what is the nature of the new course which it has set for British cinema. The television executives in control of the money have their own ideas of the sorts of films that they want their companies to be involved in. Taking into account economic and political considerations which affect their station, they have a different perception of their responsibilities to the film producers. Furthermore, many people believe that television is aesthetically, as well as technically, a different medium from that of the big screen. How important these two factors are remains a matter of some controversy and varies with the commissioning editor and the director. No one could say, for instance, that *Paris, Texas*, or *Company of Wolves*, both partly backed by Channel 4, are not cinematic. But many people, including film-makers like Alan Parker, believe

[14] English language films under £4 million can go to television immediately after release. Those over £4 million generally take two years although this requirement can be waived by the Films on Television Committee representing distributors, producers, and exhibitors.
[15] 6–7% of the Channel's programme monies go on film production.

that the sorts of films supported by the television executives are just not *real* films; that their subject matter and budgets are small scale and that this whole development has been detrimental. Many of the directors, however, once they have the money for their film, really make it primarily for the cinema—even if television is where the money originates and where most people will ultimately watch the film. This is either because they see no aesthetic differences, or because their priorities remain with the cinema rather than television. Puttnam commented on this issue, and specifically on Channel 4's *Experience Preferred But Not Essential* (which he produced and which was theatrically successful in America),

It's not a film in British terms. It may be a film in American terms because American television is genuine pulp 99.9% of the time. British television is only 85% pulp . . . I think *Angel* (directed by Neil Jordan, 1982) was a movie. *Another Time, Another Place* (directed by Michael Radford, 1983) was a film for television. The jury is out for the moment [on this issue].[16]

Mamoun Hassan while in charge of the National Film Finance Corporation reacted strongly to the way film-makers tended to ignore the nature of any differences: 'One director told me that the only difference was that the TV image was smaller. If that's the level of thinking forget it, because we're not going to get anywhere. As Kurosowa said, why should people go to the cinema when, if they do, they see what they see at home.'[17] Whatever the exact divergence ought to be in the content or aesthetically, the primary differences between seeing a film at home or at the cinema now really lie in the marketing and press coverage, the viewing in a public place so that film-going is a social event, and the higher technical quality. These factors will always remain important for the minority of enthusiasts. The impact of television for British cinema-going as a whole has been compared by Puttnam to the pop music business where groups go on tours which lose a great deal of money, but which are done because the group will recoup the money many fold in record sales. Likewise very few of these films funded by television break even, let alone make a profit, at the cinema. They are released to promote their future television transmission and the profile of the station. The long-term consequences of television's involvement in film-making still remain to be seen.

It was in the face of these constraints—the strength of Hollywood (culturally and financially) for the export of its films, the transfer of the audience to television, the lack of government support or any contributions by critics to help guide the British audience back to the fold—that

---

[16] John Walker, *The Once and Future Film*.
[17] Ibid.

the larger British companies like London Films, Rank, Thorn, Cannon, or ITC all had to operate. They were comparable in their constitution, albeit on a much smaller scale, to the Hollywood studio system, and arguably the only ones in a position to handle the more risky indigenous product, or to win back the UK market for British product. Instead, they took their risks backing American-type films aiming at the world market. Likewise, they failed to support talented young directors at their doorsteps who were snapped up instead by the big American companies.[18] However, because the market was not capable of supporting them, and, to a degree, because of some unwise, over-ambitious investments, these companies have slowly gone under or withdrawn from film.

In the 1930s and early 1940s London Films under Korda were involved in around one hundred films (eight were directed by Korda himself) which in scope and grandeur rivalled America[19] but which remained generally British. Korda himself built the large studios at Denham to provide a home for these productions. The bubble burst with the war and he never regained the same power. Rank dominated British film in the 1940s but declined in the 1950s and 1960s having also over-extended itself in the process of trying to make films to break into the American market, and moved into other business ventures, although maintaining its important distribution interest.[20] In the early 1970s EMI under Bryan Forbes had a policy of making medium budget British films with a broad appeal. Under Barry Spikings (1976–83), however, the company made high budget films aimed specifically at America, often using American directors or producers, and even moved the head offices to Los Angeles. His extremely ambitious aim was to raise the company to the level of an American studio. Several expensive failures led him to retire in 1983.[21] Ironically it was less its low profitability than the need for cash by its parent company, Thorn, which had taken it over in 1979, which led to the 1985/6 sale of the Screen Entertainment division. Cannon, who bought it, have made generally formulaic American films, and have had to sell off many of their British concerns to America to head off insolvency. They seem to be following the path which that most brilliant

[18] A third of American studio films have first time directors. They are not uncommonly British, but the larger British companies have traditionally been far less willing to take this sort of risk.

[19] Korda's two H. G. Wells films, *Things to Come* and *The Man Who Could Work Miracles*, cost £374,000.

[20] It continues to make an occasional film investment like *Bad Timing* and distribution guarantees such as for *Educating Rita*.

[21] Films he rejected included *Gandhi*, *Chariots of Fire*, and *Local Hero*. He took the risk with *Britannia Hospital*, which flopped. His backing for John Braborne's lavish Agatha Christie cycle was only initially successful. His three most profitable productions occurred in 1978—Walter Hill's *The Driver*, Pekinpah's *Convoy*, and *The Deer Hunter*, but the failures of Schlesinger's £25 million *Honky Tonk Freeway*, and *Can't Stop the Music* bit hard.

of salesmen, Lew Grade, followed with Associated Communications Corporation in the late 1970s. Films like *Green Ice* or *Escape to Athena* were generally rootless — attempting to be international and having big stars, but bereft of any passion. Several successes such as *The Muppet Movie*, along with his ability to pre-sell distribution deals, kept him in the running making fourteen films a year for several years. But he finally fell with the $35 million *Raise the Titanic* (1980), and his company was taken over by the Australian Robert Holmes a'Court. The problem is that as the various formulas — stars, special effects, locations, action, etc. — were put together to make a 'safe', 'commercial' movie the costs escalated to the extent that it became highly risky, but with little artistic or creative merit to compensate.

Although this succession of companies has done little to foster indigenous British films, they played an active role in instilling a British audience with an American film culture. Today, Cannon and Rank, like their predecessors, control a virtual duopoly of film exhibition[22] and so, to a great extent, dictate what films the public see. To Attenborough this has been 'the greatest problem that the industry has had to encounter for as long as I can remember — for almost fifty years'.[23] The power of these two companies rests partly in their ownership of all the 'first-run' cinemas and partly in their distribution deals with the major American studios. Cannon until recently guaranteed outlets for Columbia and Warner, as well as their own productions.[24] Rank has a similar arrangement with Disney, Twentieth Century Fox, United Artists, and Universal. In any one week, these deals account for as many as two-thirds of the films being shown in Britain.[25] In return, Cannon and Rank are assured a regular supply of American films which, due to the lack of competition from other exhibitors, should have been extremely profitable.[26] To take full advantage of their duopoly of exhibition, Cannon and Rank, like their predecessors, pursue a policy of 'barring'. This is done by granting exclusive rights to show films to the major cinemas in each city, which they own. Only after these films have been well and

[22] The duopoly arose during and after World War II when Rank built up an exhibition empire which was able to compete with the large ABPC circuit. Both companies maintained a long-standing commitment to American films, but the development of the duopoly made this even more rigid. ABPC was taken over by EMI in 1969, which in turn was taken over by Cannon in 1985/6 (see above). In 1935 the duopoly owned only 10% of all cinemas, in 1944 it was 22%, in 1965 29%, in 1972 32%, in 1980 40%.

[23] Interview with the authors.

[24] Through Columbia-Cannon-Warner Distributors Ltd. In October 1987, Cannon broke away from Columbia and Warner. However, Warner owns a significant minority share in the Cannon Group, which owns the exhibition circuit.

[25] *Films and Filming*, May 1987.

[26] According to a report in the *AIP and Co.*, Apr. 1985, the exhibitor's profit is as much as 60% of the box office takings; compared with the producer, who after paying the distributor and sales agent, is rarely left with more than 10%.

truly milked of profit are independent exhibitors allowed to rent them.[27] As the Boulting brothers[28] observed, 'They do not compete. They divide. And what they divide is the spoils.'[29] Such practices led to investigations by the Monopolies Commission in 1966 and 1983 and the Office of Fair Trading in 1987. Twice the Monopolies Commission concluded that a monopoly did exist. First, they did not feel the problem serious enough to demand compulsory government interference. Moreover, while condemning the practice of 'barring', they felt that the situation was not necessarily against the public interest. In an industry which, by the early 1980s, was rapidly declining, two large companies were naturally more resilient than numerous smaller ones, and it was felt that they should not be weakened. It was also argued that their size and relative security put them in a better position to take risks in exhibiting different types of films, compared to a smaller company. This made much sense, but it would take a long time before Rank and Cannon would actually broaden their programming horizons.

The problem still remains; as Stephen Woolley, joint-head of the independent British production company Palace Pictures, comments,

We have survived by ducking and weaving between the majors, releasing when we know they are not. It is getting more difficult to get theatres, they have just got too many films lined up and we have to fight tooth and nail to keep our cinemas. Even some of the independent cinemas are forced by declining audiences for foreign and subtitled films to put themselves increasingly at the call of the majors by taking more American product.[30]

The growing number of new, American owned, multi-screen cinemas does not offer much hope to such independent producers. Although these outlets have aided the recent upturn in cinema admissions they have shown a marked reluctance to screen independently produced films. Instead they have concentrated on predominantly American product.

The late seventies and early eighties saw the emergence and rapid growth of a variety of small, dynamic independent British production companies, such as Handmade Films, Palace Pictures, Working Title, Goldcrest, and those run by such entrepreneurial producers as David Puttnam, Jeremy Thomas, Don Boyd, or Simon Relph. They represent one of the largest groups of film producers in the country, and have become the bedrock of the British film industry. Several of these companies became involved in film production rather indirectly. Handmade

---

[27] Although in 1988, under the Fair Trading Act, the government introduced a draft order restricting exclusive runs to four weeks.
[28] A major British director/producer team of the 1950s and 1960s.
[29] John Walker, *Once and Future Film*.
[30] *BFI Film and Television Handbook* (BFI, 1990).

was set up in 1978 by Denis O'Brien and ex-Beatle George Harrison to back *Life of Brian*, then got involved in distributing films such as *The Long Good Friday*, *The Burning*, and *Venom*. But it was the success of *Time Bandits* that tempted it further into production. By 1988 it had a budget of £40–42 million to finance some eight features. Palace Pictures, formed by Nick Powell and Stephen Woolley, also started out in distribution. They moved into production to secure economic and creative control over their films. By 1987 they were spending £20–25 million per year on feature film production. Not all have been so successful. Goldcrest was founded by Jake Eberts in 1976 to provide modest levels of investment to films in development. The success of first *Chariots of Fire* and then *Gandhi* led one critic to write in 1985 that 'they probably represent the British film industry's last hope of independent survival'.[31] Over-expansion (their 1985 expenditure on film production was £46.5 million) amongst other causes led to the company going bust by 1986.

The emergence of the increased cable, video, and television markets for small, indigenous films has greatly helped the independents. But all these companies face certain problems because of their scale. In the first place there is the issue of cash-flow; overheads have to be met, but income is irregular and unpredictable. Producer Rick McCallum (*Pennies From Heaven*, *Track 29*, *Castaway*) comments, 'I think the curse of most of our English producers are their companies. They get locked into their overheads.'[32] Puttnam experienced how difficult this state of affairs could be while working at Goldcrest. 'Sandy Lieberson and I twice made films because we couldn't have paid the payroll without making them. In at least one case, and probably two, we knew we shouldn't have been making the film at all. But we had to make the payroll.'[33]

The fact that so few films are on the slate for an independent production company means that one major failure can spell disaster. The failure of *Revolution* was in large part to blame for Goldcrest's collapse. There are rarely sufficient assets and no predictable income to help bridge over such a loss (which a studio with a large number of films in production might be able to cope with). Moreover, because they do not have a large enough turnover to plough back their profits into future projects, the independents have to be constantly on the look out for more finance. But the financial sector in Britain has grown more and more reluctant to be involved. Rick McCallum is pessimistic. 'The city has dried up to all intents and purposes. The competition's getting

[31] John Walker, *Once and Future Film*.
[32] *Producer*, AIP Magazine, Aug. 1989.
[33] John Walker, *Once and Future Film*.

greater, there are more of us but there's no place to get the money. Its very tough.'[34] Independents therefore have increasingly turned to America, although it looks as though Japan and especially Europe are starting to offer more hopeful alternatives. Lynda Myles, first commissioning editor for independent drama at the BBC, is more cautiously optimistic. 'The tradition has been to look to the USA for money but now I think our film-makers will have to look more to Europe.[35] And that brings with it its own problems. . .'.[36] It has also meant that the independents have frequently co-produced projects with the television companies. The various consequences of that position have already been discussed in this introduction.

The independent producers have explored various solutions. One option is to keep the budgets for their films as low as possible. This is especially attractive if a television company will pay a large enough percentage of the budget. Puttnam, talking about his experiences working with Goldcrest on the series of films for television called *First Loves*, stated, 'at £400,000 no one's going to get really damaged. They might take a bath on one, but the company will limp a week or two and carry on.'[37] The problem with this approach is that the lower the budget the less the potential return. Low production value, few well-known actors, and little money to spend on marketing the film have their effects, and for a very small company the losses made by even the lowest budget film can be crippling. As such many of these independents have frequently looked towards very specific audiences. The hope for many of these companies was that films based on artistically or politically radical ideas would create their own publicity, and not demand stars or production value. Handmade, for instance, chose to make off-beat comedies, Working Title to make films with a political edge and even films which might be able to push back the boundaries of film convention. Films like *My Beautiful Laundrette* proved that films could be low budget and a success. But it has always been hard, and is getting harder, to persuade financiers that the project for them to back has any credibility if its budget is below a certain amount. Even though the risk is less, the maximum profit which low budget films can make is always going to be limited, and the perception is that the investment remains 'high risk'.

Another solution has been to secure a regular source of steady income

[34] *Producer*, AIP Magazine, Aug. 1989.
[35] According to the European Commission, the European market for film, television, and video is currently worth £11.4 billion. This could well double in the years after 1992. Britain's common language with America and common heritage with Europe places her in a good position to benefit from this. But some producers feel that the industry is not going to be economically strong enough to be able to take advantage of these opportunities.
[36] *BFI Film and Television Handbook* (BFI, 1990).
[37] John Walker, *Once and Future Film*.

to cover overheads through exploiting other film and television markets. Working Title always had another source of income from its pop video subsidiary, Aldabra, but in 1988 Tim Bevin and Sarah Radclyffe, who had set up the company four years earlier, moved into television production as well.

Palace Pictures' approach has been to keep their economic base as broad as possible. They make films on a variety of scales, and for a variety of markets. Stephen Woolley comments,

With our distribution, we have had a varied balance of releases which kept us going while other companies went bust. For the future, we will continue to tailor our production to very distinct markets. We will make a film every year or so in the USA for the American market. We'll continue with lower budget dramas like *Mona Lisa* and *Scandal* and we'll also make bigger budget films like *The Pope Must Die* for the international market. We will also go in for much smaller budget films too and will use our muscle to get theatres for films that people have said 'no' to.[38]

In the mid-1980s Goldcrest had tried something similar—but less cautiously. Under the helm of James Lee, they also tried to create a situation where the company could be run as a normal business instead of the high-risk venture that film production so often is. They went into television production, and raised further capital for expansion. But over-expansion, internal dissensions in the company, mismanagement, and a limited production slate of three large films (*Revolution*, *The Mission*, and *Absolute Beginners*) which were all rushed into production, led to disaster.

British directors and their colleagues have had to make their films in the face of this array of self-perpetuating problems, and have found few identifiable traditions to derive support from. The producer/director Don Boyd pointed out, 'We've had periods of great value, the Ealing films, the early sixties, but we've never had a tradition of cinema in the way the Italians and French have.'[39] Michael Radford (*Another Time, Another Place*, 1984, *White Mischief*) described contemporary British cinema as a collection of random individuals who have all found ways of making movies: 'I found no real tradition with which I could identify in British cinema.'[40] Attenborough commented, 'since the late forties and early fifties, the British film, as a genre has—with the notable exception of Ealing films—suffered a tragic loss of identity . . . we gave up our birth-right by a process of compromise in ever increasing and ultimately

[38] *BFI Film and Television Handbook* (BFI, 1990).
[39] John Walker, *Once and Future Film*.
[40] ICA Conference, Dec. 1984.

self defeating attempts to please the tastes of a world market.'[41] Many of the newer generation of film-makers and producers were brought up exclusively on Hollywood films. Puttnam commented, 'Cinema, to me, has always been synonymous with entertainment. My own fixation started with Disney and then bounced along through the fifties between the dramatics of Kazan, Zinneman, Rossen, and Kramer, and the sheer joy of the technicolour Hollywood musicals. No mention here you will notice of British Cinema, even Ealing. I'm afraid it was sadly neglected in favour of the glossier stuff.'[42] Many film-makers such as Hugh Hudson no longer see any reason at all to make inherently British films, or for there to be a British film culture. He followed his very English *Chariots of Fire* with the American *Revolution* and *Lost Angels*. 'To want to make a "British" film is neither here nor there . . . I don't mind where I work. I will go anywhere that the story tells me to go. The world shouldn't be considered in terms of boundaries, especially in relation to creative work. Everything else is internationalised, and while there is an argument that says it is a shame, it is a fact of modern life.'[43]

Within this context, the work of British directors since the late sixties has become polarized, directors working within their own isolated orbits, often oblivious (or just antagonistic) to the ideas in each others' work. Based on both the financial and aesthetic forces so far outlined, there are three broad poles around which the films of British directors loosely cluster: films emulating Hollywood cinema and appealing to the 'international' (i.e. American) market; those that aim to have a wide appeal but remain, in some sense, British—one might call them 'traditional'; and those that have rebelled aesthetically or politically (or both) against either the 'Hollywood' or 'traditional' films—these might be called 'oppositional'. This classification is intended to be a conceptual, rather than an all embracing, framework. In fact, most British directors have made films for more than one group, and there are of course some films which fall uncertainly between groups, while others avoid these issues altogether. Moreover, many foreign directors, especially Americans such as Joseph Losey, Stanley Kubrick, Terry Gilliam, or Richard Lester, have spent much of their careers in Britain making often quite British films from their own American perspectives.

Many British film-makers have settled well into the American context (so well indeed that they often choose to live there permanently). Alan Parker, looking back on the start of his film career, describes it well: 'I'd written five screen plays, all of them very English, very London, very angry working class. If I'd just walked up and down Wardour Street

[41] *British Films* (Kodak/British film and TV Producers' Association, 1985).
[42] George Perry, *Great British Picture Show*.
[43] Fenella Greenfield (ed.), *A Night at the Pictures* (Columbus Books, 1985).

trying to get those films made, I'd still be walking . . . I had to think—
what did I know about? And what I *did* know about was American
movies.'[44] It is not surprising that those who have made the transition
with most skill were directors who learnt their craft making commercials—like Alan Parker himself, Adrian Lyne (*Flashdance, 9½ weeks,
Fatal Attraction*), the brothers Tony Scott (*Top Gun, Beverly Hills Cop
2*) and Ridley Scott (*Blade Runner, Alien, Someone to Watch Over Me*),
and Hugh Hudson (*Lost Angels, Revolution*). There have, of course,
been other backgrounds—Michael Apted (*Coal Miner's Daughter, Continental Divide, Gorky Park, Gorillas in the Mist*) came from television.
Peter Yates (*Hot Rock, Mother, Jugs and Speed, The Deep, The Janitor,
Krull*) with the exception of *The Dresser* in 1983, has been making
Hollywood-style films since the success of *Bullitt* in 1968. Other directors who went out to America in the 1960s include Guy Hamilton
(*Diamonds Are Forever, Goldfinger, Live and Let Die*), John Guillermin
(*Death on the Nile, The Towering Inferno*), and J. Lee Thompson (*Con-
quest of the Planet of the Apes, Deathwish 4, Murphy's Law*). Robert
Stevenson (*Bedknobs and Broomsticks, One of Our Dinosaurs Is Missing, The Shaggy DA*) went to Hollywood as a screen-writer in the 1940s
and has worked as a director for Disney since the 1950s. John
Schlesinger is an example of the way many directors have found it
necessary to ply their trade between Hollywood and more English films.
Even Richard Attenborough turned his hand to making a Hollywood
film with *A Chorus Line*. Many less distinguished directors have joined
this list, either because they find America genuinely more stimulating,
or, more often, because they have been unable to work in Britain at all.
The success some of them have had, however, has been quite astonishing—resulting in some of the world's all time box office successes, such
as *Fatal Attraction* or *Top Gun*. The result for film-making in Britain
was described by Mamoun Hassan while at the NFFC as 'not so much a
brain drain, but a soul drain . . . film-makers here all seem to be working
to get that Hollywood visiting card.' If they cannot manage that in the
short term, many are prepared to make English productions in a Hollywood style, such as Steve Barron's glossy *Electric Dreams*.

The 'traditional' British films have been generally characterized by a
sense of conservatism—or more accurately 'liberal conservatism'. These
films are frequently literary adaptations, or true stories with a strong
narrative thrust and larger than life characters—often outstandingly
played by actors with a theatrical background. Frequently they display a
strong nostalgia for the past (in particular the imperial past). Even if the
dominant tone is critical it is not angry, bitter, or particularly proselytiz-

---

[44] Fenella Greenfield (ed.), *A Night at the Pictures* (Columbia Books, 1985).

ing. What comes first is the narrative. In this sense, the position of many of them could be characterized as mid-Atlantic following in the grand romantic tradition of Hollywood but with roots held firmly in British film history as far back as Korda. Deeply influenced by American styles of film-making, frequently they try to mix Britishness with the ingredients for international appeal. The films of Attenborough and Forbes (*Whistle Down The Wind, The L Shaped Room, Seance on a Wet Afternoon, The Raging Moon, The Wrong Box, The Mad Woman of Chaillot*) are some of the best examples. Younger directors in this area include Roland Joffe (*Killing Fields, The Mission*), Bill Forsyth with his wry and often whimsical, but never sentimental, view of Scotland, and Charles Sturridge (*Runners, Brideshead Revisited, A Handful of Dust*). The British films of the American director James Ivory (*Heat and Dust, Room with a View, Maurice*) seem to show some of these influences working in reverse. Even Boorman, who has been making films like *Deliverance*, or *Exorcist 2* in America since the mid-sixties, returned to make a recollection of childhood in the Blitz, *Hope and Glory*. The desire to make discernibly British films which will succeed particularly in America finds a variety of solutions. But the producer who has most characterized this whole approach in the 1970s and 1980s is Puttnam, who produced *Chariots of Fire* and whose self-declared aim is to make films which 'reflect the desire to see Britain and the world through British eyes and attitudes, and to communicate what we see in an entertaining and comprehensive manner which will appeal to audiences around the world; where cinematic images and stereo types "haunt the unconscious" . . . I see myself rightly or wrongly as an international film producer.'

The 'oppositional' films are part of a counter-culture reacting in various ways against the influences of Hollywood or what they see as the staidness of the 'traditional' British films. They might be termed 'independent' but that word is open to misinterpretation. Unable to get finance for feature films, some of these sorts of directors found refuge for large parts of their careers in television. Many of their films have strong political, often left wing undertones, and are highly critical of Britain: for example, the films of Loach, the satires of Anderson, the films of Ron Peck (*Nighthawks, Empire State*), Mike Leigh (*Bleak Moments, Meantime, High Hopes*), some of those of Frears, and films like *Ploughman's Lunch, Babylon, Business as Usual, Comrades*, or *Letter to Brezhnev*. Kureishi, writer of *My Beautiful Laundrette* and *Sammy and Rosie Get Laid*, commented, 'Whenever a right wing paper calls one of our films sick, Stephen (Frears) and I know that we must be doing the right thing . . . you could make grim, realistic films about gay people, black people, unemployment, and racism in Britain, but the age is different. What we need is irony, ambiguity and humour in Britain.'[45] The establishment is represented perhaps by the Oxford historian, Norman Stone, who was asked by the *Sunday Times*[46] to comment on a group of these films, and criticized them for portraying Britain as a 'cess pit of decay and intolerance'—a remark reflecting his own lack of understanding.

Many other films are less vehemently critical but just want to break free of the sense of restraint and respectability in British cinema—Ken Russell's idiosyncratic films (*The Music Lovers, The Devils, Tommy, Lair of the White Worm*); a variety of films which have a strong vein of bawdiness running through them (*Wish You Were Here, Personal Services, Rita Sue and Bob Too, The Fruit Machine*); a range of aggressive thrillers often with veiled social-political comments within them (*Boy Soldier, The Long Good Friday, Defence of the Realm, The Hit, Angel, Mona Lisa, Bellman and True, Empire State, For Queen and Country*); or the 'Alternative Comedy' of the Comic Strip films. Chris Petit (*Radio On, An Unsuitable Job for a Woman, Flight to Berlin*) commented, 'there's nothing coherent about British film, so what to say? Theatrical, literary, documentary, hand to mouth, a consensus of good taste, a deadly respectableness.'[47] Several other film-makers (Chris Petit, under the influence of Wim Wenders included) have reacted strongly to the influences of the American style of film-making with its fast active camera movement, narrative traditions, its ideas of pace in film editing, and its style of acting. Many look to the Continent for inspiration. Some like Mike Leigh, improvising the plot with his actors, or Ken Loach, using non-professional actors and a quiet camera, are aiming at a more realistic style. Others like Peter Greenaway or Derek Jarman, both of whom had been to art school, or Nicolas Roeg with a background as a cameraman, and many working through the London Film-Makers Co-operative and other workshops, have attempted to develop the medium as a visual art form.

This book itself is not just a study of British creative directorial talent. By putting side by side a range of British film-makers we hope that one is able to become aware of a cross-section of the problems, financial and creative, that exist in British film. The nature of any limited selection for a book of this length is necessarily personal. But our broad criteria of selection were that the directors were British (even if they no longer choose to live here), that they were well established with a significant body of highly talented work behind them, and finally that they were still working, and still totally committed to film. For all these directors film has been and remains not just a profession but a way of life. In a

---

[45] *Film Comment*, July–Aug. 1988.
[46] *Sunday Times*, 10 Jan. 1988.
[47] John Walker, *The Once and Future Film*.

sense one could say that these directors are not a totally representative sample of British film-makers because they are each, in their different ways, outstanding. But even if that is the case what remains is still a cross-section covering commercial to arts cinema, directors specializing in comedy to highly political film-makers. All of them have had to face a broadly similar range of problems throughout their careers: the lack of a strong creative identity in British film; the problems of raising finance; the influence of television; the balancing of commercial films with the desire to make personal films; and the perennial pull of Hollywood. But all of these directors have chosen, or have been forced to choose, very individual paths to pursue their careers. It is up to the reader, if he will, to decide which is most successful, laudable, or valuable.

1 LINDSAY ANDERSON

Lindsay Anderson during the filming of *O Lucky Man!*, 1973. © 1973 Warner Bros. Inc.

LINDSAY ANDERSON

LINDSAY ANDERSON's films have ranged from the romantic and idealized, to social realism, and on to caustic and anarchistic satire. They are all informed by his passionate and probing, if often didactic, interest in people's relation to society—especially British society. But his films have rarely been mainstream. Largely because he is such an intellectual film-maker, he has been virtually shunned by the British public. However, his outspoken and forceful work has attracted international attention. Of all the 'Angry Young Men' of the late fifties and early sixties he is really the only one to have resolutely stuck with his beliefs. He is impatient and short-tempered with the ignorant and narrow-minded (and often those who just disagree with him), which has frequently earned him the epithets 'abrasive' and even 'savage'. Yet with his friends, especially the close group of actors with whom he works, he is exuberant and loyal. His character is epitomized by Glenda Jackson, who affectionately refers to him as 'the old growler'.

Anderson has had a prolific career, not so much in terms of the feature films he has made, but in the variety of work he has done. He has had parallel and equally distinguished careers as author,[1] film critic,[2] and major theatre director,[3] and has even done some occasional acting[4] and production work.[5] As well as feature films, he has made documentaries, episodes of *Robin Hood* for television, television commercials, a satirical mini-series, as well as a pop video for Carmel, and a documentary record of the pop group Wham! on their tour of China. Throughout his major work there is a guiding personal philosophy based on an obstinate sense of commitment. 'No art is worth much which doesn't aim to

---

[1] Author of *Making a Film* (Allen & Unwin, 1952), and *About John Ford* (Plexus, 1981).

[2] In 1947, while at Oxford University, Anderson was a founder editor of the quarterly film magazine *Sequence*, which voiced many ideas considered radical at the time. To Gavin Lambert, who joined the editorial board in the third issue, '*Sequence* was partly a series of love letters to directors we admired, partly a succession of hate mail against work we despised' (*Cinema* (USA), Spring 1971). The editors wrote much of it under pseudonyms (taken from characters in films) to give the impression of a larger number of contributors. Anderson also wrote extensively for *Sight and Sound* in the fifties.

[3] Especially at the Royal Court in the sixties and the early seventies.

[4] Parts in *The Pleasure Garden*, *Martyrs of Love*, *Inadmissible Evidence*, and *Chariots of Fire*.

[5] Produced *The Pleasure Garden*, and co-produced *If . . .* and *O Lucky Man!*.

change the world. Of course, no artist can be judged by his success or failure to change the world, since none of us ever succeeds. We can only hope to change or to influence like-minded spirits or hearts, by telling the truth.'[6] This strong moral element can be seen burgeoning in some of his early writings. In 1956 he urged film-makers to commit themselves whole-heartedly to their beliefs. He concluded by saying, 'Fighting means commitment, means believing what you say, and saying what you believe. It will also mean being called sentimental, irresponsible, self-righteous, extremist and out of date by those who equate maturity with scepticism, art with amusement, and responsibility with romantic excess.'[7] Such words were more prophetic than he could have known, as throughout his career Anderson has been called all of these, such is *his* commitment.

Anderson has cut his own individual furrow. 'All my work, for good or bad, has been extremely subjective, embodying a very personal re-action to the world, and to people.'[8] His analysis of John Ford, whose work he much admires, reflects his own beliefs. 'There is a strong, profoundly sympathetic human quality in all his [Ford's] work. I sympathise with his temperament, which was responsive to tradition, to an idealistic view of the world, yet at the same time, markedly individual, even anarchistic. In Ford there are tensions between tradition and independence which I recognise in myself.'[9] For Anderson, independence has always been paramount, but tradition can sometimes offer, as in *Every Day Except Christmas* (1957),[10] a sense of belonging and common purpose. Nevertheless, tradition must never encroach on personal autonomy and independence. Anderson has constantly urged the individual to have dignity, to think for himself, not to be led by others or bound by what he considers to be manipulative organizations. It is this distrust of institutions and his concurrent insistence on individual responsibility which has led him to accept the title of 'anarchist'.

In his films, Anderson's aim is to penetrate beneath surface realities to portray the larger implications evident to him, to find 'essences'— even if they are in the form of ambiguities. But he is not aiming at naturalism or social-realism, which he calls merely 'making sociology'[11] and describes as 'too easy, over-explored, and not eloquent enough'.[12]

[6] *British Cinema; A Personal View*. Thames Television 1986.
[7] *Sight and Sound*, Autumn 1956, 63, 'Stand Up, Stand Up'. Article on film criticism by Anderson.
[8] Interview with the authors.
[9] *Cinema* (USA), Spring 1971, 23–36.
[10] It was a documentary about a day in the life of Covent Garden market, focusing on the community of people who worked there.
[11] *Films and Filming*, Feb. 1963.
[12] *The Times*, 21 Apr. 1973.

For Anderson, realism is not a mirror to the world. He likes to quote Brecht—'realism is not a matter of showing real things, but of showing how things really are'. The quest for a deeper meaning is always paramount in his work.

Anderson's approach to these aims has ranged from the biting satire of *If . . .* (1968), *O Lucky Man!* (1973), *Britannia Hospital* (1982), and *Glory! Glory!* (1989); films which show little hope for man, to the lyrical optimism of *Every Day Except Christmas* (1957) and *The Whales of August* (1987). Throughout, Anderson has felt strongly that 'If your indignation is valid, it has got to be compensated by warmth of feeling.'[13] Certainly there is warmth when he concentrates on human details; people's mannerisms and their interaction with each other. He is extremely sensitive in many of his more optimistic works: *Thursdays Children* (1953), a documentary about teaching a class of deaf children to read and speak, as well as his most recent feature film, *The Whales of August*. However, this contrasts with the coldness of many of his satirical films which tend to be dominated by cerebral issues, with a heavy emphasis on symbolism and metaphors at the expense of emotional ones.

For Anderson, intellectual concerns always dominate aesthetic considerations. Visually, he aspires to follow Ford in the supposition that one should let the action come to the camera, rather than the camera creating the action.[14] In 1969, he described his technique as 'extremely sober'. It is something stemming naturally from his own personality, but also a conscious, determined policy.

The more, what we might call, 'trendy', or eccentric, or showy technique has tended to become in the last few years, the more I have wanted to try and make films with as much simplicity and directness as possible. Of course, simplicity and directness are actually the most difficult things, and maybe, sometimes, one fails to bring it off, and it is simply dull. But this is the direction in which I try and work . . . qualities of rhythm, and balance, and composition, inside a very straightforward and sober technique, are the problems that interest me most.[15]

Ten years earlier, when the Free Cinema Manifesto of 1959 declared that 'an attitude means a style; a style an attitude,'[16] it was attempting to counter what Anderson even then had long considered to be the excessive attention to 'form' in cinema. To Anderson, what he says is ultimately much more important than how he says it. He dismisses all of what he terms 'mannerism' in film and is perfectly prepared to admit, 'I've never really been able to shoot anything purely in terms of style

[13] *Sunday Telegraph*, 27 Mar. 1988.
[14] As a tribute to Ford he has sometimes copied compositions directly from his films.
[15] Elizabeth Sussex, *Lindsay Anderson* (Praeger, 1969).
[16] From the Free Cinema Manifesto (1959).

or technique.'[17] Often, his minimalist, usually static, visual style suits the mood of his subject, as in his essentially narrative, dramatic films: *This Sporting Life* (1963) and *In Celebration* (1974), with their claustrophobic atmosphere and their emphasis on the text and actors. However, some of his satirical films such as *O Lucky Man!* and *Britannia Hospital*, which are extremely restrained visually, bordering on the sterile, would undoubtedly have benefited from a more dynamic style which could have provided another dimension for the articulation of the underlying irony and satire. By refusing to take advantage of the full range of cinematic resources available, Anderson is effectively turning a blind eye to the fact that imaginative, visual craftsmanship or technique is a powerful ally for any cause.

Anderson prefers to turn his attention to the actors with whom he has a good reputation as a director (for stage and screen alike). He aims for a balance between the actor's exploration of character and his style—'a psychological awareness balanced by a feeling for presentation'.[18] To Malcolm McDowell, 'Lindsay is one of the very few directors that actually likes actors. He is like a psychiatrist, in that he directs every actor in a totally individual style.'[19] Anderson is aware that 'all actors need praise,' and as Richard Tombleson, production assistant on *Britannia Hospital*, commented, 'He makes actors feel that what they do is of value, he makes them feel they're engaged together in a valuable activity.'[20] Anderson's success with actors derives largely from the contagiousness of his own self-confidence and sense of conviction. Jill Bennett has commented, 'We all liked Lindsay so much—I turned down a West End play in order to do the film [*The Old Crowd*, 1978] with him. None of us read the script much in advance . . . But that was Lindsay all over. He had confidence in himself, so we had confidence in him.'[21]

Anderson's confidence and sense of conviction ensured that even at the beginning of his career, his views of what film should be were outspoken, representing a challenge to the orthodoxy of British cinema, which he saw as class-bound and self-satisfied. In these early years, Anderson was inspired by the strong socialist sympathies of British politics after World War II, and became involved in the 'New Left' movement of the mid-1950s. Imbued with the optimism of the time, which he later described as a 'blissful false dawn', Anderson, together with a group of friends and fellow film-makers—Karel Reisz, Tony

[17] *Creative Review*, July 1982.
[18] *Transatlantic Review*, Spring 1962.
[19] Ann Lloyd and David Robinson (eds.), *Movies of the Sixties* (Orbis, 1983).
[20] *Creative Review*, July 1982.
[21] Alexander Walker (ed.), *No Bells on Sunday: The Journals of Rachel Roberts* (Pavilion, 1984).

Richardson, and the Italian Lorenza Mazzetti—founded a movement, called 'Free Cinema',[22] which they hoped would carry the spirit of the 'New Left' into the cinema. They saw themselves as challenging contemporary documentary-makers who, unlike their predecessors, were no longer seen as critical of the establishment. The aim of the 'Free Cinema' movement was in part to capture 'the poetry of the everyday'. Certainly all the directors produced highly personal works, reflecting the influence of the great British documentary-maker Humphrey Jennings.[23] But shrewdly, they also realized that the establishment of a movement, with its film programmes at the National Film Theatre, would provide a platform for the directors' previous documentary work.[24]

Anderson's first documentary film, *Meet the Pioneers* (1948), was commissioned to celebrate the fiftieth birthday of a firm in Wakefield, which had been the first to install conveyors in coal mines. According to the owners' wishes it specifically highlighted the family nature of the firm. The owners proceeded to commission Anderson for three further documentaries,[25] and recommended him to the local Wakefield newspaper, for whom he made *Wakefield Express* (1952). This was a documentary about the paper and its role within the town. Looking at the lives of a close-knit group of people was the focus of one of the best Free Cinema films, Anderson's *Every Day Except Christmas* (1957), about a day in the life of Covent Garden fruit and vegetable market. Both these films are lyric pieces about busy, hard working people who live in humane, friendly communities. They portray an extremely positive, purposefully romanticized view of a community, full of smiling people, apparently incapable of exchanging an angry word with each other. Anderson accepts that they are idealized, but believes that 'something can be idealised without being untrue . . . it becomes an image'. If such images were a little too sentimental, a perfect balance was achieved with *Thursday's Children* (1953). It was a simple documentary, narrated by Richard Burton, about a school for deaf children who learn

[22] After the title Anderson had given to an article in *Sequence* magazine. The article was actually written by Alan Cooke, on his first two films at Oxford, *Black Legend* and *The Starfish*, which he co-directed with John Schlesinger.

[23] Jennings is, alongside Ford, Anderson's other great cinematic hero. Jennings (1907–50) was a documentary film-maker who, in the early 1940s, made some of the warmest, most lyrical documentaries ever produced in Britain. During the war years, his films accurately reflected the mood of London during the blitz. His work includes: *The Birth of a Robot* (1936), *London Can Take It* (1940), and *Fires Were Started* (1943).

[24] In particular three films: *O Dreamland*, made by Anderson in 1953, a study of a cheap funfair that showed the pathetic way people can be debased by the demands of a popularist culture; *Momma Don't Allow*, the first film made by Tony Richardson and Karel Reisz, in 1956, about jazz enthusiasts in a dance hall; *Together* (1956), directed by Lorenza Mazzetti and edited by Anderson. It is about a pair of deaf mutes living in London's East End, and how they are treated by the surrounding community.

[25] *Idlers that Work* (1949), *Three Installations* (1952), and *Trunk Conveyor* (1954).

both to read and speak under the patient guidance of their teachers. The film looked at the learning process from the children's point of view, their sense of belonging, and their achievements. It let the children's personalities and feelings of joy and discovery come to the fore, and the film well deserved its Academy Award. While he was making this, Anderson also made the quite different *O Dreamland*—a bitter mockery of a cheap funfair in Margate. Most of these films were not shown publicly until the first Free Cinema programme, in 1956. Altogether there were six programmes of such films at the NFT in three years, and they were greeted with considerable enthusiasm. However, in the absence of any long-term sponsorship (the initial grant given by the Ford Motor Company was never repeated), the movement could not extend beyond being simply a group of friends making films.

This Sporting Life (1963), Anderson's feature film debut, was made fifteen years after his first documentary, and a full five years after his previous film, *Every Day Except Christmas*. He had been working at the Royal Court theatre directing several notable plays, in particular *The Long and the Short and the Tall*, *Sergeant Musgrave's Dance*, and *The Lily White Boys*. He was finally given a chance to direct a version of David Storey's novel, *This Sporting Life*, by his Free Cinema colleague, Karel Reisz. The backers, Independent Artists, had initially asked Reisz to direct, but after directing *Saturday Night and Sunday Morning* he was keen to get some experience producing and so offered it to Anderson.

The film was shot in Yorkshire around Wakefield, an area familiar to Anderson, where he had made his first four documentaries. It is a tale of a fatally mismatched relationship between the young Frank Machin (Richard Harris) and his widowed landlady, Mrs Hammond (Rachel Roberts), against the background of Rugby League Football—'the professional working class version of the game'. Anderson filmed the rugby scenes with unremitting violence. The effect is a series of brutal encounters reflecting Machin's own inner torment. His philosophy of life is simple—get what you want, by force if necessary. Anderson describes the film as 'primarily a study of temperament'. For all Machin's physical power and aggressiveness, he is emotionally inarticulate, even subhuman, like the ape he resembles[26] as he leans against the beams of the kitchen. He may well be equipped for the rugby pitch, but not for the human relationship that he longs for. He is unable to communicate his tenderness for Mrs Hammond, who is herself unable to express her repressed sexuality. The film ends with Mrs Hammond dying from cancer, and Machin finally discovering the tenderness which he had

---

[26] An idea perhaps picked up by Karel Reisz in *Morgan: A Suitable Case for Treatment* (1966). On several occasions in the film Morgan is shown swinging around like an ape.

Apprenticeship in the North Country. Anderson directs one of his early documentaries, *Wakefield Express* (1952), about Northern community life as seen through the work of the local newspaper. © Wakefield Express.

never been able to extend to her. For Anderson, *This Sporting Life* represented a powerful and ambitious entrance into feature films, and was amongst the best of the diverse generation of social-realist films of the sixties. It reflected Anderson's own brand of subjective realism; as he said, 'We were very aware that we were not making a film about anything representative; we were making a film about something that was unique; not about a "worker" but about an extraordinary (and therefore more deeply significant) man, and about an extraordinary relationship.'[27] Despite its critical success, the film was too grim to attract a mass audience. Karel Reisz noted that 'people came away having been pained rather than cheered'.

This Sporting Life may have been black, but much of Anderson's work thereafter was even darker. He moved away from this kind of dramatic realism towards satire, being first prompted in this development by his success in directing Max Frisch's play, *The Fire Raisers*, at the Royal Court. Through absurdist comedy he hopes to make films which have a greater universal human relevance. 'The follies that you read about every day when you open the paper are so absurd, that the only way to comment on them, is through laughing, because if you try to be serious about them they dwarf you. I am sure I don't have to quote Byron to you: "And if I laugh at any mortal thing, 'tis that I may not weep".'[28] Anderson's trio of films scripted by David Sherwin are all in this vein. They employ interesting metaphorical structures for commenting on the chaos, absurdity and misguided values of British society—the school in *If . . .*, the hospital in *Britannia Hospital*, and the classical odyssey in *O Lucky Man!* The style of this social satire has led Anderson to be called 'The English Buñuel'. Parts of these films are purposefully obscure, aiming to encourage thought and debate. 'People always ask me what I mean in certain sequences, and I can't always tell them. But surely that's the poetry of the thing.'[29] Likewise he uses alienation devices to invite people 'to use their intelligence. It's a very dangerous invitation because it will almost invariably be rejected.'[30] These include the use of musical interludes, shots of camera and crew, black segments of film between or even in the middle of scenes, or randomly cutting from colour to black and white, or vice versa. Some of these techniques have been more successful than others, but much to Anderson's frustration, they have often distracted the audience away from the very issues he was trying to get them to reflect on.

The short work which started this particular style of film-making was

[27] *Films and Filming*, Feb. 1963.
[28] Interview with the authors.
[29] *Roughcut Film*, Summer 1976.
[30] Interview with the authors.

the rarely seen *The White Bus* (1966)[31] (Anderson calls it 'my lost film'), based on the story by Shelagh Delaney. In it Anderson sardonically exposes bourgeois bigotry and philistinism. A party of local citizens and dignitaries are taken on a 'See Your City' bus tour of Manchester and Salford. Amongst them is a young girl, through whose innocent eyes we see a series of incidents and hear comments which illustrate 'civic conceit and middle class bigotry'.[32] The film abounds in ambiguities and metaphors; the 'white bus' itself is a clever metaphor for the blinkered ignorance of its passengers and their engrained prejudices. Unlike many other protagonists in Anderson's films, the girl is merely a passive observer. That she makes no attempt to change anything serves to heighten the audience's sense of frustration at seeing clearly the hypocrisy, yet remaining helpless in their vicarious desire to act through the girl.

Anderson's next film is his best known work. *If . . .* (1968) was the first of Anderson's trilogy of biting social satire to be scripted by David Sherwin. The film starts out as a realistic drama, but the public school setting, representing a microcosm of British society, develops into a powerful metaphor. The film reflected many of Anderson's preoccupations—'the basic tensions, between hierarchy and anarchy, independence and tradition, liberty and law, which are always with us'.[33] The school is effectively run by 'whips' (prefects) who revel in their power and privileges. The teachers are pathetic, eccentric, and perverted or at best bland, while the pupils conform unthinkingly to the 'system' with its repression and senseless rituals. But there are a few exceptions, led by Mick (Malcolm McDowell), who violently rebel. The film, almost apocalyptically, ends with their defiant if ultimately futile last stand. They proceed to gun down not only the staff and student collaborators, but also, turning a gun towards the camera, the audience—partly for their complicity, partly, no doubt, just to make them feel uncomfortable. As Anderson said,

Far from filling me with dread, I find the last sequence of the film exhilarating, funny (its violence is so plainly metaphorical), a bit shocking and finally sad. It doesn't look to me as though Mick can win. The world rallies as it always will, and brings its overwhelming fire power to bear on the man who says 'No'.[34]

Although their cause may be defeated, their spirit is not.

To Anderson *If . . .* is a romantic and 'deeply anarchistic' reaction against the destructive aspects of the class hierarchy, traditions, and hallowed British institutions with their notions of responsibility, which

[31] Anderson's 45-minute contribution to a project originally entitled *Red, White and Zero.*
[32] David Robinson, *Financial Times*, 17 Jan. 1968.
[33] Anderson's preface to the script of *If . . .* (Lorimar, 1969).
[34] Preface to Lorimar edition of the post-production script.

are so at odds with his own. But the indistinct border between what is reality and what is Mick's fantasy tended to divert the audience, and so work against the film. The contrast between the colour and mono-chrome sequences in particular caused much debate. It was a distancing device which Anderson had used in *The White Bus*, but its immediate cause was that it was the only practical solution to the problems of lighting the chapel for colour photography on a limited budget; as Anderson said at the time, 'why not?' Moreover, at certain points, fantasy confusingly degenerates into farce, breaking some of the mounting tension and losing the sense of unease which the audience feels as it tries to work out what side of farce the film really lies on. The scene where the rebels have to apologize to the chaplain, who is produced from the drawer in the headmaster's office, is one example.

Despite this, *If . . .* is a film magnificent in its anger, and it effectively exploits the public school metaphor to mirror British society. It was, moreover, commercially successful, largely due to its fortuitous co-incidence with the prevailing temperament of 1968, with its student revolts and political upheavals.

Anderson's next film, *O Lucky Man!* (1973), was the second of his satirical British trilogy. It took the form of a contemporary morality play, chronicling the journey of Mick Travis (Malcolm McDowell), full of the greedy ambition of youth, from illusion to self-realization. Its structure is the classical one of an epic odyssey showing the absurdities of life. 'It is not just one story but five.'[35] Mick passes through numerous adventures and encounters with all manner of people, from rich tycoon to the down-and-outs, with many of the different roles being cleverly played by the same actors. Mick is abused, exploited, and misunderstood. Corruption and callousness are found at every stage, and at each Mick shows a different form of innocence. In contrast to *If . . .*, open revolt is not even considered worth trying; Mick passes a slogan on the wall which reads 'Revolution is the Opium of the Intellectuals'. At the end of the film, Mick, virtually destitute, sees a notice for an audition. Standing in front of 'the director'—Anderson himself—he is asked to smile, but refuses, demanding to know why he should. The director strikes him over the head with a script and he smiles, a smile of instant understanding—certain things one can accept without 'conforming'. Acceptance does not necessarily imply surrender; indeed, Anderson seems to be saying, it is often important if one is to make anything of one's life in a real world. His early romanticism was being displaced by pragmatism.

He intended every minute of the film to be packed and intense, which it is, but not overly so, largely due to the musical interludes by Alan

---

[35] *Cinema TV Today*, 26 May 1973.

Anderson's anarchism in *If . . .*, 1968. Malcolm McDowell guns down members of his school in a desperate and ultimately futile act of rebellion. Courtesy of Paramount Pictures.

Price. These effectively break the film up into sections and provide time for reflecting on what has just been seen, while listening, as to a Greek Chorus, to an overview given by the words of the song.[36]

Although to a lesser degree than *If . . .*, the weakness of *O Lucky Man!* lies in the failure to distinguish clearly between what is realism satirized and what is fantasy. There is nothing which is stricty unbelievable, nor are there any puzzling black and white sequences. But because it begins, as *If . . .* does, within the terms of conventional narrative naturalism, the use of farce later in the film is less successful than it might have been; arguably the audience should be aware of a guiding logic and consistency in a film.

In Celebration (1974) marked a temporary withdrawal from the world of social satire, and back to the kind of realism of *This Sporting Life*. Both films were written by David Storey, and Anderson had in fact directed the stage play of *In Celebration* at the Royal Court in 1969. Feeling that they share the same way of looking at the world, Anderson has worked constantly with David Storey, directing eight of his stage plays, as well as transferring three of his works to the screen. *In Celebration* was made as part of the British Film Theatre's series of film versions of contemporary plays. It is the story of three well-educated sons, a teacher (Brian Cox), a factory negotiator (James Bolam), and a rebellious artist (Alan Bates), who return for a weekend to their cramped home in a coal mining town to celebrate their parents' fortieth wedding anniversary. Looking beneath the surface appearance of family love and unity, the film explores the diverse forces which have alienated members of the family from each other—the father's determination to educate his children, yet his hatred of the fact that they now live 'middle class' lives; the guilt the children feel about satisfying their parents' expectations of them; as well as the mother's own false sense of virtue. Alan Bates observed, 'It is the story of the crumbling façade of family life. It is the story of all our lives.'

Anderson used the same actors who had played in the stage version five years earlier. Consequently, he feels, with characteristic confidence, that they achieved 'a deepening and maturing of understanding, which I believe results in one of the finest examples of ensemble playing to be preserved anywhere on film'. He kept the film very faithful to the play and rightly avoided the temptation to 'open it up'. The play's sense of claustrophobia is maintained by the absence of many exterior scenes and by the use of wide, static camera shots. Here his sparse camera style is particularly effective in focusing the audience's attention on the power and tension of the acting, and on the text, which unlike that of his

[36] Anderson himself remarked, *post facto*, on the similarity of this to Brecht's use of a singer in *The Threepenny Opera*.

Richard Harris and Rachel Roberts in Anderson's first feature film, the social-realist *This Sporting Life*, 1966. Courtesy of the Rank Organisation plc.

previous films, contains long dialogue scenes. Consequently, when there is the occasional close up, it has a powerful visual effect.

The Old Crowd (1982), was Anderson's first BBC television film, and of all his work to date received the most hostile reviews. It was a joint project with playwright Alan Bennett, but the criticism was directed exclusively at Anderson, despite comments made by Bennett himself, the producer Stephen Frears, and the actors, about how fruitful and enjoyable they had felt the whole experience to be. Bennett even commented, 'If it were to be shown again, I think people would begin to see what Lindsay was getting at. It came in for a lot of criticism, but I think that Lindsay was simply ahead of his time.'[37] The play was about the gathering of a middle class family in an empty house. It poked fun at them, and in particular at their narrow-mindedness and the frustrations of their women. What really alienated the critics (and one must assume a significant proportion of the audience as well) was the use of distancing techniques, such as a shot of the technicians watching the film being made at the point where the host asks one of the guests, 'You're not overlooked?' or the use of a table which is eighteen inches too high for the guests and which sarcastically makes them look like children. This attempt at making a non-naturalistic drama angered those who were not interested in anything so out of the ordinary for British television.

Britannia Hospital (1982), the most accomplished of Anderson's satirical trilogy, is also his most scathing work—a caustic attack on virtually every aspect of British social and political life. Although based on actual events which Anderson had read about in the newspapers, the hospital, in which the film is set, is also a metaphor for what he sees as a country which has undergone a nervous collapse, having ceased to believe in itself. Little escapes his onslaught. Health service workers leave the sick for the sake of a strike to defend their tea break. Kitchen staff refuse to serve private patients. They are egged on by union leaders who subsequently sell them down the river for a seat on the table with the members of the royal family, who are visiting on that day. But even the royal visitors have to be smuggled into the hospital in bandages to avoid rioters at the gate who are protesting against one of the private patients, a ruthless and corrupt Black African leader. Behind all this surface mayhem prowls the insane medical professor, first seen in *O Lucky Man!* Now, however, his experiments have become more advanced and he has in fact become a terrible Frankenstein figure, working on his vision of a mix-and-match 'new man'. In this pessimistic work no hope is offered for the future. In *If . . .* the rebels' cause may have been defeated, but they themselves were unbowed. *O Lucky Man!*

---

[37] *Western Mail*, 9 Nov. 1982.

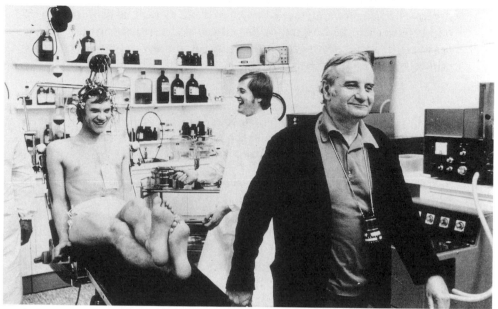

Anderson shares a joke with McDowell and James Bolam on the set of *O Lucky Man!*, 1973. Anderson's strong rapport with his actors ensures their recurrent appearances in his films. McDowell starred throughout Anderson's satirical British trilogy. © 1973 Warner Bros. Inc. Photograph: BFI.

had the elixir of self-realization at the end of the odyssey. But at the end of *Britannia Hospital* no such hopes remain. The final scene is of the professor telling an audience of royal visitors as well as protestors who had broken into the hospital of his vision of the future for man—as he uncovers a grotesque mechanical brain. A hollow disembodied voice quotes from *Hamlet*:

> What a piece of work is man.
> How noble in reason, how infinite in faculty,
> in form and movement how express and admirable,
> in action how like an angel,
> in apprehension how like a God . . .

This is indeed harsh irony. Throughout the film Anderson has shown how pathetic man has become. Yet here is the aspiration for what man is capable of, great aims which are all too often not even acknowledged, but which here are to be incarnated in a machine! None of his previous work had cut so deep. But for the jingoism that swept over the British public at the time of its release, the same year as the Falklands crisis, it might have been much more popular and successful.

LEEDS METROPOLITAN UNIVERSITY LIBRARY

The Whales of August (1987), his most recent feature film, had many critics wondering how such an apparently abrasive satirist could have so changed his vision of humanity. It is a story of two sisters confronting old age together on a quiet rural island off the American mainland. Certainly the film's absorbing character studies mark it out from his previous work and make it one of his most appealing films. But this warm vision is not wholly new, it is a return to the lyricism of his earlier documentaries such as *Every Day Except Christmas*, a return to a film-making which expresses all his hopes for mankind. The real sense in which it is different is superficial—it has an American, not British, cast and location.

The two sisters are very different characters. Libby (Bette Davis), adopting the attitude often expected of the old, feels that life is effectively over, whereas Sarah (Lillian Gish, who had not acted in a major role for thirty years) believes that life is still for living to its full. If one were looking for Anderson's conception of the ideal individual, one would no doubt come close to finding it in Sarah—honest and kind, ignoring irrelevant traditions and social mores, seeing people for what they really are and accepting them as such. In this mature work Anderson has found the perfect vehicle for the romantic and optimistic aspects of his view of life.

Since then Anderson has directed another American piece, *Glory! Glory!* (1989). At a loose end at the time, he was asked to direct it, and shot it—within the short thirty-five-day shooting schedule. He made few changes to the script of the 3¼-hour mini-series, produced for the cable station Home Box Office. On the surface it is a satirical look at the rapacious, fundamentalist television preaching industry. But Anderson saw in it universal themes applicable to the whole of society. The series follows the fortunes of the change in leadership of the television empire of an evangelical church, which passes from the fiery, faith-healing father to his bland but earnest son. Donations fall dramatically until, with the ministry near to collapse, the son recruits a boozing, cocaine-addicted female rock singer to instil new life into the TV services. Through the rawness and sexual overtones in her singing, she rejuvenates the finances of the ministry—as well as experiencing her own spiritual revival. The timing of the work was particularly appropriate given the recent flurry of actual disgraced television preachers in America. The film has much humour and is probably Anderson's funniest work to date. But there is, once again, a deep pessimism behind it. Everyone is viewed as ultimately corruptible. Although the characters may occupy the moral high ground, it is only temporary. They will always succumb to temptation.

The mood of Anderson's work has generally fluctuated between optimistic, romantic lyricism and bleak, caustic satire. But his satirical works, dense with ideas, are often much less coherent than some of his other films, such as *This Sporting Life* and *The Whales of August*, with their more conventional dramatic structure and emphasis on characterization. It is unfortunate that although Anderson's social comment is so sharp and incisive, the effectiveness of many of his films for a mass audience has often been blunted by excessive complexity and ambiguity. As Pauline Kael harshly wrote, 'even his best sequences are sometimes baffling—heavy with multiple meanings that he doesn't appear to think need sorting out'.[38] But, by the same token, Anderson's great contribution to British cinema is that he has made stimulating and challenging films, which, because they are complex, require the audience to make an effort and think for themselves.

[38] *Going Steady* (Temple Smith, 1970).

LINDSAY ANDERSON

You have a reputation for being an anti-conformist, yet your upbringing was entirely traditional.[1] When and why did you rebel against these values?

Well, I would like to say that it's because I was born with a certain intelligence. I don't think one pinpoints one's development with that kind of precision, unless there's been some kind of crucial experience in one's life which has changed one completely, but that certainly never happened in my life. I think that many people born into an upper or upper middle class environment disassociate themselves from the conventional values of their background—that's quite a common phenomenon. But I think it is really a question of temperament—you see, although much of my background was quite characteristically English upper middle class, my blood is Scots. This definitely makes a difference, but one which English people don't particularly credit because to them England and Britain are the same thing. I've become aware of a difference between myself and many of the English people I've found myself among, and I've wondered why I seemed to them to be as you said 'anti-conformist', which is a very English kind of reaction. I don't think you'd necessarily find the same reaction amongst Scots people, because they tend to be abrasive and they have faith in their own intelligence; they enjoy argument or dispute and, unlike the English, they don't think it's strange or rude to behave in this way.

Certainly you've always spoken your mind.

I've never been cautious in what I've said, which may or may not be a good thing. I've mostly got away with it, but not entirely. I remember an anecdote with *This Sporting Life* which illustrates this. Julian Wintle, the head of Independent Artists, who were making the film, panicked. He decided to show the film to David Lean to get his opinion on whether it would work or not. Well, unfortunately I'd previously done a little bit of film criticism, writing the column for the New Statesman when *Bridge on the River Kwai*[2] came out. I had disliked the film and rather than go

[1] Born in India, father was an army officer. Went to Cheltenham College, followed by Wadham College, Oxford, where he studied Classics.
[2] Directed by David Lean in 1957.

…n directing Lilian Gish in *The Whales of August*, 1987. © Curzon Film Distributors Limited.

into detail I had given it only two or three lines calling it a 'chocolate-box war film' and devoted the rest of the column to reviewing Wajda's first film *A Generation*, which I'd seen in Cannes and which wasn't on release in Britain. I later heard that David Lean was absolutely furious about this. So, when I heard that my film was going to be shown to David Lean for him to say whether it worked or not, I was exceedingly alarmed. But fortunately he was out of the country at the time. Still, a little later I was actually dismissed from the columns of the *New Statesman* because of what I had written on *Bridge on the River Kwai*.

How much control do you demand as a director?

I want total control working with extremely good, skilled, and sympathetic collaborators, which is of course difficult; and, of course, 'control' means 'responsibility'.

Have you consistently used the same crew?

I love to if I can, but I haven't always been able to, partly because I don't make films very often. For example, take cameramen. Three pictures were shot by Miroslav Ondricek, whom I met in Czechoslovakia in about '63 or '64, when the Czech cinema was really blossoming; then I used Mike Fash who did *Britannia Hospital*. He then emigrated to the States: by great good luck he was available to shoot *The Whales of August* and the mini series, *Glory! Glory!* for American television. So generally I haven't been able to work with a faithful group of adherents as crew. Although this, of course, doesn't apply to the actors.[3]

As a director how do you try to help actors in their performance?

That is a very difficult question to answer, because I don't proceed with them according to a worked out approach. One gets the best out of actors by an intuitive sympathy, or understanding, without trying to work it out too much, or too cautiously. One just has to know, feel, and understand their problems. The really important thing in the actor/director relationship is the casting. If you offer an actor the part, then in a sense, you have made the first gesture. Then, if they accept, they come forward offering the interpretation of the part, and I respond to that.

Do you find that some actors are particularly hard to work with?

It can happen. I wouldn't wish anyone the fate of having to work with Bette Davis. I had to use all my instinct and all my guile in *The Whales of August*, to get past her. She is totally egocentric, and she needs to be, demands to be, the centre of attention.

Do you surrender to this?

---

[3] For example Malcolm McDowell or Arthur Lowe, with whom he has worked many times.

Well, you surrender up to the point where she starts making demands that you cannot go with. She would say; 'Oh, I want this to be there, and that over there'—you know. In the end you have to say, 'I'm sorry, this isn't your role; you're not taking over,' and she walked out! We all waited around, and she came back again. Bette is a star, and thinks of herself as a star. Actors are generally better, but things do happen.

Do you storyboard your films?

No I don't, although I've often wanted someone to do it for me. I'm not very good at working things out beforehand. I tend to work a sequence out with the actors and cameraman, although this takes a bit more time.

How did you come to direct Meet the Pioneers *even though you had no previous experience?*

It was totally by chance. Having edited the film magazine *Sequence* at Oxford, I was approached by Lois Sutcliffe, the wife of the managing director of a factory in Wakefield in Yorkshire, who made conveyors. She came and asked me to make it for them after getting boring treatments from two London documentary firms. At the time I didn't really have an idea of exactly what I was going to do—in a way that's very typical of me, I'm afraid. When I began, the cinema and theatre were totally mysterious. My first reaction was, 'This is crazy because I know nothing about making films,' but she said, 'Well you've always got to start somewhere' and so I said 'All right, I'll have a try.'

In retrospect how do you think you coped?

Amazingly well. This was a marvellous way to begin to make films, which probably wouldn't be possible today—everyone today is much more professional. I don't think anyone would be able to go out with a friend and a 16 mm. camera and make a film. But I didn't know anything when I made that first film. I didn't know how to edit. I thought you just wrote a script, wrote the shots down, took the shots and then joined them all together. So on the schedule I gave a couple of days to edit what was a 40-minute film. There was a naïvety there which has vanished from the world of film.

Your creation with Tony Richardson and Karel Reisz of the Free Cinema 'Movement', with its manifesto of principles, was rather less naïve wasn't it?

Yes, we got the idea to form a movement to get our films shown, and the Free Cinema manifesto was made up. But this didn't mean that it was exactly untrue, although of course it tended to give the impression that here was a group of people who'd got together, believed these things, and had gone out and made some pictures—which *was* quite untrue. In the subsequent films I've made I don't think I have, and I don't think any of us have, been in any way affected by that manifesto—except, of course, in so far as the manifesto was a reflection of ourselves.

Looking at the films you've made throughout your career, would you agree that you do seem to have a personal manifesto in terms of recurrent themes?

One doesn't wish to be one's own analyst, but today one is continually forced to be. What seems to emerge is a mistrust of institutions, a mistrust of authority, and a corresponding belief in, or liking for individualism, and individual responsibility. 'Humanism' is another word that might crop up.

There is certainly a feeling for community, which in *Every Day Except Christmas*, for instance, is seen as an ideal, a kind of poetic expression of community at its best. In a sense, *The Whales of August* is a return to the warmth of this earlier approach (I found it ridiculous that certain critics found it out of character). But I think this type of lyrical expression of humanity is difficult to hold onto as one gets older, and a lot of my work since then has tended to veer more towards satire and the comedy of human folly, which can be seen in various ways in *If . . .*, *O, Lucky Man!*, and *Britannia Hospital*.

Humour has always been important in your work hasn't it?

I suppose so. But my comedy is apt to be edged; horrifying and satirical as well as funny. For example, in *O Lucky Man!* in the scene at the clinic when Mick, seeing what he thinks is a man shivering in bed, pulls back the sheets to discover a man's head on a pig's body. And again in *Britannia Hospital* when Professor Miller liquidizes the brain and drinks it. Unfortunately, audiences, particularly English audiences, want to be amused and not disturbed. Monty Python is in that great British tradition of facetiousness. But it is not satire and it does not aim to change anything.

At the beginning of If . . . *there is an epigraph which says that 'wisdom is the principal thing, therefore get wisdom, and with all thy getting, get understanding.' That attitude seems very central to the ideas and aims of your films.*

Well actually, those words owe their origin to when we were shooting an earlier film *The White Bus* in the library in Manchester and these words were written on the ceiling. It probably does say quite a lot about me, because being a bit critical, I think I have a rather pedagogic attitude which many people don't like. They don't like the idea that you're telling them. They instinctively respond, 'Oh, you think you know better do you? Well, my ideas are as good as yours!' But an ignorant idea is never as good as an intelligent one. In fact, from my point of view, anything that can make people think is good. So for example, with *If . . .*, *O Lucky Man!*, and *Britannia Hospital*, they are not strictly a trilogy although I would say they have common and developing themes. Mick in *O Lucky Man!* plainly isn't the Mick in *If . . .*, except that again it's Malcolm McDowell. When he's interrogated he's asked, 'Was your headmaster right to expel you from school?' Well, either you're amused by that or you find it irritating, but if you think about that, you also think about the related themes.

But it is not always pedagogic—you don't try and resolve all the issues in the film . . .

I also think it's very important to balance the intelligence and the intuition. So that when you're working out something, you should preserve what is intuitive or a free exercise of personality, humour, or imagination, as well as maintaining inherent ambiguities. Particularly with *Britannia Hospital* people said 'Oh, it's a muddle,' but I think it is a very clear film—if you can accept the fact that clarity does not eliminate ambiguity.

Can you give an example of this ambiguity?

Yes, towards the end of the film, there is a scene when the demonstrators are at the gates screaming and yelling like fools and those inside the hospital say, 'We'd better do the ceremony now.' They walk across the courtyard watched by the demonstrators who look through the gate. I held the shot, and the protestors are not absurd any more. The people looking through the gate have a certain dignity. You might say, 'Well that's not consistent,' or else, 'Now we're seeing another way of looking at them,' which is of course a challenge to the intellect of the viewer—a dangerous thing to do. I think that the attempt to preserve this ambiguity is one of the reasons why my films are not viewed with any consistency or consensus by the critics, as they don't know how to handle them.

You mentioned a mistrust of institutions and authority. How does that relate to the traditionalism evident in some of your earlier films?

I've got nothing against traditionalism unless it is distorted and made absurd. But often this is misunderstood. In *Every Day Except Christmas* there is a little prologue when a truck is loaded and driven off through the dark streets to the sound of 'God save the Queen'. Now when that film was shown to the National Film Theatre the audience laughed because that was the trendy London audience who thought that if you played 'God save the Queen' you must be sending it up. But this wasn't the intention. In fact *Every Day Except Christmas* is a very strongly traditional film.

What is interesting is that in those early days when we did 'Free Cinema', we called two films — *Every Day Except Christmas* and *We're the Lambeth Boys* (which Karel did) — 'Look at Britain'. Well that was just us really, trying to be 'patriotic'. Of course the irony was that *Every Day Except Christmas* was turned down by the British selection committee to be shown at the Venice Film Festival because they thought that this isn't the way to show Britain, not with all those mucky workers. The kind of thing they wanted to show was Benjamin Britten or something like that. So we only got it to Venice by putting it in the category of films for television, which it wasn't. And then it won the Grand Prix and everyone was quite annoyed. In other words, one was an outsider then for being 'patriotic', having a feeling for tradition, as one is an outsider now for being, I don't know what, just out of touch I suppose.

Why do you think the audience has failed to respond to what you're saying?

I'd say that since the mid-seventies the British, including the young, are suffering from a creeping and now galloping conformism.

How did you react to Punk in the seventies — Britain's only really anarchistic movement?

I think it was quite anarchistic, but it was devoid of any thought or ideas. It couldn't amount to anything, and it could easily be gobbled up by commercial interests. The British don't think anymore, do they? You can't make people think in the cinema, it's impossible. That's why my films have been failures. I've asked the audience to think, which is fatal really. Yet there is a certain naïve quality in me, or has been, where even with something like *Britannia Hospital* I thought that there would be sufficient people who would recognize what it was saying, which after all are the kind of things they grumble, laugh, or get angry about in the pub every night. Not that the film would be a huge success, but that it would find an audience. Well, when that film found no audience whatever I thought, 'Well, that's it.' I suppose it was much too complex for the audience to understand. Everyone was mocked in it; the workers, the demonstrators, the establishment. I don't believe that it failed because of the Falklands incident. It failed becaue the British are just too conservative.

You see, an artist should be truthful and that means an artist is liable to make people feel uncomfortable. People are generally weak, and you either feel sorry for them or you want to kick them. It's a matter of temperament. In most of the films I've made, I tended to want to kick

Britain on the verge of a nervous breakdown. Anderson directing the riot police in *Britannia Hospital*, 1982. © Weintraub Screen Entertainment. Photograph: BFI.

them, which naturally doesn't get a very favourable response. It's very frustrating. I see all the mess in the world and I say, 'Serves you right! I told you but you wouldn't listen!' The popular audience has been totally corrupted, and television has hastened the process. We have a public which is totally deformed by American pictures.

In what way?

I don't think that the British public is at all interested in British films. That is the way the culture has gone, especially Thatcherite culture. Even *Time Out*,[4] which is supposed to be radical in some way, puts Tom Cruise on the cover. The whole taste of the British public for film has been formed by American cinema. Go and look outside the Museum of the Moving Image in London. There are large blown up stills of the stars, and personalities of the cinema—how many photos of British talents do you think there are?

Sir Laurence Olivier?

No, not one—all American or classic European. You have an immense advantage working in a language of your own country, and not that of Hollywood. In Britain we have never had a chance. Puttnam is characteristic, in that he wants to have it both ways. He talks a lot about British cinema, but actually his films are completely American in appeal. The primacy of American cinema is an established fact, and then you have to think why; it's partly because of the market and finance there, but partly because, unlike our society, it is relatively classless. I did have a propagandist impulse for British cinema which to some extent I've stayed with. But I must be completely realistic. We, or I, completely failed in the attempts to make British cinema prized or valued. You can't defeat history, and that's why one feels completely on one's own. You might describe this as bitter, but I think I'm being realistic.

But much of your work has been American financed?

Yes, it is ironic, isn't it? But I have not made Hollywood films—they are something completely different, and I have never set my cap at Hollywood, and never gone out there looking for subjects.

Both Whales of August, *and* Glory! Glory! *have American settings and actors. Why do them if you feel this strong desire to make films about Britain?*

It isn't that I feel a necessity to make films about Britain, but that British subjects hold a strong appeal for me. But I'm just not asked to make

---

[4] A fashionable London listings magazine.

films in Britain anymore. Anyway when I choose what films I want to make, I do it on the instinctive immediate appeal of the subject. But both those scripts had that. The script for *Glory! Glory*, for instance, was very interesting, and I'd never done anything like it before. I have this reputation for the most boring of virtues, integrity, so I thought I'd see if I could sell out. It also had the virtue of being immediate, so I didn't have to develop a script or think about it, just go in and see what happened. I've always been like that. It was the same at the very beginning when I shot *Meet the Pioneers*.

Why do you think television has been so harmful? You say it is corrupting.

It drugs the audience into not thinking. Television's finished everything really. It's a disaster.

Is this why you've done so little work for television?

I haven't been asked to do much work because I'm temperamentally not attuned to working in an environment which has to be conformist. I did once apply to the BBC television course, but I was turned down.

   Unfortunately I am not, in any way, what John Ford called a 'career man'. I just am not, and it's not to my credit. It ultimately meant that when the Ford motor company[5] didn't want to finance documentaries anymore, I didn't know how to get commissions. By a great stroke of luck Tony Richardson was working at the Royal Court and so I got involved in the direction of theatre. And I didn't do a feature film until *This Sporting Life*, which was after the revolution caused by *Look Back in Anger*[6] and *Saturday Night and Sunday Morning*.[7] But it is also why, when that movement collapsed, I didn't go off to work in Hollywood like all the other 'career men'. It is interesting to see what has happened to all those people. Tony Richardson took off, because he can't stand the English. Karel Reisz has become partly a British and partly an American director. So many others for financial, or whatever, reasons went out there. Peter Yates remained out there. Alan Parker has always wanted to be an American director . . . John Schlesinger . . . Ken Russell . . . Of the ideals of the late fifties, and early sixties, for good or ill, I am the only one who has stuck with them—although in a rather different form.

What do you feel about the actual quality of British cinema directors at the moment?

----

[5] Financed a series of 'Free Cinema' films with no advertising in them, of which *Every Day Except Christmas* was the first.
   [6] 1958, directed by Tony Richardson.
   [7] 1960, directed by Karel Reisz. The success of these films encouraged American studios to take on less well-known directors to tackle new material.

One thing that is missing from British film now is any real sense of excitement. I think that British cinema tends to be unambitious in a funny way. There is a little anger, but you have to judge the quality of the anger. Ken Loach should have developed as a film-maker but has become stuck in other people's half-baked Trotskyist ideas. He's a sentimental leftist. I think something like Frears's *Sammy and Rosie Get Laid* looks like a sort of dissident and ambitious film, but is in fact muddled and opportunist. Even its dissident quality seems to me to have a *Time Out* quality to it. I feel the same with *My Beautiful Laundrette*. They both lack conviction. They are half-baked. They are clever, but there isn't really anything positive about them. I feel there's a certain glossy shallowness that you'd get from the *Sunday Times* magazine. It will have all those good pictures of the East End and all that, but in the end it just says, 'This is how things are.' There is no thrust for change in them. They are essentially conventional. Stephen (Frears)[8] is a really interesting case of somebody who is intelligent without being exactly committed. He is not one for sticking his neck out in any particular direction. But one just gets resigned. Think of the early days—of the hopeful radicalism of the New Left of the late fifties and early sixties. Kureishi[9] may be a Royal Court writer, but one of the latter days of the Royal Court—which can now be regarded as the safety valve of the Establishment. Nobody is really shocked by anything that happens there. The moment they got a really controversial subject like, for example, *Perdition*,[10] they ran away from it. Everything has slid back into Thatcherism. But it's not possible to say 'lets kick her out,' because the opposition is so inept. What on earth would she be replaced with?

Do you feel there is enough contact between British film directors?

Behind the camaraderie which you may think exists, there are tensions. Artists are always jealous of each other. I've often tried to explain to young film-makers what we did with 'Free Cinema', and tried to convince them to get together, because when we did, we got a lot of publicity, which helped. But of course the English don't do that. They don't get together.

Besides, we often don't have much in common. My work has always tended to have a social element to it, whereas someone like Nicolas Roeg's has not. You see, as you'll probably tell from the things I've

---

[8] Stephen Frears directed *My Beautiful Laundrette* and *Sammy and Rosie Get Laid*. He was assistant to Lindsay Anderson on *If . . .*, and in 1978 asked Anderson to direct the BBC play *The Old Crowd*.

[9] Script-writer of *My Beautiful Laundrette* and *Sammy and Rosie Get Laid*.

[10] Written by Jim Allen, it was to be directed by Ken Loach, but under considerable public protest it was cancelled by the Royal Court.

made, I'm not very keen or good at sophistication of technique and camera, unlike many other directors. I think there's a great deal too much camera in films today, and this may be the result of the whole disastrous movement of film culture, film appreciation, and film schools, where the emphasis is more on manner rather than content. I think British cinema lost its way with the romantic neo-baroque of Roeg or Ken Russell. People all adopt academic attitudes, or are influenced by 'auteurism' and all that rubbish. But they are much more aware of what they are doing, whereas in the fifties we were just discovering.

How then do you think your films fit into the context of British cinema?

I would say all my films have been sore thumbs as far as the British film industry is concerned. I don't know how to make a British film which the British want to see. I've never felt part of the film industry here, and I don't think I am. Read the film books—I'm rarely there. David Lean is a very 'English' film-maker. But I don't understand his situation, because although he makes films extremely well from the technical point of view and from the point of view of 'form', what else he has I don't know, and I don't know to what his great reputation is owed.

Perhaps as a story-teller?

Well, yes, that's a good way out. But when I see, say, *A Passage to India*, I don't think that it is a story well told. It's certainly not as good as the book . . . I don't exist anymore as a British film-maker. I have never had a nomination, not that I give a damn, from the British Film Academy. That is perfectly OK because I know what I do is not to the English taste—Fuck 'em.

Don't you think for example If . . . is an important British film?

I don't think it was influential at all. *Another Country*[11] is to me much more like a British film, it's glossy, and photographed in that seductive way. There's all the difference in the world between *If . . .* and *Another Country. If . . .* is a sort of sore thumb.

How do you feel about your work to date? You sound both resigned and satisfied at the same time?

Well, there's not much point in not being is there? I feel in a way, 'Well, that's it.' It would be nice to do something more, and I probably will,[12] but there is a sort of body of work there which will do and anything else

---

[11] 1984, directed by Marek Kanievska. A much more romantic film, also set in a public school, about the anti-establishment activities of a young Guy Burgess.

[12] At the end of the interview David Sherwin arrived to talk to Anderson about a possible sequel to *If . . .*

will be a plus . . . It's called 'age'. It's also recognizing, when looking around, what I can and cannot take on. I can try and survive and do something, fine, but the idea that one will change anyone's thinking has gone.

FILMOGRAPHY

BRIEF BIOGRAPHICAL DETAILS

Born in 1923 in Bangalore, South India, where he spent his first two years. Educated at Cheltenham College, and then at Wadham College, Oxford, where he read Classics. Education interrupted by the war in which he served in the Intelligence Corps of the British Army as a cryptographer. He returned to Oxford to take a degree in English.

FILMS DIRECTED BY LINDSAY ANDERSON

1948   *Meet the Pioneers*. 33 mins. Richard Sutcliffe Ltd. PRODUCERS: Desmond and Lois Sutcliffe. PHOTOGRAPHY: John Jones and Edward Brendon. EDITORS: Lindsay Anderson and Edward Brendon. MUSIC ARRANGED BY: Len Scott. COMMENTARY: Lindsay Anderson.

1949   *Idlers that Work*. 17 mins. Richard Sutcliffe Ltd. PRODUCER: Richard O'Brien. PHOTOGRAPHY: George Levy. CONTINUITY: Lois Sutcliffe. COMMENTARY: Lindsay Anderson.

1952   *Three Installations*. 28 mins. Richard Sutcliffe Ltd. PRODUCER: Desmond Sutcliffe. PHOTOGRAPHY: Walter Lassally. ADDITIONAL PHOTOGRAPHY: John Jones. EDITOR: Derek York. COMMENTARY: Lindsay Anderson.

       Wakefield Express. 33 mins. The Wakefield Express Series Ltd. PRODUCER: Michael Robinson. PHOTOGRAPHY: Walter Lassally. PRODUCTION ASSISTANT: John Fletcher. SONGS: Snapethorpe and Horbury Secondary Modern Schools. COMMENTARY: George Potts.

1953   *Thursday's Children*. 20 mins. World Wide Pictures (A Morse Production). WRITTEN AND DIRECTED BY: Guy Brenton and Lindsay Anderson. PHOTOGRAPHY: Walter Lassally. MUSIC: Geoffrey Wright. COMMENTARY: Richard Burton. With children from the Royal School for the Deaf, Margate.

       O Dreamland. 12 mins. A Sequence Film. PHOTOGRAPHY: John Fletcher.

1954   *Trunk Conveyor*. 38 mins. Richard Sutcliffe Ltd./National Coal Board. PRODUCER: Desmond Sutcliffe. PHOTOGRAPHY: John Reid. EDITOR: Bill Megarry. COMMENTARY: Lindsay Anderson.

1955   *Green and Pleasant Land.* 4 mins. *Henry.* 5½ mins. *The Children Upstairs.* 4 mins. *A Hundred Thousand Children.* 4 mins. National Society for the Prevention of Cruelty to Children (Basic Film Productions). SCRIPT: Lindsay Anderson. PRODUCER: Leon Clore. PHOTOGRAPHY: Walter Lassally.

20 a Ton. 5 mins. *Energy First.* 5 mins. National Industrial Fuel Efficiency Service (Basic Film Productions). PRODUCER: Leon Clore. PHOTOGRAPHY: Larry Pizer. PRODUCTION MANAGER: John Fletcher.

Foot and Mouth. 20 mins. Central Office of Information for the Ministry of Agriculture, Fisheries and Food (A Basic Film Production). SCRIPT: Lindsay Anderson. PHOTOGRAPHY: Walter Lassally. PRODUCER: Leon Clore. EDITOR: Bill Megarry. COMMENTARY: Lindsay Anderson.

1955–6 Five episodes in *The Adventures of Robin Hood* series. Incorporated Television Programme Company (Weinstein Productions for Sapphire Films). EXECUTIVE PRODUCER: Hannah Weinstein. ASSOCIATE PRODUCER: Sidney Cole. PHOTOGRAPHY: Ken Hodges. ART SUPERVISOR: William Kellner. SUPERVISING EDITOR: Thelma Connell. CAST: Richard Greene, Alan Wheatley, Bernardette O'Farrell, Archie Duncan, Donald Pleasance.

1955   *Secret Mission.* 25 mins. SCRIPT: Ralph Smart. MUSIC: Edwin Astley.

The Impostors. 25 mins. SCRIPT: Norman Best. MUSIC: Edwin Astley.

1956   *Ambush.* 25 mins. SCRIPT: Ernest Borneman, Ralph Smart. MUSIC: Albert Elms.

The Haunted Mill. 25 mins. SCRIPT: Paul Symonds. MUSIC: Edwin Astley. CAST: Includes John Schlesinger.

Isabella. 25 mins. SCRIPT: Neil Collins. MUSIC: Edwin Astley.

1957   *Every Day Except Christmas.* 40 mins. Ford of Britain (A Graphic Production). PRODUCERS: Leon Clore and Karel Reisz. PHOTOGRAPHY: Walter Lassally. MUSIC: Daniel Paris. RECORDING AND SOUND EDITING BY: John Fletcher. COMMENTARY: Alun Owen.

1963   *This Sporting Life.* 134 mins. Independent Artists (A Julian Wintle/Leslie Parkyn Production). SCRIPT: David Storey (based on his novel). PRODUCER: Karel Reisz. PHOTOGRAPHY: Denys Coop. PRODUCTION DESIGNER: Alan Withy. EDITOR: Peter Taylor. MUSIC: Roberto Gerhard. CAST: Richard Harris, Rachel Roberts, Alan Badel, Colin Blakely, Arthur Lowe.

1966   *The White Bus.* 46 mins. (One of three episodes for a projected film which was never realized, *Red, White and Zero.*) United Artists (A Woodfall Film Presentation). SCRIPT: Shelagh Delaney (from an original story in her book *Sweetly Sings the Donkey*). EXECUTIVE PRODUCER: Oscar Lewenstein. ASSOCIATE PRODUCER: Michael Deeley. PHOTOGRAPHY: Miroslav Ondricek. PRODUCTION DESIGNER: David Marshall. SOUND EDITOR: John Fletcher. EDITOR: Kevin Brownlow. MUSIC: Misha Donat. CAST: Patricia Healey, Arthur Lowe, John Sharp, Anthony Hopkins.

1967   *The Singing Lesson/Raz, Dwa, Trzy.* 20 mins. Contemporary Films (Warsaw Documentary Studios). PHOTOGRAPHY: Zygmunt Samosiuk. EDITOR: Barbara Kosidowska. ASSISTANT DIRECTOR: Joanna Nawrocka.

1968   *If . . . .* 112 mins. Paramount (A Memorial Enterprises Film). SCRIPT: David Sherwin (from the original script *Crusaders* by David Sherwin and John Howlett, written 1958–60). PRODUCERS: Michael Medwin and Lindsay Anderson. PHOTOGRAPHY: Miroslav Ondricek. CAMERAMAN: Chris Menges. PRODUCTION DESIGNER: Jocelyn Herbert. EDITOR: David Gladwell. ASSISTANTS TO THE DIRECTOR: Stephen Frears, Stuart Baird. MUSIC: Marc Wilkinson. CAST: Malcolm McDowell, David Wood, Richard Warwick, Robert Swann, Christine Noonan, Arthur Lowe, Robin Askwith, Peter Jeffrey, Graham Crowden, Brian Pettifer.

Home. 90 mins. New York Public Television (Channel 13). As produced at The Royal Court, Apollo Theatre, and Morosco Theatre, New York. SCRIPT: David Storey (from his play). EXECUTIVE PRODUCER: Jac Venza. MUSIC: Alan Price. CAST: John Gielgud, Ralph Richardson, Dandy Nichols, Mona Washbourne, Warren Clarke.

1973   *O Lucky Man!* 174 mins. Warner Bros. (A Memorial-Sam Film). SCRIPT: David Sherwin (based on an idea by Malcolm McDowell). PRODUCERS: Michael Medwin, Lindsay Anderson. PHOTOGRAPHY: Miroslav Ondricek. PRODUCTION DESIGNER: Jocelyn Herbert. EDITOR: David Gladwell. MUSIC AND SONGS: Alan Price. CAST: Malcolm McDowell, Ralph Richardson, Rachel Roberts, Arthur Lowe, Helen Mirren, Vivian Pickles.

1974   *In Celebration.* 131 mins. Ely Landau's Organisation (London)/Cinevision (Montreal). SCRIPT: David Storey (based on his play). PRODUCER: David Landau. EXECUTIVE PRODUCER: Otto Plaschkes. PHOTOGRAPHY: Dick Bush. PRODUCTION DESIGNER: Alan Witny. EDITOR: Russell Lloyd. MUSIC: Christopher Gunning. CAST: Alan Bates, James Bolam, Brian Cox, Bill Owen, Constance Chapman.

1978   *The Old Crowd.* 90 mins. LWT. SCRIPT: Alan Bennett. PRODUCER: Stephen Frears. CAST: Isabel Dean, Peter Jeffrey, Jill Bennett, Rachel Roberts, Valentine Dyall, Frank Grimes.

1982   *Look Back in Anger.* 100 mins. Taped record of the stage production by Ted Craig at Roundabout Theatre, New York. PRODUCER: Church Braverman, and Don Boyd. CAST: Malcolm McDowell, Lisa

Barnes, Fran Brill.

Britannia Hospital. 116 mins. EMI Films and General Productions. SCRIPT: David Sherwin. PRODUCERS: Davina Belling, Clive Parsons. PHOTOGRAPHY: Mike Fash. PRODUCTION DESIGNER: Morris Spencer. EDITOR: Michael Ellis. MUSIC: Alan Price. CAST: Malcolm McDowell, Leonard Rossiter, Graham Crowden, Joan Plowright, Jill Bennett, Marsha Hunt, Brian Pettifer, Vivian Pickles.

1986   *Foreign Skies.* Documentary/Pop video for *Wham!* on their tour of China. PRODUCER: Martin Lewis. Never completed: it was argued that there were too many shots of China and too few of the group. Director Andy Morahan, and Producer Strath Hamilton, took over.

Free Cinema 1956–? An Essay on Film. 60 mins. In *British Cinema: A Personal View* (Thames). Producers: David Gill, Kevin Brownlow.

1987   *The Whales of August.* 91 mins. Alive Films/Circle Associates/In association with Nelson Entertainment. SCRIPT: David Berry (based on his play). PRODUCERS: Carolyn Pfeiffer, Mike Kaplan. PHOTOGRAPHY: Mike Fash. PRODUCTION DESIGNER: Jocelyn Herbert. EDITOR: Nicolas Gaster. MUSIC: Alan Price. CAST: Bette Davis, Lillian Gish, Vincent Price, Ann Southern, Harry Carey Jnr.

1988   *Cinema Masterclass.* 60 mins. Section on *My Darling Clementine* for Channel 4.

1989   *Glory! Glory!.* 195 mins. (Mini-series.) Home Box Office. SCRIPT: Stan Daniels. EXECUTIVE PRODUCERS: Bonny Dore, Leslie Greif. PRODUCERS: Stan Daniels, Seaton McLean. PHOTOGRAPHY: Mike Fash. EDITOR: Ruth Foster. MUSIC: Steve Tyrell. CAST: Barry Morse, Richard Thomas, Ellen Greene, James Whitmore.

STAGE PRODUCTIONS DIRECTED BY LINDSAY ANDERSON

1957   *The Waiting of Lester Abbs,* by Kathleen Sully. Royal Court Theatre (Sunday night production without décor).

1959   *The Long and the Short and the Tall,* by Willis Hall. Royal Court Theatre.

Progress to the Park, by Alun Owen. Royal Court Theatre (Sunday night production without décor).

Sergeant Musgrave's Dance, by John Arden. Royal Court Theatre.

1960   *The Lily White Boys,* by Harry Cookson with songs by Christopher Logue. Royal Court Theatre.

Billy Liar, by Keith Waterhouse and Willis Hall. Cambridge Theatre.

Trials by Logue, by Christopher Logue. Royal Court Theatre.

1961   *The Fire Raisers,* by Max Frisch. Royal Court Theatre.

1963   *The Diary of a Madman,* by Gogol (adapted by Richard Harris and Lindsay Anderson). Royal Court Theatre.

1964   *Andorra*, by Max Frisch. National Theatre.

Julius Caesar, by William Shakespeare. Royal Court Theatre.

1966   *The Cherry Orchard*, by Anton Chekhov. Chichester Festival Theatre.

Inadmissible Evidence, by John Osborne. Contemporary Theatre, Warsaw.

1969   Took over artistic direction at the Royal Court Theatre with William Gaskill and Anthony Page.

The Contractor, by David Storey. Royal Court Theatre.

In Celebration, by David Storey. Royal Court Theatre.

1970   *Home*, by David Storey. Royal Court Theatre.

1973   *The Changing Room*, by David Storey. Royal Court Theatre.

1974   *The Farm*, by David Storey. Royal Court Theatre.

Life Class, by David Storey. Royal Court Theatre.

Early Days, by David Storey. National Theatre.

1975   *What the Butler Saw*, by Joe Orton. Royal Court Theatre.

The Bed before Yesterday, by Ben Travers. The Lyric.

The Seagull, by Anton Chekhov. The Lyric.

1977   *The Kingfisher*, by William Douglas Home. The Lyric.

1981   *Hamlet*, by William Shakespeare. Theatre Royal, Stratford East.

1985   *The Playboy of the Western World*, by Sean O'Casey. Edinburgh Festival, Riverside.

The Cherry Orchard, by Anton Chekhov. Haymarket Theatre.

1987   *Holiday*, by Philip Barry. Old Vic.

1989   *The March on Russia*, by David Storey. National Theatre.

OTHER WORK

1952   Producer of, and acted in, James Broughton's film *The Pleasure Garden*.

1955   Organized a season of John Ford's films at the National Film Theatre.

1956   Supervising editor of Lorenza Mazzetti's *Together*.

1956–9 One of the originating members of Free Cinema.

1958   Worked on the Nuclear Disarmament film *March to Aldermaston*.

He has made commercials for Mackeson's Stout, Kellogg's Cornflakes, Alcan Foil, Rowntree's Fruit Gums, amongst others; but none recently. He has also been narrator on numerous programmes — most recently the following.

1987   *The Arts and Glasnost Cinema*, one of the 'Omnibus' series.

Buster Keaton — A Hard Act to Follow, a three-part documentary by

Kevin Brownlow and David Gill.
1988   *Ingmar Bergman: The Magic Lantern*. Thames TV.

AS ACTOR

1952   *The Pleasure Garden*. Director: James Broughton.
1966   *Martyrs of Love*. Director: Jan Nemec (Czech film).
1967   *About 'The White Bus'*. Director: John Fletcher (as himself).
1968   *Inadmissible Evidence*. Director: Anthony Page (for TV).
1981   *Chariots of Fire*. Director: Hugh Hudson.

BIBLIOGRAPHY

Sequence, Winter 1947. 'Angles of Approach'.

Sequence, Spring 1948. 'A Possible Solution'. Anderson surveys contemporary film production.

Sequence, Autumn 1948. 'Creative Elements'.

Sequence, Spring 1949. 'British Cinema: The Descending Spiral'.

Sequence, Autumn 1949. 'The Films of Alfred Hitchcock'.

Sequence, Summer 1950. 'They were Expendable and John Ford', by Anderson.

Sequence, Autumn 1950. 'The Director's Cinema?', by Anderson.

Sequence, New Year 1951. 'Goldwyn at Claridges'. Interview with Samuel Goldwyn.

Sequence, New Year 1952. 'The Quiet Man'. Anderson interviews John Ford.

Lindsay Anderson, *Making a Film: Secret People* (Allen & Unwin, 1952).

Sight and Sound, Apr.–June 1952. Review of Anderson's book *Making a Film: Secret People*.

Sight and Sound, Apr.–June 1954. 'Only Connect: Some Aspects of the Work of Humphrey Jennings', by Anderson.

Sight and Sound, Oct.–Dec. 1954. Anderson reviews French critical magazines.

Sight and Sound, Jan.–Mar. 1955. Anderson attacks the last sequence of *On the Waterfront* for being hypocritical and unrealistic.

Sight and Sound, Summer 1955. Article by Anderson on the Cannes Film Festival.

Sight and Sound, Summer 1956. Anderson reports on the Cannes Festival.

Sight and Sound, Autumn 1956. 'Stand Up, Stand Up'. Article on film criticism by Anderson.

Sight and Sound, Winter 1956–7. 'Notes from Sherwood'. Anderson reports on the filming of his episodes for the *Robin Hood* series for television.

Declaration, 1957. 'Get Out and Push', by Anderson.

Sight and Sound, Summer 1957. Anderson reports on the Cannes Film Festival, and on American films made by TV directors.

Sight and Sound, Winter 1957–8. 'Two Inches off the Ground'. Article by Anderson on Japanese cinema.

Sight and Sound, Autumn 1958. 'The Critical Issue'. Discussion between Anderson, Rotha, Wright, and Houston on films.

The Times, 15 Apr. 1959. 'English Theatre in a Philistine Society', by Anderson.

International Theatre Annual, 1961. 'Pre Renaissance', by Anderson.

Transatlantic Review, Spring 1962.

Films and Filming, Feb. 1963. 'Sport, Life and Art'. Anderson interviewed on the making of *This Sporting Life*, and his career to date.

MFB (351). Credits of Anderson; dates of all film and stage productions, critical writings, and other work.

Film Quarterly, Summer 1964. Interview with Anderson.

The Times, 15 Dec. 1964. 'No Nonsense about Shakespeare'. Anderson on directing Julius Caesar at the Royal Court, and the pros and cons of cutting and adding dialogue to Shakespeare.

Tulane Drama Review, Autumn 1966. Interview by Paul Gray.

Financial Times, 17 Jan. 1968. 'The Anderson Eye'. David Robinson on Anderson's documentaries to date including *The White Bus* and *The Singing Lesson*.

Lindsay Anderson, Preface to the script of *If . . .* (Lorimar, 1969).

Elizabeth Sussex, *Lindsay Anderson* (Praeger, 1969). Anderson's biography. Interesting but dated.

Time Out, 22 Aug.–5 Sept. 1970. Interview with Anderson, and biographical details.

Interview, No. 1, 1970. Anderson talks to Glenn O'Brien and Paul Morrissey about his career, films, and theatre work.

Cinema (USA), Spring 1971. 'John Ford, by Lindsay Anderson'.

Cinema Journal, Spring 1972. A survey of the British cinema and documentary movement with specific reference to Anderson.

Film Dope, Dec. 1972. Biofilmography.

The Times, 21 Apr. 1973. 'Stripping the Veils Away'. David Robinson on *O Lucky Man!*

Cinema TV Today, 26 May 1973. Report from Cannes: Anderson gives his views on British directors and film production.

Sight and Sound, Summer 1973. Anderson's change of attitude over the last sixteen years is examined with particular reference to *O Lucky Man!*

Cinema TV Today, 19 Oct. 1974. Anderson refers to the gloomy state of the British film industry.

Millimeter, Jan. 1975. Interview in which Anderson discusses his long and varied career.

Montage, Winter 1975. Articles on Anderson's career, films, and his work as a film critic.

The Times, 11 June 1976. David Robinson on *In Celebration*.

Sunday Times, 13 June 1976. Dilys Powell praises *In Celebration*.

Sunday Times, 31 July 1976. 'Our Film Disgrace'. Anderson on the need for an investigation of the British Film Institute.

Roughcut Film, Summer 1976. Interview in which Anderson gives his views on auteurism, film critics, the BFI, new directors, *If . . .*, *O Lucky Man!*, and playwrights.

Screen International, 13 Aug. 1977. Pictorial coverage of reception for Lindsay Anderson at the NFT which was holding a season of his works.

NFT Booklet, Aug. 1977. 'British Cinema. Part 3: Lindsay Anderson'. On screening a season of his films.

Moving Target, Winter 1977–8. Interview with Anderson on working in cinema and theatre, post-war British cinema, the relationship between the director and the actor, and the progression of his films. Transcript of BBC Radio Brighton Arts Interview.

Show Biz, 25 Apr. 1980. Brief note on his role in the film *Chariots of Fire*, and his directorial activities at the National Theatre.

Lindsay Anderson, *About John Ford* (Plexus, 1981). Anderson's seminal work on John Ford.

Guardian, 2 Mar. 1981. 'At the Root of the Cinema's Problems is a Critical Betrayal'. Anderson on film critics.

Radio Times, 28 Mar.–3 Apr. 1981. Anderson talks about O *Lucky Man!*

Allison Graham, *Lindsay Anderson* (Twayne Publishers, 1981). Biography.

Daily Telegraph, 15 May 1982. 'O Plucky Dissident', by Robin Stringer.

Screen International, 29 May–5 June 1982. Interview on *Britannia Hospital*.

Time Out, 4–10 June 1982. Profile of Anderson with his own comments on his work, the British reaction to it, and *Britannia Hospital*.

Television Today, 10 June 1982. Details of a planned documentary about Anderson and his making of *Britannia Hospital*.

Creative Review, July 1982.

Jeffrey Richards and Anthony Aldgate, *Best of British* (Basil Blackwell, 1983). Including an essay on *If . . .*

Ann Lloyd and David Robinson (eds.), *Movies of the Sixties* (Orbis, 1983).

Cineaste, No. 4, 1983. Interview on *Britannia Hospital*, the Free Cinema movement, and Anderson's position with the critics.

Alexander Walker (ed.), *No Bells on Sunday: The Journals of Rachel Roberts* (Pavilion, 1984). Especially 55–9, 101–2, 131–2, on her involvement in *This Sporting Life*, O *Lucky Man!*, and *The Old Crowd*.

Film News (Canada), No. 15, 1984. Report on an evening hosted by Anderson at the Ontario Film Theatre, in which he talked about his films and introduced extracts.

Guardian, 7 May 1984. Anderson discusses the problems of balancing independence with conformity with special reference to Eastern European Cinema. He also discusses *A Singing Lesson*.

Cinema Canada, June 1985. Interview with Anderson about his films.

Mail on Sunday, 1 Dec. 1985. 'The Great Wail of China'. Fiasco surrounding Anderson's documentary video of Wham! on tour in China.

Televisual, Mar. 1986. Comment on Anderson's contribution to the Thames TV series *British Cinema: Personal View*.

American Cinematographer, Oct. 1987. Interview on *The Whales of August*.

Films and Filming, Mar. 1988. Interview on the making of *The Whales of August*.

Sunday Telegraph, 27 Mar. 1988. 'Odd Man out in British Cinema'. Daniel Farson on *The Whales of August*.

Listener, 19 May 1988. Letter from Anderson answering recent criticisms made about the BBC2 Film Club series.

Time Out, 25 May–1 June 1988. Anderson, Davis, and Gish talk about working with each other on *The Whales of August*.

Sight and Sound, Spring 1989. '35 Days in Toronto'. Gerald Pratley interviews

Anderson on his first mini-series *Glory! Glory!*, and the differences between shooting for TV and the cinema.

BFI Special Collection, n.d., on Anderson.

2 Sir Richard Attenborough

Portrait of Sir Richard Attenborough. Photo by Terry O'Neill.

SIR RICHARD
ATTENBOROUGH

IT is hard to find anyone who has committed more of himself to the British film industry than Richard Attenborough. He was one of Britain's first truly cinematic actors, for many years an enterprising producer, and then director of seven medium to high budget feature films, three of which he co-produced or produced himself. As well as all this he has been deeply committed to numerous film industry organizations; he is chairman of the BFI, Channel 4, the British Screen Advisory Council, the European Script Fund, vice-president of BAFTA, and from 1970 to 1981 governor of the National Film and Television School. Moreover, his involvement stretches well into other media. He is chairman of Capital Radio, the Royal Academy of Dramatic Art, and the Duke of York Theatre; he was even a director of Chelsea Football Club from 1969 to 1982; has been Pro-Chancellor of Sussex University since 1970; from 1976 to 1982 a trustee of the Tate Gallery; has had significant involvement in a wide variety of charities; and was active in the founding of the Social Democratic Party. He was awarded the CBE in 1967, knighted in 1976, and has also received great, international honours including the Martin Luther King Jr. Peace Prize and the Padma Bhusan in India (the highest honour for a non-Indian). He has been appointed a Chevalier Légion d'Honneur and Commandeur des Arts et des Lettres in France, and has been made a Goodwill Ambassador by UNICEF. Not surprisingly, what has enabled him to do all these things is an insatiable energy—'I'm bored with leisure. I'm only really happy working,'[1]—combined with tenacity, persistence, and an inability to resist a challenge—'I love them. They're part of my life. Professionally I've never ever wanted to play safe and indeed I don't believe I have.' His brother, the famous naturalist Sir David Attenborough, describes how 'he gets totally absorbed in his projects; eating, breathing and sleeping them to an extent that is claustrophobic.'[2] But with the help of four secretaries and a very tight diary he somehow manages to fulfil his manifold responsibilities. Through all this he hopes to maintain his determination while avoiding becoming ruthless. 'I am uncomfortable in the company of sharp elbows, but that doesn't mean that I personally

[1] *Daily Express*, 1 Apr. 1965.
[2] *The Times*, 22 Nov. 1982.

lack ambition. I don't really know if I am ruthless, but I'm hugely determined in achieving ends and objectives which I have set myself or have been set for me.'[3] In fact, many in the film industry have been surprised by the degree to which he has led a long and highly fruitful career without being in any way corrupted. John Boulting[4] commented, 'Everyone is very fond of Dickie, and he really is as decent as he appears to be. He has managed to survive in a profession that can be very cynical and ruthless and has done so without becoming defiled.'[5]

Attenborough has the extrovert manner of an actor able to play any role. But he is aware of this: 'I know I'm flamboyant. I express my feelings vehemently. They're usually expressed in emotional terms and a number of people find that somewhat boring, even nauseating.'[6] He has even described himself as 'an old male Mary Poppins . . . hopelessly optimistic'. But many people—certainly those who know him—find him totally charming. Jake Eberts[7] describes him as being 'really like a vicar—you go to him in a crisis, or to make "confession",' and to William Goldman,[8] 'Richard Attenborough is by far the finest, most decent human being I've met in the picture business.'[9]

Attenborough's interest in drama actually arose at a very early age. He recalls how, in 1935, when he was 12, he tried to get his younger brother David to help him stage his first show.

Unscrupulously, I cashed in on his fondness for animals . . . Naturally I was to be the script-writer, producer and star but Dave was needed to run errands, bring up the curtain and play supporting parts. He refused absolutely until I played my trump card: all the proceeds were to go to the RSPCA.[10] We bought penny notebooks (haggling for them in the market), and then resold them with pencils attached by ribbon for threepence a piece. This earned a profit of ten shillings which we paid as a deposit for the hire of the Hall.[11]

Determined to act, Attenborough won the Leverhulme scholarship and

[3] *Guardian*, 6 Apr. 1987.

[4] John Boulting, and his brother Roy, were the producers and directors of many films in the 1940s and 1950s. John directed *Journey Together* (1945), *Brighton Rock* (1947), and *I'm All Right Jack* (1959). Attenborough acted in many of their films.

[5] *The Times*, 22 Nov. 1982.

[6] *Guardian*, 6 Apr. 1987.

[7] The founder of Goldcrest Films in 1977, and Chief Executive until December 1983. He returned to the post in 1985.

[8] Established American script-writer. His work includes Attenborough's *A Bridge Too Far* (1977) and *Magic* (1978); Schlesinger's *Marathon Man* (1977); *Butch Cassidy and the Sundance Kid* (1969, dir. George Roy Hill); and *All the President's Men* (1983, dir. Alan Pakula).

[9] William Goldman, *Adventures in the Screen Trade* (Futura, 1985).

[10] Royal Society for the Prevention of Cruelty to Animals.

[11] Interview with the authors.

so went to RADA[12] at the age of 17, rather than to university. It has become one of his greatest regrets that he never pursued the opportunity for further learning—heightened because both his brothers went to Cambridge and his father was, in fact, the principal of Leicester University College.

Attenborough's parents were both highly dynamic people, with strong social consciences. His mother took under her wing fifty Basque refugees from the Spanish Civil War. His father was chairman of a Midlands committee for getting Jewish refugees out of Germany, and the family looked after two Jewish girls for the duration of the war. Both parents were active in the Labour Party and he remembers the book shelves of their house being filled with the orange jackets of the Left Book Club.

For Attenborough, these various influences were to combine to create desires which have found release through directing films.

I am anything but an intellectual, but I have an enormous desire to communicate. I've always wanted to, and I can't write, I can't paint, I can't compose, but what I can do is act a bit. However, if you're directing you have the opportunity to say, 'That is what I believe. This is my credo'.[13]

But, having got into RADA, it was still twenty-five years before he was to make his own films, and so satisfy this desire.

He began his acting career while he was at RADA. He was spotted while working in a local repertory company[14] and was put straight into Noel Coward's film, co-directed by David Lean,[15] *In Which We Serve* (1942) as a young sailor whose nerve fails under fire. The following year he joined the RAF to train as a pilot, but before gaining his wings, was seconded to the Royal Air Force Film Unit at Pinewood. There, in 1945, he acted with Edward G. Robinson in John Boulting's *Journey Together*, which proved a pivotal influence on the formation of his acting style. While working on the film he also watched Humphrey Jennings[16] edit a number of documentaries—'He unquestionably influenced everyone working at Pinewood at that time.' Attenborough's first really great role was as Pinky in John Boulting's *Brighton Rock* (1947). There was a story that when the film was launched in the States, the New York distributor wired frantically, 'Attenborough's name too big for banners.' The Boulting brothers, 'British to the core wired back, "Get bigger

---

[12] The Royal Academy of Dramatic Art, the premier acting academy in Britain.
[13] *Movie Maker*, Sept. 1972.
[14] At the Intimate Theatre, Palmers Green, London.
[15] This was Lean's directorial début. The partnership with Noel Coward continued with *This Brief Life* (1944), *Blithe Spirit* (1945), and *Brief Encounter* (1945).
[16] Distinguished British documentarist. Responsible for a series of World War II films which sensitively recorded the mood of the times. Films included: *London Can Take It* (1940), *Listen to Britain* (1941), and *Fires Were Started* (1943).

banners".'[17] But as his film acting progressed throughout the 1950s he felt increasingly typecast: 'OK they were fun days alright—no denying that. But from a *creative* point of view, not all that satisfying, I'm afraid.'[18] So in the late 1950s Attenborough turned producer. Not only would he win independence for himself as an actor, but he would also move his career in a new and creatively exciting direction.

While sheltering from a sandstorm making *Sea of Sand* (1958), Michael Craig,[19] Guy Green,[20] and Attenborough decided to make *The Angry Silence* (1960). Attenborough starred in and co-produced it with his friend and writer, Bryan Forbes.[21] 'I discovered that I adored producing and planning; the administration required to call people together and solve problems.'[22] His role as a producer was characterized by great imagination. To get *The Angry Silence* off the ground, he got the budget down to a practical level by persuading most of those involved to take a percentage of the profits rather than payment. Similar creative thinking was applied to *The League of Gentlemen* (1960). In a clever attempt to gain independence on a project which was not immediately commercial, Attenborough joined with Forbes, Basil Dearden,[23] Michael Relph,[24] Jack Hawkins,[25] and Guy Green to form Allied Film-Makers, their own independent production/distribution company. Over the next few years, Attenborough continued to work closely with Forbes, producing or co-producing *Whistle Down the Wind* (1961), *The L-Shaped Room* (1962), and *Seance on a Wet Afternoon* (1964).[26]

It was at this stage that he was offered films to direct himself. He became less interested in producing,[27] and even acting became increasingly unimportant to him. He has not appeared on stage since 1958 and has not acted in a film since 1979. But his desire to direct was not inspired by a fascination with the actual medium—'The cinema as an

[17] Attenborough in *Sunday Telegraph*, 22 Aug. 1965.

[18] Interview with the authors.

[19] British actor of the 1950s–1970s. Starred in *The Angry Silence* (1960), *The Royal Hunt of the Sun* (1969).

[20] British director of photography and later director. Photographed Lean's *Great Expectations* (1946) and directed *The Angry Silence* (1960). He later produced extremely disappointing larger budgeted work in Hollywood, such as *The Magus* (1968).

[21] Director, producer, screen-writer, actor. Worked with Attenborough on a series of films in the early 1960s. From 1969 to 1971 he was head of EMI. Since then he has worked frequently in America.

[22] *Sunday Telegraph Magazine*, 13 Nov. 1983.

[23] British director who, between the late 1940s and late 1960s, formed a writing-producing-directing partnership with Michael Relph. Films include: *Dead of Night* (1945) and *Victim* (1961).

[24] British producer-director. Worked in collaboration with Dearden.

[25] Respected British actor who has starred in numerous films by British directors, including those of Reed, Lean, and Attenborough.

[26] Attenborough considers his performance in this film to be the best of his career.

[27] He has, in fact, only produced three films since then: *Oh! What a Lovely War* (1969), *Gandhi* (1982), and *Cry Freedom* (1987).

Attenborough as producer, with director Bryan Forbes, on the set of *Whistle Down the Wind*, 1961. Courtesy of the Rank Organisation plc.

end unto itself, is not enough. It cannot afford to be narcissistic'[28] . . . 'I don't make movies because I get sexual satisfaction from pulling film through my fingers or because I am mesmerized by the pyrotechnics of it'[29]—nor is it an urge to intellectualize. There are however, three discernible desires which seem to drive Attenborough to want to direct films.

The first relates back to the values engendered in him by his parents.

The whole atmosphere, and ambience of my house was of social awareness and social concern. It became part of my very being—like breathing. That anyone isn't concerned with these sorts of issues or doesn't want to do something about it, seems very odd to me. By making movies, I can express the feelings which I've felt since I was a kid. I wish to make a plea to the strong for the weak. I found it partially possible as an actor, but then you're in large measure an interpreter, not an originator. As director it goes much deeper.[30]

He has even admitted that, 'I can't think of one illiberal idea I have. What a pain in the arse I am.'[31] He feels that making a movie is 'like going into a monastery. It's like taking up a religion. You have to have an absolutely sacred dedication to the job in hand.'[32]

Secondly and related to this is his desire to communicate to a wide audience. 'In some quarters "popularise" is a dirty word. But it is a banner under which I will gladly sail.'[33] 'Cinema is the greatest means of communication ever invented and I am a compulsive communicator. I love conveying ideas and stimulating reactions from them. There's no point making a statement in the cinema if no one's watching.'[34] He feels the importance of reaching a large audience is intensified by economic reasons:

We are involved with vast sums of other people's money. If you write or paint then you involve your own time and hence money, and it doesn't matter whether people like it or not. But this doesn't apply to cinema. For the French and Italians things are different; but here we have to compete within the American market.[35]

His third reason for wanting to direct, was his twenty-year obsession with making a film of the life of Gandhi. In 1979 he confessed, 'actually

---

[28] *Sunday Telegraph*, Aug. 1965.
[29] Interview with the authors.
[30] Interview with the authors.
[31] *Sunday Express Magazine*, 15 Nov. 1987.
[32] *Films and Filming*, June 1969.
[33] *Sunday Telegraph*, 29 Aug. 1965.
[34] Interview with the authors.
[35] *Films and Filming*, June 1969.

the truth is that I don't really want to be a director at all. I just want to direct that film.'[36] In the 1950s David Lean had his eyes on the project which had been brought to him by an Indian, Motilal Kothari, but then he got involved with *Lawrence of Arabia* (1962), and Kothari took it to Attenborough. On reading about Gandhi,

I gradually felt, agnostic though I am, that somebody was talking to me about spiritual values—certainly moral values—in a manner which seemed to me accessible, and I felt for the first time deeply touched by what a man had to say, and what a man was prepared to do . . . I knew nothing of India, nothing of Gandhi other than a school-boy would know. But although I had never direc-ted, when I read this biography, I was absolutely bowled over.[37]

Not surprisingly it was a subject which had interested many other direc-tors. Besides Lean and Attenborough, there had been Pascal,[38] Preminger,[39] and Michael Powell.[40] However, it was only Attenborough who maintained the struggle to make this apparently uncommercial movie, turning down many other offers of work, both as a director and actor, to do so.

Attenborough's principal contribution to his films is in terms of the vision he brings to the project—in the choice of subject matter, and broad approach. All his films have a strong conventional narrative thread which he hopes will get the audience hooked:

Narrative is what interests me. The dramatic dynamic thrusts you forward. You must be teased into wanting to know what will happen next. I want the suspen-sion of disbelief in the films to be total. If the tension of the story-telling is right, you can present the most ravishing, evocative moments in the cinema.[41]

Naturally he has chosen to focus on topics of social concern.

Gandhi, and *Oh! What a Lovely War* are the two films most precious to me. The latter was like a first child . . . But *Cry Freedom* is also of enormous importance . . . *Oh! What a Lovely War* was a massive anti-war statement, and so was *A Bridge Too Far* in terms of the stupidities, the arrogance and the inefficiencies of the Second World War. *Gandhi* was what it was, and had its

---

[36] *Guardian,* 5 Feb. 1979.

[37] *Stills,* No. 5, 1982.

[38] Hungarian producer-director who came to Britain in the 1930s, won the esteem of Bernard Shaw, and was entrusted with filming several of his plays. He produced *Pygmalion* (1938), and directed *Caesar and Cleopatra* (1945).

[39] Films include: *Fallen Angel* (1945), *Angel Face* (1952), and *Rosebud* (1974).

[40] Films include: *The Life and Death of Colonel Blimp* (1943), *A Matter of Life and Death* (1946), and *Black Narcissus* (1946). Frequently worked in partnership with Emeric Pressburger.

[41] Interview with the authors.

own political content, but I suppose *Cry Freedom* is the most overt, and uncompromising political statement I've made yet.[42]

Biography has been another fascination. It is there in *Young Winston* (1972), *Gandhi*, *Cry Freedom* to an extent, as well as his current two projects on Charlie Chaplin and the revolutionary thinker Tom Paine. Yet it is an interest which had shaky beginnings. Speaking of *Young Winston*, about the early life of Churchill, he commented, 'I think the very content, the very biographical subject has such inherent, built in restrictions that I'm not at all sure that in cinematic terms, they're ever totally successful. The truth is biographies don't work very well in cinema.'[43] But the success of *Gandhi* altered his perspective, and soon after he said, 'I'm not interested in detective stories or love stories. I'm fascinated by figures who have changed, and who still are changing, our society.'[44] The biographical film is a genre which complements his interest in narrative cinema and his desire to convey a message. His biographies are not complex psychological explorations and are in some ways rather two dimensional. But, being based on fact, he sees them as excellent vehicles for his aims:

I don't actually enjoy fictional reading nearly as much as I do factual subject matter. I love biography with a background of political and social circumstances. I feel now that if there are things that I want to say, questions which I want to illuminate, doing it on a fictional level is nothing like as effective as biography. Biography is a means to an end.[45]

During the making of his films, Attenborough, naturally enough, addresses most of his attention to working with the actors. 'What I love in the actual execution of a film is really working with the actors. They are what I really care about.'[46] But as he points out, 'First and foremost, the vital question arises in terms of casting. The chances are that, if they're good at their job and have done their homework, it is total folly to restrict them by telling them precisely how the scene must be played.'[47] His films have been characterized by some very daring casting of lead actors; Simon Ward in *Young Winston*, who previously had little cinematic experience and was hitherto associated with vaguely effeminate juvenile roles; Anthony Hopkins in *Magic*, despite pressures on Attenborough to use a star like Hoffman, De Niro, or Jack Nicholson; or Ben Kingsley as Gandhi, who was largely unknown and had primarily

[42] *London Evening Standard Magazine*, Nov. 1987, 'My Office and I: Sir Richard Attenborough', by Steve Clarke.
[43] *Dialogue on Film*, Feb. 1973.
[44] *The Times*, 20 Apr. 1983.
[45] Interview with the authors.
[46] Interview with the authors.
[47] *Dialogue on Film*, Feb. 1975.

Attenborough's 'intimate epic', *Young Winston*, 1972. © Courtesy of Columbia Pictures Industries Inc.

a theatrical background. When directing the actors he feels that he has a natural advantage. 'There is, in a way, an extraordinary bond between actors. I think they are quite proud of the fact that an actor is directing them.'[48] Anne Bancroft who played Lady Churchill in *Young Winston* agrees, 'Being an actor himself, he is the soul of understanding. I had such confidence in him and he gave me such confidence, that I was able to play scenes which at one time I never believed I could master.'[49] Attenborough feels that it is crucial to build up a rapport with his cast. 'You must have an affair with your actors. You really must. They must adore you, and you must adore them, and you must want to work with them, and want to create with them.'[50] His philosophy is that no actor should ever be left floundering, or be made to feel foolish. 'You must never place them in a position where they turn to you in an embarrassed way and say "What do you want me to do now?"'[51] Simon Ward believes that 'a director has to be a sounding board for ideas, and Dickie is the nearest I have ever come to having myself out there, looking at me.'[52]

Visually, Attenborough's films are simply shot, lacking great flair although they are meticulously planned. Something of his approach is shown in his statement that all art forms 'demand "order", although there is a vast spectrum into which you may choose'.[53] He sticks very closely to the script and admits that his style is over-disciplined. 'I'm over-ordered—that's been my handicap as a film-maker. I'd be a more exciting director if I could hang loose a bit more, grab for more than I can comfortably hold.'[54] There is no real consistency of style in his films. It varies according to who he is working with, especially the script-writer and cameraman, and range from his most dynamic film, *Oh! What a Lovely War*, to the sober *Magic*. All, however, show a fondness for camera movement—tracking shots and the use of the crane— although he hopes to keep these unobtrusive.

Attenborough also has a reputation as the maker of epic films. Indeed, in terms of budget, scale, and ambition, he follows in David Lean's footsteps. To achieve this, he needs a dedicated crew working in close collaboration, and relies heavily on their creative input. He also needs to retain a firm control, and always praises the first assistant director, whose job it is to ensure that this is achieved.[55] In terms of the scale of

[48] Interview with the authors.
[49] *Films Illustrated*, July 1972.
[50] *Dialogue on Film*, Feb. 1975.
[51] Ibid.
[52] *Films Illustrated*, July 1972.
[53] *Dialogue on Film*, Feb. 1975.
[54] *Sunday Telegraph Magazine*, 18 Nov. 1979.
[55] Especially David Tomlin who has worked on Attenborough's three biggest films.

the project, and the subsequent need for organization, Attenborough's first film could have been a baptism of fire, but for the support of his talented crew and cast.

Having failed to get *Gandhi* off the ground since 1962, Attenborough unexpectedly received a screenplay of the stage musical *Oh! What a Lovely War.*[56] It had been sent to him by his friend and actor John Mills who was working on it with Len Deighton. Attenborough accepted the project because he felt it would be a good learning experience if he was ever going to direct *Gandhi*; and, moreover, 'The pacifism that attracted me to *Gandhi* was epitomized in the screenplay. It rang the same bells for me.'[57] But his immediate problems were not as director but as co-producer; he was expected to raise the funding for the film. To do this it was necessary to assemble an all star British cast prepared to work for low fees. Typically his success was absolute; Laurence Olivier, Dirk Bogarde, John Gielgud, Jack Hawkins, Kenneth More, Michael Redgrave, Ralph Richardson, Maggie Smith, Vanessa Redgrave, and John Mills all agreed. To help him in directing the film, Attenborough was equally successful in obtaining a top crew; Claude Watson as first assistant director, Ann Skinner, a 'brilliant' continuity girl,[58] Gerry Turpin as cameraman, Kevin Connor as editor, and 'the best art director in England at the time', Don Ashton. 'All these people got me through. It was wonderful that they accepted what I asked for.'[59] However, despite all this, and despite the fact that 'I'd been in our business for nearly thirty years, I had no conception of just how great the total involvement would be once I accepted the post.'[60] It was a hugely ambitious film, described as an obituary to the lost generation of the First World War, with a large cast and numerous musical numbers. Its style is flamboyant, capturing the sense of bravado and gaiety with which so many people senselessly entered the war. Attenborough's next film, *Young Winston*, made two years later, was a challenge in a different way.

Churchill, having seen *The Guns of Navarone*,[61] had asked Carl Foreman to make a film of the first part of his autobiography, *My Early Life*, saying, 'show me, warts and all'. Foreman, who scripted and produced

---

[56] The original stage version was written by Joan Littlewood and Charles Chilton.

[57] *American Film*, Mar. 1983.

[58] Ann Skinner has worked extensively with Attenborough. She was the production secretary on *The L-Shaped Room*, *Whistle Down the Wind*, and *Seance on a Wet Afternoon*, and continuity on *Oh! What a Lovely War*, *Young Winston*, and *Magic*. At the same time she worked, in the same capacities, on several of Schlesinger's films. In 1978 she founded Skreba Productions, with the producer Simon Relph and director Zelda Barron. Since then she has co-produced *The Return of the Soldier* (1982), *Secret Places* (1984), and produced *The Kitchen Toto* (1987).

[59] Interview with the authors.

[60] *Action*, Jan.–Feb. 1970.

[61] 1961; directed by J. Lee-Thomson from a script by Carl Foreman.

the film, wanted an English director—'a director with the right sensitivity. As soon as I saw *Oh! What a Lovely War*, I knew that Dickie was my man.'[62] Their approach was not to make 'an adventure film in the accepted sense. If it's not a contradiction in terms, what Carl and I set out to make was an *intimate* epic.'[63] But Attenborough faced a lot of interference from Foreman, and this has soured his feelings about the film. He also found a lot of practical difficulties in balancing the desire for a truthful biography with the demands of cinematic drama. He did, however, tackle the epic scenes showing Churchill fighting in India and the Sudan with daring. But the film failed to get its money back; an important blow for Attenborough's bankability as a prospective director of *Gandhi*.

Attenborough agreed to direct Joseph Levine's[64] *A Bridge Too Far* (1977), after further frustrating and fruitless attempts to get *Gandhi* financed. His long and ultimately bitter involvement with Levine centred around the mogul's frequent, although ultimately hollow, assurances that he was interested in producing Attenborough's great dream. Already he had pulled out from financing *Gandhi* three times, but Attenborough believed that by accepting *A Bridge Too Far* as well as his next film, *Magic*, also produced by Levine, he might change his mind. But to Levine Attenborough was just the right man to make the $25 million *A Bridge Too Far*, with an all star cast.[65] Scripted by William Goldman, it is an account of Operation Market Garden, one of the most ambitious, if ill-fated, campaigns of the Second World War. It was an attempt by Montgomery to force an early German surrender, by dropping Allied forces behind enemy lines and so cutting off their withdrawal from occupied Europe. The film was an immense undertaking, and it is probably true to say that no other director would have had the energy to complete the twenty-three-week shoot:

I was scared stiff nearly all the way through with the logistics of the film; up at 6.30 am; shooting from 8 am to 6 or 7 pm; then I saw the rushes; then I cut the film for about an hour; then an hour at the production office; I would eat at 11 pm, and got to bed at 12–12.30 . . . at the end of filming I just literally collapsed for four days.[66]

Commenting on the military props for the film, 'by the time shooting

[62] *Films Illustrated*, July 1972.
[63] Ibid.
[64] A film distributor turned producer. By the 1960s he had become the sole showman tycoon in an industry run by big corporations. Films produced include: *8½* (1963, dir. Fellini), *The Graduate* (1967, dir. Mike Nichols), *Carnal Knowledge* (1971, dir. Mike Nichols).
[65] Including Dirk Bogarde, James Caan, Michael Caine, Sean Connery, Edward Fox, Elliott Gould, Gene Hackman, Anthony Hopkins, Laurence Olivier, Robert Redford, and Liv Ullman. The Americans received $1 million each except Redford who received $2 million.
[66] *Photoplay*, July 1977.

A logistical nightmare. Attenborough's military epic, *A Bridge Too Far*, 1977. © MGM/UA.

began, we were so well equipped we could have fought Russia,—and won.'[67] Although commercially successful, the film was not as powerful as *Oh! What a Lovely War*. It had some of the same sorts of criticisms of military incompetence, but its impact was reduced partly by its emphasis on action rather than analysis, and partly by its length—two hours fifty minutes.

Magic (1978) suffered from a script, also by Goldman, which was distinctly uncinematic; although it could have made a interesting piece of theatre or a one-hour television film. It explores the disintegrating mind of a young magician and ventriloquist, who imbues his dummy with his sinister, subconscious personality. The idea itself was derivative, first used in *The Great Gabbo* (1929), and later in *The Devil Doll* (1936), episodes of *The Twilight Zone*, and the Redgrave episode of *Dead of Night* (1945). Despite the rather sterile atmosphere of the film, it was successful in America.

Over the years, Attenborough had approached every major film company to try and finance *Gandhi*.[68] He had constantly avoided turning it into a television mini-series or casting an inappropriate star to raise the money. The film was to be made on his own terms. Ultimately, that the principal backer was British—Goldcrest—was fortuitous.[69] Before the production began, a large amount of the money had to go to Levine who had, in 1978, purchased the rights to the project from AVCO Embassy, for $100,000. Levine now demanded $2 million, and 7.5 per cent of the profits or 2.5 per cent of the gross, far more than Attenborough had made from *Magic* and *A Bridge Too Far*. But despite this, Attenborough was overjoyed that the project was finally going ahead when for so many years 'nobody had thought it was a viable proposition. Nobody thought that we could ever gather an audience to see "A little brown man dressed in a sheet, carrying a bean pole," as one American studio head called the Mahatma. Nobody had wanted to know about it.'[70] During the long gestation period, Attenborough's approach to the film, particularly relating to the writer and lead actor, had changed radically. The first script was by Gerald Hanley but Attenborough did not consider it sufficiently cinematic. He then took it to Robert Bolt who wrote two scripts, but he became so unwell that he could not undertake a third. Jack Briley, who wrote the final 187-minute screenplay, became

[67] *Films Illustrated*, July 1977.

[68] Including many in Britain, such as Rank under John Davis, and Thorn-EMI under Barry Spikings.

[69] Jake Eberts raised the $11 million budget through a variety of sources, including the Post Office Super-Annuation Fund, the Miners' Pension Fund, International Film Investors, as well as $3 million from the Indian Film Development Corporation who stepped in after an Indian Maharajah pulled out.

[70] *American Film*, Mar. 1983.

involved in 1979–80. Attenborough's ideas about casting had also changed considerably. He believed that the days when Western actors could be accepted as Indians by using brown make-up were over. But whoever played Gandhi had to be able to age from youth to extreme old age; to play private scenes with delicacy; as well as having a charismatic presence. The problem seemed insuperable. Attenborough could find no Indian actor with the experience and skill. Over the years Alec Guinness, Albert Finney, Peter Finch, Tom Courtenay, John Hurt, even De Niro and Dustin Hoffman were all considered. Eventually Attenborough's son, who worked in the theatre,[71] suggested the relatively unknown Ben Kingsley, who was actually half Indian. He was perfect, and seemed to take on the very appearance of Gandhi himself. The script-writer John Briley was determined on the idea of John Hurt in the lead but later admitted that Kingsley's was the best screen performance he had ever seen. The portrayal of Gandhi was a simple one. Attenborough preferred to concentrate on his achievements rather than his complexities. One of the two cameramen, Billy Williams, commenting on his view of the film said,

I can't do a picture unless I really care about what it is trying to say. One of the most satisfying films I've ever done was *Gandhi*, because I think what Gandhi stood for, and what he tried to teach, should be known by millions of people. So I was absolutely thrilled when Richard Attenborough asked me to photograph it. The film didn't preach, but it put all that across and I'm sure it can stand as a great piece of entertainment as well. Now that was something that I was very proud to be associated with.[72]

Attenborough had to face numerous production difficulties. Besides the logistics of the film and the Indian bureaucracy, two-thirds of the way through shooting Billy Williams slipped a disc and Ronnie Taylor had to take over. Attenborough felt that the style of the film was 'in the tradition of British movies . . . in the genre of Lean, Carol Reed and Fred Zinneman, who works here all the time, in the sense that it was a very simply shot film. There may be those who criticise it as being old fashioned.'[73] But it was, despite almost two decades of struggle, an incredible critical and commercial success. An amount equal to a third of the budget was wisely risked on publicity, and the film in the end took over $130 million at the box office. Attenborough had climbed his Everest. His obsessiveness was vindicated.

Attenborough had several reasons for being involved in the $20

[71] Now the Executive Producer of the Royal Shakespeare Company.
[72] Dennis Schaefer and Larry Salvato, *Masters of Light* (University of California Press, 1984).
[73] *Stills*, No. 5, 1982.

million film of the hit American stage musical *A Chorus Line*,[74] when a
handful of top directors, including Mike Nichols[75] and Sidney Lumet,[76]
had already turned it down for being an impractical project. His experience on *Oh! What a Lovely War* had given him exposure to directing
musicals, and he adored their theatricality, and entertainment value.
'The idea of being part of the world of the broadway musical was
irresistible to me. I absolutely adored it.'[77] But more significantly, his
agent felt that for pragmatic reasons he should avoid being over-associ-
ated with films which had social connotations.

To make this hit musical about the audition of dancers for a show into
a viable, visual film, it really needed to be moved away from its major
location of a theatre. But Attenborough probably rightly felt that this
destroyed the premiss of the musical. He hoped to make it more cinematic by using close-ups of the actors; 'Sitting way back in the stalls you
never get to see the look in these kids' eyes as they succeed or fail, the
beads of sweat, the trembling lips, the fear, as well as the joy of getting
the job.'[78] Yet, Attenborough should really have waited for another
vehicle to prove his diversity of interests because the film was rather
staid, and it is probably fair to say that it was impossible to convert it
into little more than a filmed version of the staged performance.

Attenborough's most recent film, the $22 million *Cry Freedom*, a
passionate attack on South African apartheid, was a return to more
familiar and successful territory. 'I was taught by my father that if you
think something is wrong you don't keep quiet . . . you shout about it.
And racism has got to be one of the world's greatest evils.'[79] In the late
1950s, even before *Gandhi*, Attenborough had worked on a screenplay
about apartheid, *God is a Bad Policeman*, and had, over the years
optioned various scripts on South Africa. Like *Gandhi*, *Cry Freedom*
was scripted by John Briley (from two books by the South African
journalist Donald Woods), photographed by Ronnie Taylor, designed
by Stuart Craig, and produced and directed by Attenborough. Typically,
Attenborough wanted to reach the widest possible audience and wanted
a story which 'won't be depressing, even though it ended in tragedy—
because it is about black people and white people working together
towards the same ends. It is intended to be a message of hope.'[80]
Although the film did have a great impact, and succeeded in heightening

[74] Which was already at an advanced stage of production when Attenborough got involved.
[75] Films include: *The Graduate* (1967), *Catch 22* (1970), *Silkwood* (1983), and *Working Girl* (1988).
[76] Films include: *12 Angry Men* (1957), *Equus* (1977), and *The Verdict* (1982).
[77] *Sunday Times*, 21 Jan. 1986.
[78] Interview with the authors.
[79] *Guardian*, 11 Jan. 1986.
[80] Ibid.

After nearly two decades of struggling, Attenborough finally raised the money to make *Gandhi*, 1985. The different faces of Attenborough as director: above, his work with Ben Kingsley, and below, his concern for the shot. © Courtesy of Columbia Pictures Industries Inc,

people's awareness, it also received criticism from those who wanted a more direct and hard hitting movie.

This is the problem which Attenborough has had to face in all his films; how to balance the urge to get his messages and beliefs across to the widest possible audience with the need to be faithful to the complexities of the issues and people he is focusing on. He has recently secured a three-picture deal with Universal,[81] each film having a budget of $25 million. As the first two projects he has in the pipeline are biographies, with strong social and moral themes, this dilemma will continue to face him in the future. The first is about the great *auteur* Charlie Chaplin,[82] and the second, scripted by Trevor Griffiths,[83] about Tom Paine, the great eighteenth-century egalitarian thinker who wrote *The Rights of Man* and was one of the leaders of the American Revolution. Yet, this problem does not seem immediately soluble. Although no one can doubt the importance of heightening the social conscience of the audience, there is an inherent conflict between the necessity for entertaining popularist films (a consequence of the very high budgets which Attenborough works with, largely dictated by his desire to make period dramas), and the demands of the factual subject matter for sophistication and a strict adherence to the truth.

[81] In return for Universal's acceptance of a screenplay and one principal actor, Attenborough has total control over the 'final cut' of the film.

[82] From an original storyline by Diana Hawkins, scripted by Bryan Forbes.

[83] British socialist writer for stage, film, and television. He scripted *Reds* (1981, dir. Warren Beatty), and *Fatherland* (1986, dir. Ken Loach).

SIR RICHARD
ATTENBOROUGH

Bearing in mind your active involvement in film-making in Britain since the 1940s, we should possibly start our discussion with the British Film Industry.

Well, the difficulty is that I don't think there is a British 'cinema' or film industry as such. I don't think it exists any more. It is a somewhat homogeneous group of people who perhaps make movies in Britain and about Britain.

Do you think it makes sense to talk of a British film culture?

It's very hard to know what a film culture is. Is Parker part of British film culture? I don't know. *The Mission*[1] is a very English film. *Cry Freedom* is a very English film because of its attitudes (its traditional British understatement), its style, and manner. The films of Stephen Frears, some of which I like, some of which I don't like, are essentially of here, of our time. I think it would be sad if we lost that. Tragic. I desperately believe in internationalism, but if it erodes the indigenous moral concepts, or manners, or tastes of a nation it fails. You see it's very difficult with film. It's strange because it's different with British graphic art. You can't think of Blake, Moore, Sutherland, Smith, or Lowry as anything other than British.

The most important influence on British cinema in the 1980s has been Channel 4, hasn't it?

Yes, television in the last few years has largely kept British film alive. When I accepted my position at Channel 4[2] it was only on the condition that the channel would help pay television's debt back to the film industry. Look what's come out of Film on Four. Marvellous.

Do you see any danger of an over-domination of television?

Yes. Film on Four is made with a certain eye to cinema as well as straight television drama. But it is my opinion that this whole move is now in danger of becoming counter-productive. Film-makers are starting to realize that the only way to get a film made is to interest Channel 4,

---

[1] Made in 1986. Directed by Roland Joffe and produced by David Puttnam.
[2] Initially as deputy chairman, and since 1987 as chairman.

British Screen (who depend greatly on Channel 4), and the BFI, and therefore only set out to make something which falls within their budget and their concept. Young film-makers, almost subconsciously, only contemplate material with a view to that television market. We are no longer fighting for the larger production which is so essential to maintain a cinema audience. It erodes the fundamentals of what cinema is about—going out to the 'cinema'. Those films have a place; they have a market; but if they are the only apples on the stall then we are in trouble.

What do you think that television drama can do better than cinema?

At the end of the day one is comparing apples and pears even if the artistic input is very similar. There are requirements for the cinema, particularly if you move into Panavision, of subtlety of lighting, make-up, and performance which are necessary if your eye is six foot across, but which is not needed if your eye is a pin-prick on a twenty-eight-inch screen. This, of course, makes the budget much cheaper for television. Fiction on television also has an immediacy which is hard to achieve in cinema. *Boys from the Blackstuff*[3] is a very good example. Its very lack of supposed polish, its resemblance to the newsreel and the documentary make it much more effective for television than it could ever be for the cinema.

How would you describe the difference between seeing a film at the cinema and on television or video where most people watch it?

Well, high definition television, larger screens, and better technology will eventually make television and cinema technically comparable although it certainly isn't at the moment. But more importantly there is something inherently exciting about enjoying the experience of seeing a film in the cinema with other people. When you see it on television auntie always wants a cup of coffee, or the cat wants to be let out. I'm currently putting my money where my mouth is by backing the setting-up of a small, 175-seat cinema here in Richmond[4] with a group of other people, including David Puttnam.

How far do you feel that there is a battle between television and cinema drama?

Well, I think they must coexist. They can offer each other facilities and talent. I was involved, for instance, in the evolution of the old British Film Academy and the Guild of Television Producers and Directors into the British Academy of Film and Television Arts. To a large extent the

---

[3] BBC, 1982. A highly acclaimed drama series by Alan Bleasdale, based on his play *The Black Stuff*. Directed by Phillip Saville.
[4] An area of south-west London, where Attenborough lives.

demarcation is impossible because of the mixed funding, and because they both represent the moving image as an art form. But I do not think that they will ever be totally compatible or interchangeable. In the end one will vie for an audience with the other.

Do you think it's possible to attract people from television back to the cinema?

Well, the battle is being won. Cinema audiences have risen consistently from the 1984 all time low of fifty three million, to over ninety three million this year [1989]. People are coming back into the cinema.

How far do you identify the problems of the British Film Industry with the exhibition duopoly?

It is the greatest problem that the industry has had to face for as long as I can remember—for almost fifty years. It's a problem that exceeds even that of American involvement. In the old days a group of us, Launder and Gilliat,[5] the Boultings, Carol Reed, Mickey Powell, and Emeric Pressburger and so on, left the British Film Producers' Association to form the Federation of British Film-Makers because the BFPA involved the exhibitors and we couldn't mount the attack on the duopoly that we wanted to. Sadly, we eventually folded and returned to the BFPA. The problem was that two men, the chief film bookers for Thorn-EMI and Rank, had become almost the sole arbitrators of what was shown at the cinema; their tastes prevailed; they decided on the advertising that went with a film; on how long the film would run. Any movies outside the beaten track were often suffocated, and to a large measure this still exists with Cannon and Rank, although there are other groups coming in. The government should have intervened. It is to the shame of both major parties that they did not. It's stultifying.

You are a strong advocate of government involvement aren't you?

Yes. The government has to accept that cinema as well as television is worthy of investment concessions. It is worthy of the national facilities which would be granted to any other major performing art form whether theatre, ballet, opera, or whatever. The rest of the world look at us as a joke. If you set foot in an American city and say, 'Might I make a film here?', you have twenty people saying, 'Of course you can. What do you want?' In England the government hasn't even appointed a true Minister for film. We bounce between the DTI, the Home Office, Arts and Libraries, Foreign Office, and so on, with no one department taking us seriously. The industry doesn't need a begging bowl. It is a

---

[5] British comedy script-writer and director team. Films include: *Millions Like Us* (1943), and *Green for Danger* (1946).

commercial business, but we're entitled to moves to decrease the volatility of financing especially from the City. This is particularly important given the influence the exhibition duopoly has on the financing of a film. Capital allowances[6] were granted for a while by Geoffrey Howe, and as a result, we saw the greatest City investment in film since Korda. But when he left for the Foreign Office, he was replaced by Nigel Lawson who withdrew them. Government support is necessary if we are going to compete on an even footing with other countries.

Do you think this inability to raise government or alternative sources of finance is the reason for many directors turning to America?

Yes, of course. Because of the current circumstances of making a film in the UK, each time we find an amazing talent, where do they go? They go out to Los Angeles and buy a mansion on Sunset Boulevard. I don't blame them. They want to make films on a particular scale and in a particular manner which is impossible in this country. But it shouldn't be just like Film on Four. We need a more catholic identity in British film. I want to make larger films like *Cry Freedom* or *Tom Paine* which need $25 to $35 million. So I myself have no alternative but to go elsewhere for funding. I'm one of the very few producer/directors who, although I use American money, make the films for my own company here in the UK. I just wish that there were one or two more people around here who would also engage British writers, crews, actors, and actresses, with large American financed films.

Do you find that any compromises are necessary in order to make these sorts of high budget films?

I don't compromise—but it's a fine line. I've never put in a character, or dialogue, or changed anything that I've wanted to say as a concession merely to obtain funds. But I envisage the concept of the picture, right from the word go, to appeal to a world market so as to recoup the budget. I could make the most marvellously self-indulgent film for $30 million. But because it would only gross $2 million, I would never work again. Since there are subjects that I'm bursting to make, I have to try and make viable films.

But in truth, I didn't make *Cry Freedom* simply to make large returns for Universal or myself. However, I'm not interested in appealing to two men and a dog in a barn. I don't want to make films which preach to the converted. I made it because I feel passionately about apartheid and want to find a large audience, whether they're interested in apartheid or not. Then I want them to come out of the film angry and wanting to do something. Film is after all a *mass* medium. I hope that I find in my own

[6] See Introduction.

Attenborough on the set of his most political film to date, *Cry Freedom*, 1987. Copyright © by Universal Pictures, a Division of Universal City Studios, Inc. Courtesy of MCA Publishing Rights, a Division of MCA Inc.

way, and with my own integrity, the means to attract an audience to hear what I want to say. Whatever happens, you must not destroy that integrity. The moment the audience feels that something is phoney, you place the entire picture in jeopardy. They question everything and why shouldn't they?

Well, do you feel that the logistics of big budget movies imply a different style of film-making—especially for a producer/director like yourself?

Yes, I'm very much aware of the fact that I'm involved in major budget movies—often out of necessity not choice. But I must ensure that it's up on the screen and not wasted; that the film comes in on budget. If it doesn't, you won't get money so easily the next time.

The result for me is that the script becomes a bible and everything is planned meticulously. I believe that, if I'm absolutely honest and put my hand on my heart, I must say that I tend to stick precisely to what is planned, in order to stay on budget. Certain people who don't feel that constraint, or, without being pompous, sense of responsibility to the financiers, sometimes come up with more exciting material than I do. My style is preconceived and, therefore, I think it's sometimes a bit mundane. If I am self-critical, I have to say that I think my work tends perhaps to be over-formulated. I sometimes wish that I'd been a bit more unconventional—flamboyant almost.

Can you give an example?

I think that the freedom of style in the opening scenes and the Soweto sequence of *Cry Freedom* had a panache, which I wish had occurred more through the rest of the narrative. David Lean is self-evidently a hero of mine but he has some of these constraints. *Lawrence of Arabia* is not really flamboyant. Much of it is conventionally shot. Part of this is caused by shooting in Panavision which technically constrains the volatility of the sort of film one can shoot.

You have talked of your respect for a very different kind of film-maker, Ken Loach.

When I think of Ken, I think that a number of us, perhaps, attempt too much in some ways. We attempt to reach out in too many directions. Ken also has something to say which he believes in passionately and I respect his total lack of concession to anything and anybody. In so doing he has broken rules which most of us believed were almost sacrosanct. He can work without a script for sequences; he marries professional and amateur; he is a fundamentally creative, unique film-maker. But the British director whom I admired more than any other was Carol Reed, although I suppose his great trilogy *Odd Man Out* (1947), *The Fallen Idol* (1948), and *The Third Man* (1949) would be considered somewhat staid now. Carol adored actors unlike David [Lean], and as a result gained fantastic performances. Carol also knew how to move actors within the frame and that contributed to an extraordinary degree to the drama of the moment. Then, of course, there's Alan Parker. I would wish to gain his admiration as much as anybody I know currently making films—although I question his taste a bit here and there. He's wonderful. I'm devoted to him. One of my most treasured possessions is a letter he wrote to me after seeing *Cry Freedom*.

You have referred to film generally as a semi-art form, rather than an art form. What did you mean by that?

Well, I think there are the great figures like Chaplin, Satyajit Ray, Jacques Tati, Welles, and so on. These are people who virtually write, direct, compose, edit, design, operate the camera, and photograph the whole movie. They do 90 per cent of the creative work themselves. It's their inspiration and vision. But film rarely achieves such fine art status. I'm not a great movie director. I'm more a co-operative director. I work with a band of people both on the floor and in the production office, which operates under my permanent executive producer, Terry Clegg. I hold the strings, but they all have their elements and contributions. The assembly of talent is diverse and for me to say that too much of it is my creation is profoundly dishonest.

You have talked about your strength as a producer. Have you ever thought of producing other people's films with your own company?

I have, and quite a large number of people have attempted to persuade me to do it. My American agent said, 'You could have six pictures going like Puttnam. There are so many directors who would adore to do a picture under you.' The problem is that as a director I loathe some executive producer telling me what to do. In truth I don't think I could keep my hands off it. I work out of a burning passion in a blinkered atmosphere of sacred dedication. I don't know any other way of doing it. I'm too inextricably committed. That's what drives me. As a result, it really is a 'hands-on' job for me.

FILMOGRAPHY

BRIEF BIOGRAPHICAL DETAILS

Born in Cambridge in 1923. Educated at Wyggeston Grammar School, Leicester. Won the Leverhulme Scholarship to the Royal Academy of Dramatic Art. Joined the RAF in 1943, and seconded to the RAF Film Unit for *Journey Together* in 1945. Awarded the CBE in 1967, and knighted in 1976.

FILMS DIRECTED BY SIR RICHARD ATTENBOROUGH

1969   *Oh! What a Lovely War*. 144 mins. Paramount/Accord. SCRIPT: Len Deighton (from the stage play by Joan Littlewood and Charles Chilton). PRODUCERS: Brian Duffy and Richard Attenborough. ASSOCIATE PRODUCER: Mack Davidson. PHOTOGRAPHY: Gerry Turpin. CAMERA OPERATOR: Ronnie Taylor. PRODUCTION DESIGNER: Don Ashton. EDITOR: Kevin Connor. STAGE SHOW: Joan Littlewood and Charles Chilton. SONGS ORCHESTRATED AND INCIDENTAL MUSIC COMPOSED BY: Alfred Ralston. CAST: Dirk Bogarde, Phyllis Calvert, Jean Pierre Cassel, John Clements, John Gielgud, Jack Hawkins, John Mills, Kenneth More, Laurence Olivier, Michael Redgrave, Ralph Richardson, Maggie Smith, Susannah York.

1972   *Young Winston*. 157 mins. Columbia/Open Road/Hugh French. SCRIPT: Carl Foreman (based on Churchill's autobiography *My Early Life*). PRODUCER: Carl Foreman. ASSOCIATE PRODUCER: Harold Buck. PHOTOGRAPHY: Gerry Turpin. CAMERA OPERATOR: Ronnie Taylor. PRODUCTION DESIGNERS: Don Ashton, Geoffrey Drake. EDITOR: Kevin Connor. MUSIC: Alfred Ralston. CAST: Simon Ward, Robert Shaw, Anne Bancroft, John Mills, Jack Hawkins, Ian Holm, Anthony Hopkins, Patrick Magee, Edward Woodward.

1977   *A Bridge Too Far*. 175 mins. United Artists/Joseph E. Levine. SCRIPT: William Goldman (from the book by Cornelius Ryan). PRODUCERS: Joseph Levine and Richard Levine. CO-PRODUCER: Michael Stanley-Evans. PHOTOGRAPHY: Geoffrey Unsworth. PRODUCTION DESIGNER: Terence Marsh. ART DIRECTORS: Stuart Craig, Alan Tomkins, Roy Stannard. EDITOR: Anthony Gibbs. MUSIC: John Addison. PRODUCTION MANAGER: Terry Clegg. CAST: Dirk Bogarde, James Caan, Michael Caine, Sean Connery, Edward Fox, Elliot Gould, Gene Hackman, Anthony Hopkins, Hardy Kruger, Laurence Olivier, Ryan O'Neal, Robert Redford, Maximilian Schell, Liv Ullman.

1978   *Magic.* 107 mins. TCF/Joseph E. Levine. SCRIPT: William Goldman (based on his novel of the same name). PRODUCERS: Joseph Levine and Richard Levine. PHOTOGRAPHY: Victor Kemper. PRODUCTION DESIGNER: Terence Marsh. EDITOR: John Bloom. MUSIC: Jerry Goldsmith. CAST: Anthony Hopkins, Ann-Margret, Burgess Meredith, Ed Lauter.

1982   *Gandhi.* 188 mins. Columbia/Goldcrest/Indo-British/International Film Investors/National Film Development Corporation of India. SCRIPT: John Briley. PRODUCER: Richard Attenborough. EXECUTIVE PRODUCER: Michael Stanley-Evans. PHOTOGRAPHY: Billy Williams, Ronnie Taylor. PRODUCTION DESIGNER: Stuart Craig. EDITOR: John Bloom. MUSIC: Orchestral score by Ravi Shankar and additional music by George Fenton. IN CHARGE OF PRODUCTION: Terence Clegg. CAST: Ben Kingsley, Candice Bergen, Edward Fox, John Gielgud, Trevor Howard, John Mills, Martin Sheen, Rohini Hattangady, Roshan Seth.

1985   *A Chorus Line.* 118 mins. Embassy-Polygram/Columbia/Rank. SCRIPT: Arnold Schulman (based on the play by James Kirkwood and Nicholas Dante). PRODUCERS: Cy Feuer and Ernest H. Martin. PHOTOGRAPHY: Ronnie Taylor. PRODUCTION DESIGNER: Patrizia von Brandenstein. EDITOR: John Bloom. MUSIC/LYRICS: Marvin Hamlisch, Edward Kleban. CHOREOGRAPHY: Jeffrey Hornaday. CAST: Michael Douglas, Terence Mann, Alyson Reed.

1987   *Cry Freedom.* 159 mins. SCRIPT: John Briley (based on Donald Wood's books *Biko* and *Asking for Trouble*). PRODUCER: Richard Attenborough. CO-PRODUCERS: Norman Spencer and John Briley. EXECUTIVE PRODUCER: Terry Clegg. PHOTOGRAPHY: Ronnie Taylor. PRODUCTION DESIGNER: Stuart Craig. EDITOR: Lesley Walker. MUSIC: George Fenton, Jonas Gwangwa. CAST: Kevin Klein, Penelope Wilton, Denzel Washington.

OTHER FILM WORK

1986   *A Marriage of Convenience.* 60 mins. For Thames Television series *British Cinema: A Personal View.* PRODUCER: David Gill and Kevin Brownlow. EXECUTIVE PRODUCER: Catherine Freeman. PHOTOGRAPHY: Ted Adcock.

FILM PERFORMANCES

(* indicates guest appearances by Attenborough)

1942   *In Which We Serve.* Director: David Lean/Noel Coward.

1943   *Schweik's New Adventures.* Director: Karl Lamac.

1944	*The Hundred Pound Window*. Director: Brian Desmond Hurst.
1945	*Journey Together*. Director: John Boulting.
1946	*A Matter of Life and Death*.* Director: Michael Powell.
1947	*School for Secrets*. Director: Peter Ustinov.
	The Man Within. Director: Bernard Knowles.
	Dancing With Crime. Director: John Paddy Carstairs.
	Brighton Rock. Director: John Boulting.
1948	*London Belongs to Me*. Director: Sidney Gilliatt.
	The Guinea Pig. Director: Roy Boulting.
	The Lost People. Director: Bernard Knowles.
1949	*Boys in Brown*. Director: Montgomery Tully.
	Morning Departure. Director: Roy Baker.
1950	*Hell is Sold Out*. Director: Michael Anderson.
	The Magic Box.* Director: John Boulting.
1951	*Gift Horse*. Director: Compton Burnett.
1952	*Father's Doing Fine*. Director: Henry Cass.
1953	*Eight O'Clock Walk*. Director: Lance Comfort.
1954	*The Ship That Died of Shame*. Director: Basil Dearden.
1955	*Private's Progress*. Director: John Boulting.
	The Baby and the Battleship. Director: Jay Lewis.
1956	*Brothers in Law*. Director: Roy Boulting.
1957	*The Scamp*. Director: Wolf Rilla.
	Dunkirk. Director: Leslie Norman.
1958	*The Man Upstairs*. Director: Don Chaffey.
	Sea of Sand. Director: Guy Green.
	Danger Within. Director: Don Chaffey.
1959	*I'm All Right, Jack*. Director: John Boulting.
	Jetstorm. Director: Cy Endfield.
	S.O.S. Pacific. Director: Guy Green.
1960	*The Angry Silence.* Director: Guy Green.
	The League of Gentlemen. Director: Basil Dearden.
1961	*Only Two Can Play*.* Director: Sidney Gilliatt.
	All Night Long. Director: Basil Dearden.
1962	*The Dock Brief*. Director: James Hill.
	The Great Escape. Director: John Sturges.
1963	*The Third Secret*. Director: Charles Crichton.
	Seance on a Wet Afternoon. Director: Bryan Forbes.
1964	*Guns at Batasi*. Director: John Guillermin.

1965	*Flight of the Phoenix.* Director: Robert Aldrich.
1966	*The Sand Pebbles.* Director: Robert Wise.
	Doctor Dolittle. Director: Richard Fleischer.
1967	*The Bliss of Mrs Blossom.* Director: Joe McGrath.
	Only When I Larf. Director: Basil Dearden.
	*The Magic Christian.** Director: Joseph McGrath.
1969	*The Last Grenade.* Director: Gordon Flemyng.
	A Severed Head. Director: Dick Clement.
	*David Copperfield.** Director: Delbert Mann.
	Loot. Director: Silvio Narizzano.
1970	*Ten Rillington Place.* Director: Richard Fleischer.
1974	*And Then There Were None.* Director: Peter Collinson.
	*Rosebud.** Director: Otto Preminger.
	Brannigan. Director: Douglas Hickox.
	Conduct Unbecoming. Director: Michael Anderson.
1977	*The Chess Players.* Director: Satyajit Ray.
1979	*The Human Factor.* Director: Otto Preminger.

FILMS PRODUCED OR CO-PRODUCED

(* indicates co-producer)

1960	*The Angry Silence.** Director: Guy Green.
1961	*Whistle Down the Wind.* Director: Bryan Forbes.
1962	*The L-Shaped Room.** Director: Bryan Forbes.
1964	*Seance on a Wet Afternoon.* Director: Bryan Forbes.
1969	*Oh! What a Lovely War.** Director: Attenborough.
1982	*Gandhi.* Director: Attenborough.
1987	*Cry Freedom.* Director: Attenborough.

STAGE PERFORMANCES

1941	*Ah Wilderness.* Intimate Theatre.
1942	*Awake and Sing*; *Holly Aisle*; *Twelfth Night.* Arts.
	The Little Foxes. Piccadilly.
1943	*Brighton Rock.* Garrick.
1949	*The Way Back.* Westminster.
1950	*To Dorothy a Son.* Savoy.
1952	*Sweet Madness.* Vaudeville.

1952–4   *The Mousetrap*. Ambassadors.

1956–7   *Double Image*. Savoy.

1957–8   *The Rape of the Belt*. Piccadilly.

TELEVISION

1956   *Talk of Many Things*.

1960   *They Made History*.

BIBLIOGRAPHY

London Evening Standard, 14 Apr. 1950. Attenborough discusses being a disc jockey for 'Record Rendez-vous'.

Films and Filming, Sept. 1961. Article about Attenborough's production plans.

Films and Filming, Aug. 1963. Article on his career and appreciation of his work.

Sunday Telegraph, 15 Aug. 1965. Attenborough on Levine's first promise to back *Gandhi*; his childhood; and his move into acting.

Sunday Telegraph, 22 Aug. 1965. Attenborough discusses learning the techniques of screen acting and his early film roles.

Sunday Telegraph, 29 Aug. 1965. Attenborough on acting in Hollywood, and his love of painting.

Cinema (USA), Mar. 1966. Interview with Attenborough about his acting career.

London Evening Standard, 18 Nov. 1966. Alexander Walker on *Dr Dolittle*, and Attenborough's unfulfilled plans to produce a Hollywood film *Insurance Italian Style*.

London Evening Standard, 10 Apr. 1969. Attenborough's production company Marble Arch Productions is bought by Constellation Investments—a group owned by various British stars.

Films and Filming, June 1969. Interview by Gordon Gow about Attenborough's new role as a director.

Sunday Mirror, 19 Oct. 1969. On some of the forty films on offer at the time for Attenborough to direct; including a musical on the life of Nijinsky; a musical *Roman Holiday*; a life of Nelson; a version of *Hamlet* with Richard Harris and of *The Tempest* with Gielgud; *The Feathers of Death* for Forbes; Isaac Singer's *The Slave* for Levine; and *Henderson, The Rain King* for United Artists.

Action, Jan.–Feb. 1970. Interview on 'Why I became a director'.

Sunday Times, 24 Jan. 1971. Article on Attenborough by Philip Norman.

Films in London, 24 Jan.–6 Feb. 1971. Biofilmography.

Sunday Times, 31 Jan. 1971. Letters complaining about misleading statements in Philip Norman's article.

TV Times, 17 Feb. 1972. David McGill on Attenborough's career.

Films Illustrated, July 1972. Background information on *Young Winston*.

Photoplay Film Monthly, Aug. 1972. Interview on *Young Winston*.

Movie Maker, Sept. 1972. Interview with Attenborough and Foreman at the opening of *Young Winston*.

Dialogue on Film, Feb. 1973. Transcript of a seminar Attenborough gave at the

American Institute's Centre for Advanced Film Studies.

Film Dope, Mar. 1973. Brief career article.

Daily Mail, 28 Aug. 1973. Leslie Watkins on Attenborough's involvement in Capital Radio.

The Times, 1 Sept. 1973. Barry Norman on Attenborough as chairman of Capital Radio, and being 50.

Films Illustrated, Sept. 1974. Interview about his ambitions to make a film on the life of Gandhi.

Dialogue on Film, Feb. 1975.

Sunday Express, 3 Aug. 1975. On Attenborough's struggles to do work which is creatively satisfying in his acting career.

Screen International, 4 June 1976. Interview on directing epics, in particular *A Bridge Too Far*.

Sunday Express, 21 Nov. 1976. Roderick Mann on Levine's and Attenborough's amiable working relationship.

American Cinematographer, Apr. 1977. Comments on various aspects of the production of *A Bridge Too Far*.

Sunday Telegraph, 12 June 1977. On Attenborough acting in Satyajit Ray's *The Chess Players*.

Photoplay Film Monthly, July 1977. Illustrated interview in which he discusses some of the production problems on *A Bridge Too Far*.

Films Illustrated, July 1977. Interview on *A Bridge Too Far*.

Hollywood Reporter, 1 Dec. 1978. Brief comments from Attenborough on *Magic*, and the problems of portraying Gandhi on screen, having chosen Anthony Hopkins for the lead role.

Film-maker's Monthly, Dec. 1978. Interview in which he discusses the production of *Magic*; *A Bridge Too Far*; *Gandhi*; budgeting for his films; and his working relationship with Levine.

Screen International, 27 Jan. 1979. Interview discussing the criticisms his films have received; *Magic*; and his hopes to make a film of *Gandhi*.

Films and Filming, Feb. 1979. Interview in which he discusses the pre-production preparations with William Goldman for *Magic*; working with Satyajit Ray on the *Chess Players*; and working with actors on his recent films.

Guardian, 5 Feb. 1979. Derek Malcolm on *Magic* and *The Chess Players*.

Sunday Telegraph Magazine, 18 Nov. 1979. Good article by Alexander Walker on Attenborough's career.

Daily Mail, 30 May 1980. Martin Jackson on Attenborough's involvement in the setting up of Channel 4.

Guardian, 9 Aug. 1980. Peter Miesenwand on Indian finance for *Gandhi*.

Broadcast, 1 Sept. 1980. Attenborough announces financing for *Gandhi*.

Screen International, 10–17 Oct. 1981. Note that as of January 1982, Attenborough is to take on the post of chairman and governor of the BFI.

Broadcast, 12 Oct. 1981. Note on Attenborough's appointment as BFI chairman.

BFI News, Mar. 1982. Note that he is the new chairman.

Stills, No. 5, 1982. Interview on the making of *Gandhi*, with comments from Ben Kingsley, who plays the title role.

Sunday Express, 21 Nov. 1982. Roderick Mann on finishing *Gandhi*.

The Times, 22 Nov. 1982. Profile on Attenborough's career.

Eyepiece, Nov.–Dec. 1982. Interview on why he wanted to make *Gandhi*.

Screen International, 4–11 Dec. 1982. Attenborough talks about the making of *Gandhi*; his twenty-year obsession with the subject; and his future plans.

Screen International, 11–18 Dec. 1982. Note that he has been awarded the 1982 Gold Star Award by Plitt Southern Theatres.

Vogue, Dec. 1982. On Ben Kingsley's role in *Gandhi*.

Hollywood Reporter, 23 Dec. 1982. Attenborough talks of the origin of his interest in *Gandhi*.

David Castell, *Richard Attenborough: A Pictorial History* (Bodley Head, 1982).

Richard Attenborough, *In Search of Gandhi* (Bodley Head, 1982).

Screen International, 22–9 Jan. 1983. Note that he has been named as one of the recipients of the 1983 Martin Luther King Jr. Non-Violent Peace Prize.

Screen International, 12–19 Feb. 1983. Note on Attenborough being awarded the 'Padma Bhusan', the Indian equivalent of a British knighthood.

American Film, Mar. 1983. Interview in which he talks about how he moved into directing, and his struggle to get *Gandhi* made.

Guardian, 14, 16, 20 Apr. 1983. Articles on Attenborough's possible attendance at and then withdrawal from premières of *Gandhi* in South Africa.

Observer, 17 Apr. 1983. Profile of Attenborough.

Broadcast, 25 Apr. 1983. Details of the postponement of his trip to South Africa for the première of *Gandhi*.

The Times, 14 May 1983. Attenborough replies in full to criticisms of *Gandhi* by Salman Rushdie.

Screen International, 14–21 May 1983. Attenborough's decision to cancel his visit to South Africa.

Daily Telegraph, 6 June 1983. James Allan on Attenborough's support of Jenkins and the SDP/Liberal Alliance at the Hillhead by-election.

Daily Express, 26 Aug. 1983. Victor Davis on the reaction to *Gandhi*.

Hollywood Reporter, 6 Oct. 1983. Note on *A Chorus Line*.

NFT Booklet, Oct.–Nov. 1983. 'Richard Attenborough' season.

Films and Filming, Dec. 1983. The story of his quest to film *Gandhi*.

Daily Telegraph, 11 Feb. 1984. On Attenborough's visit to South Africa to research *Cry Freedom*.

The Times, 28 Nov. 1984. Sheridan Morley's location report of *A Chorus Line*.

Hollywood Reporter, 14 Dec. 1984. Attenborough talks about his use of film stock in an advert for Eastman Kodak.

Screen International, 13–20 Apr. 1985. Text of his speech at the inauguration of British Film Year.

Screen International, 18–25 May 1985. Item on *Cry Freedom*.

Screen International, 3–10 Aug. 1985. His appointment as chairman of Goldcrest Film and Television, and deputy chairman of the holding company, Goldcrest Holdings.

Variety, 7 Aug. 1985. Attenborough appointed chairman of Goldcrest.

Hollywood Reporter, 8 Oct. 1985. Attenborough offers an explanation of British Film Year.

Daily Mail, 16 Dec. 1985. Attenborough hits at critics of *A Chorus Line*.

Hollywood Reporter, 20 Dec. 1985. He talks about his style of film-making; defends *A Chorus Line*; and talks about his next film, then titled *Biko*.

Diana Carter, *Richard Attenborough's* A Chorus Line (Bodley Head, 1985).

Films in Review, Jan. 1986. Interview about his directing career.

Guardian, 11 Jan. 1986. Derek Malcolm on *A Chorus Line* and Attenborough's plans for *Cry Freedom*.

Sunday Times, 21 Jan. 1986. Sue Summers on *A Chorus Line* and his plans to make *Cry Freedom*.

Hollywood Reporter, 21 Jan. 1986. His investment as Commander in the French Order of Arts and Sciences; the favourable European reception of *A Chorus Line*; and his next project *Biko*.

Televisual, Mar. 1986. On his contribution to the *British Cinema: Personal View* series for Thames TV.

Sunday Times, 23 Mar. 1986. Attenborough defends 'British Film Year'.

Screen International, 14–21 Mar. 1987. Attenborough's comments on his view of the future of Channel 4.

Guardian, 6 Apr. 1987. Attenborough on Channel 4 and finding a successor to Jeremy Isaacs, independent radio, and satellite broadcasting.

Screen International, 18–25 July 1987. In his capacity as the new chairman of the British Screen Advisory Council, he calls on the government to give urgent attention to the British film industry.

Screen International, 25 July–1 Aug. 1987. Awarded an honorary degree by the University of Sussex.

Screen International, 26 Sept.–3 Oct. 1987. Attenborough talks with Elem Klimov about the launch of an Anglo-Soviet film initiative.

London Evening Standard Magazine, Nov. 1987. Description of Attenborough's personality and his office.

Screen International, 7–14 Nov. 1987. Attenborough is signed by Universal Pictures to produce and direct three big budget productions.

London Evening Standard, 9 Nov. 1987. Announcement of his three-picture deal with Universal.

Sunday Telegraph Magazine, 15 Nov. 1987. By Attenborough, on his visit to South Africa to research *Cry Freedom*.

Sunday Express Magazine, 15 Nov. 1987. Career profile by Michael Rye.

Films in Review, Dec. 1987. Interview about the making of *Cry Freedom*.

Hollywood Reporter, 14 Dec. 1987. Profile of Attenborough.

Donald Woods, *Filming with Attenborough* (Penguin, 1987).

Richard Attenborough, *Richard Attenborough's Cry Freedom: A Pictorial Record* (Bodley Head, 1987).

Screen International, 1–6 Feb. 1988. Presented with the Jean Renoir Humanitarian Award.

Screen International, 2–9 July 1988. He voices his disquiet at recent government treatment of television.

Screen International, 16–23 July 1988. He pleads for government intervention over the sale of Elstree studios.

Television Today, 4 Aug. 1988. Attenborough praised by former Equity president Derek Bond, for attempting to get *Cry Freedom* released in South Africa.

Screen International, 17–23 Dec. 1988. Attenborough introduces the European script fund.

Screen International, 1–4 Feb. 1989. Attenborough elaborates on the European script fund.

Screen International, 4–11 Feb. 1989. Letter giving more details about the financing of the script fund.

3 BILL FORSYTH

Bill Forsyth on the set of *Housekeeping*, 1987. © Courtesy of Columbia Pictures Industries Inc. Photograph: BFI.

ESSAY ON

BILL FORSYTH

IN the early 1980s, Bill Forsyth received more public attention than any other British film-maker. With his first two films, *That Sinking Feeling* and *Gregory's Girl*, playing simultaneously in 1981, and two more in the pipeline,[1] he appeared to have come from nowhere.[2] Equally unusual was the fact that he was a writer-director—a rare breed in British cinema. What struck the audience and critics alike was the originality and charm of his films. The wry humour, eccentric characters, and absurdist touches prompted Mamoun Hassan[3] to remark, 'Bill is one of those rare writer-directors who are able to treat serious subjects with great warmth.'[4] His initial prolific output, five films in as many years, established him as the new hope of British film-making. These films, all set in Scotland, made him a paradigm for those hoping to create indigenous films with international appeal.

But he was never a very likely champion. He is shy, gentle, and unassuming—'the most modest of British film-makers'.[5] There is a revealing story that after a press screening of *Local Hero* in New York, Forsyth was stuck in the same elevator as Vincent Canby—one of the most powerful critics in the US. But he was reluctant to introduce himself, even though the critic had liked his previous film, *Gregory's Girl*. When Canby asked him for the time, Forsyth, fearing that his Scottish accent would give him away, could only point to his watch and mutter 'ten after one', hoping to pass as a Canadian. Even while shooting, he keeps a low profile, being reluctant to call 'action' and 'cut', leaving this to his first assistant. He admits that he is self-conscious: 'When I'm standing around on a shoot looking pre-occupied, I'm actually just whistling to myself, hoping people will think I'm looking

---

[1] *Local Hero* and *Comfort and Joy*.

[2] In fact it was only with the success of his second film *Gregory's Girl* that he was able to secure a national distribution for *That Sinking Feeling*.

[3] Managing director of the National Film Finance Corporation, 1979–84. The NFFC was dissolved as a result of the Film Act (1985), and succeeded by the British Screen Finance Corporation.

[4] *Sight and Sound*, Autumn 1981.

[5] John Pym in *Sight and Sound*, Winter 1987–8.

preoccupied.'[6] David Puttnam[7] sees him as a 'remarkably uncorrupted and unsoured man'.[8] Fulton Mackay[9] believes that 'Forsyth's films show that he is innocent. I would say that it is his greatest asset.'[10] But a certain amount of what is seen as innocence is in fact reticence. Puttnam goes on to describe Forsyth as 'sometimes taciturn so that you're a little unsure as to where you're going, or what you're doing'.[11] But for Forsyth, his silent manner is, in part, a defence mechanism to protect him from being distracted from his vision of his films. As Chris Menges[12] noted, 'Time and time again people would come up and say "We are over-shooting" or "The script is too long", and Bill would say "Yes, I agree" and do nothing about it.'[13]

Born in Glasgow in 1946, Forsyth has lived there ever since.[14] He recalls his childhood as being 'remarkable for its unremarkableness'. He had no great interest in film, although a school screening of Tati's *Monsieur Hulot's Holiday*[15] did catch his attention. 'I didn't even mind that it was in black and white and in a foreign language because almost all of the fun in it was visual.' But it was quite fortuitous that he actually went into film as a career. After leaving school at 17, he planned to work for a year in order to save enough money to go to Greece with some friends. 'I answered an ad which said "Lad required for film company," and I thought it sounded better than "Lad required for butchers".'[16] After proving what he called his 'skill at lifting heavy objects', he found himself working for an eccentric, local documentary-maker, Stanley Russell.[17] 'That was in the days of a one man documentary film company. One man and a boy. And I was the boy.'[18] Forsyth helped make sponsored industrial films about factories, industrial dams, motorways, fishing, and forestry. His decision to go freelance, after the death of Stanley Russell six months later, marked the beginning of his commitment to a film career. He started to watch films seriously and was struck

[6] *Screen International*, 10 July 1982.

[7] Producer of *Local Hero*, and backer of *Housekeeping* during his tenure as head of Columbia Pictures in 1987.

[8] Andrew Yule, *Puttnam: A Personal Biography* (Mainstream, 1988).

[9] Well-known Scottish actor who played Ben in *Local Hero*.

[10] Allan Hunter and Mark Astaire, *Local Hero: The Making of the Film* (Polygon Books, 1983).

[11] Andrew Yule, *Puttnam: A Personal Biography*.

[12] Director of photography on *Local Hero* and *Comfort and Joy*.

[13] Allan Hunter and Mark Astaire, *Local Hero: The Making of the Film*.

[14] His father was a grocer, plumber, and warehouse-manager: a working-class background.

[15] Made in 1952, it is a gentle comedy about the clumsy, well-meaning Hulot on holiday in a provincial seaside resort.

[16] *Blitz*, No. 6, 1982.

[17] Within three weeks of joining, Russell handed Forsyth his notes for a script of a sponsored film for the Bank of Scotland, and asked him to write it. As Forsyth remembers, 'Naturally, for a boy of seventeen, this was all very exciting.'

[18] *Films and Filming*, Aug. 1984.

by the New Wave directors, particularly Godard[19] and Malle.[20] 'I wanted to be Louis Malle when I was nineteen because he had made *Le Feu Follet*'[21] . . . 'I combed my hair like Maurice Ronet[22] and started to smoke Gauloise Bleues. I began to take a physical pleasure in the art of splicing film.'[23] When he was 20, he briefly left Glasgow to work in London at the BBC as an assistant editor.[24] 'That was my year breaking out, but no miracle happened. Within a year I was home making films in Glasgow, although on the train back I had promoted myself to editor.'[25] It was at this time that Forsyth made his only two experimental films—*Language* (1969) and *Waterloo* (1970). Describing these short films as 'structuralist', he was trying to transmit emotions as simply as possible by using images rather than narrative or dialogue. 'I was trying to re-invent cinema, but I was really just like a kid playing in a sandpit.'[26] One shot in *Waterloo* was ten minutes long, incorporating images of a man reading a book and a car journey through part of Glasgow to the accompaniment of a sentimental Jim Reeves record, and ending in a park where a young couple were kissing. 'I was trying to transmit feelings of memory, loss, love and the passage of time. You see, the old man and the young couple are joined in different ways—through his memory, linked through the sentimentality of the song, or physically joined because they are on one piece of film.'[27] This film was screened to a dismal reception at the Edinburgh Film Festival, most of the audience leaving before it had finished. This audience reaction had an important impact on Forsyth's subsequent approach to tackling serious issues in his feature films.

In 1971 Forsyth set up *Tree Films* with two friends to make privately funded industrial films. In the same year he was accepted into the newly founded National Film School. He hoped that by becoming a student he might catch up on some reading. But too heavily committed to *Tree Films* and unable to divide his time, he only stayed at the college for a term. Due to the amount of work they had, Forsyth soon moved from

---

[19] Jean-Luc Godard: French innovative director; a pioneer of non-linear narrative, and one of the leading members of the New Wave movement in the 1960s. Films include (as director and screen-writer): *Breathless* (1960), *A Woman is a Woman* (1961), *Alphaville* (1965), *La Chinoise* (1967).

[20] Louis Malle: French director. Films include: *Frantic* (1957), *The Fire Within* (1963), *Murmur of the Heart* (1971), *Au revoir les enfants* (1987).

[21] *Le Feu follet* (The Fire Within). Made in 1963, it is one of Malle's most mature and accomplished films. It tells the poignant story of the last few days in the life of a suicidal alcoholic, played by Maurice Ronet. Comment from *Sight and Sound*, Autumn 1981.

[22] Star of Malle's *Le Feu follet*.

[23] *American Film*, Nov. 1984.

[24] Working on *Chronicle*, *Z Cars*, and *Play for Today*.

[25] *New Yorker*, 1 Oct. 1984.

[26] *Sight and Sound*, Autumn 1981.

[27] Interview with the authors.

editing to directing the films.[28] By 1977 he had had enough, and wanted to turn his hand to features. 'You can only be so creative with a sponsored film, it's the law of diminishing returns. You cease to learn things after a while.' Over the next two frustrating years he wrote *Gregory's Girl*, had it rejected three times by the BFI; wrote *That Sinking Feeling*, persuaded his film-making friends to give up their time to help him shoot it; and at last broke through to become a feature film director.

Forsyth's films reflect his idiosyncratic view of the world. Within people's daily lives he finds all the material that he wants for his films: people's need for self-fulfilment, eccentric characters, humour, absurdity, and even a touch of fantasy. 'I wish people were more aware of the fantastic nature of ordinary life. Maybe people are and they just don't talk about it. I suspect that that's the truth. People stand at bus stops and work out the absolute wonderment of life but don't mention it to anyone else.'[29]

Of central importance to all Forsyth's work is his fascination with people. 'All you are doing when you make a film is to take two or three characters from one point to another and observe what happens to them on the way. So whether the film is about unemployed kids [*That Sinking Feeling* (1979)] or a retired sea captain [*Andrina* (1981)], it's still the same: the characters are the strength of the thing.'[30] Puttnam sees him as having 'a very wonderful vision, an innate belief in the best of people . . . Not so much life, as people.'[31] But it is his view of life, and the painful and often lonely dilemmas facing the central characters in his films, which provides the backbone of his work. There is always an underlying melancholy as they struggle to discover their own identity. Sometimes this discovery is caused by being exposed to a totally different way of life as in *Local Hero* (1982), with the American oil executive, MacIntyre, who comes into contact with village life in Scotland. In *Comfort and Joy* (1984) the disc jockey, Alan 'Dickie' Bird, faces a mid-life crisis after his long-standing girl-friend suddenly leaves him, and he begins to question his whole existence. In *Housekeeping* (1987) it is young Ruth's struggle against conformity to find herself, and similarly in *Breaking In* (1989) the lonely adolescent, Mike, searches for a sense of belonging.

It is the characters' tortuous journey to some form of self-realization which is the guiding force in all Forsyth's work and which dictates the loose, organic structure of his films. As he sees it, 'In some films it's

---

[28] One of the more ambitious projects he got involved in was *The Legend of Los Tayos*, an ill-fated documentary about an expedition into caves in the Andes. Forsyth comments that the background of the expedition was used to 'speculate about man's place in the Universe!' The project was never sold, and the final payment of £15,000 never materialized.
[29] *NME*, 27 Oct. 1984.
[30] *Radio Times*, 28 Nov.–4 Dec. 1981.
[31] Andrew Yule, *Puttnam: A Personal Biography*.

That Sinking Feeling, 1979, Forsyth's first feature film, made in Glasgow for £6,000. © Bill Forsyth. Photograph: BFI.

possible to put the beginning at the end and move it around. I don't think I've ever made a film where that's possible; there's always a structure which is difficult to play with because certain things have got to happen for the characters to change.'[32] Puttnam jokes, 'Bill's a rambler—but that's where his genius lies, in letting the story follow its nose.'[33] Traditionally, and especially in American films, structure, based on conflict, dictates character development. For Forsyth the priorities are reversed, and it is the development of characters which determines the structure of his films. He believes that 'You have to look for tension within and between the characters. Conflict should be embedded in the film but not necessarily be the guiding principle from scene to scene.'[34]

One of the main problems with this approach is that for a mass audience, accustomed to a more rigid structure and a clear knowledge of the film's direction, based on the early establishment of the dramatic conflict, there is less to hold onto. Forsyth recognizes this, 'Really, I suppose my films boil down to just a collection of moments which may give the audience a certain amount of insight into either a character, a situation or a place, or varying degrees of each of them. And that's about all that I would want anyone to get from a film that I've made.'[35] Unfortunately, this more serious side to his work is all too easily over-looked, especially given the infectiousness of his humour.

He is genuinely uncertain about the place humour holds in his films. 'I've never really thought of myself as a funny person. I've tended to use comedy as a disguise. I'm still frightened of being serious. What I dread most of all is to make something serious which people will laugh at.'[36] What he does find amusing derives largely from a fascination with the absurdity of life. 'In the areas I work, comedy sort of follows me, it belongs there because the situations demand it.'[37] It is there in *Local Hero*, in the villagers' unexpected reaction to the attempt of an American oil corporation to buy them up, namely, to sell out for as much as possible. Again in *Comfort and Joy* there is the apparent absurdity of an ice-cream war between two rival Italian families.

Forsyth delights in poking fun at eccentricities: MacIntyre's executive traits—charging the batteries of his electric briefcase; his watch alarm set to remind him of 'Conference time in Houston'; or Dickie Bird's fastidiousness about his car. Such traits set these characters up for their humorous come-uppance, and it is Dickie's car which is constantly attacked by the ice-cream mafia. But although such characters seem

[32] *Films and Filming*, Aug. 1984.
[33] *Time Out*, 11–17 Mar. 1983.
[34] Ibid.
[35] *Film Quarterly*, Spring 1985.
[36] *American Film*, Nov. 1984.
[37] NME, 27 Oct. 1984.

quite eccentric, Forsyth does not regard them as particularly abnormal. 'Strangeness is in everyone, it's just a matter of whether you choose to reveal it or not.'[38] What makes these characters interesting is the difference between the seriousness with which they view themselves and the way they are perceived by others. 'One of my pre-occupations is the idea of how everyone inhabits the world and sees themselves in a certain way and absolutely no-one else sees them in that way. In fact, the whole philosophy of *Gregory's Girl* (1980) was that, especially in adolescence, you have a very precise image of yourself but you are completely unaware of the fact that the world sees you as something different.'[39]

Parody is another of his favoured devices. He enjoys placing familiar situations in an abnormal context. There is nothing unusual about the story of *That Sinking Feeling*, but the carefully planned robbery by a group of youths is not on a bank, but a stainless steel sink factory. The opening scene of *Comfort and Joy* shows a smart woman blatantly shoplifting and being watched by a man, who, one assumes, is a store detective. As she leaves the shop, he follows her, grabs her by the arm saying, 'Maddie, you'll be the death of me,' and the couple walk off together. The same approach often applies to dialogue. In *Gregory's Girl* the school headmaster solemnly discusses with the games' teacher the problems raised by a mixed-sex football team, in the same tone as the mastermind in a bank robbery film asks the explosives expert if he can manage a particularly difficult safe—'The showers . . . Think you can handle it?' Forsyth extends this humour and makes it more absurd through repeating jokes. 'The audience has got to laugh sooner or later because the joke keeps coming back.'[40] In *Local Hero* a motor-bike passes at full speed when MacIntyre first steps out of the front door of the hotel into the otherwise deserted village street. After this, every time MacIntyre leaves his hotel it careers past, until finally he (like the audience) learns to anticipate it, and with new-found imperturbability is able to restrain an unsuspecting companion from stepping out without looking.

There are also elements of fantasy in Forsyth's films. 'You've got to open your eyes to the magical things in mundane life—you don't need to get it from *Star Wars*. That kind of cinema has taken the magic from things.'[41] There is the Aurora Borealis in *Local Hero*, which to the local inhabitants is merely a regular occurrence, but to the city-bred MacIntyre assumes the scale of a revelation. Similarly in *Housekeeping* there is a shack in a remote forest area which, due to its cold and shaded

[38] *Films and Filming*, Dec. 1987.
[39] *Sight and Sound*, Summer 1983.
[40] *Time Out*, 11–17 Mar. 1983.
[41] *Vogue*, Nov. 1981.

location, is covered with ice. Again it is not particularly unusual in such a setting, but to an eccentric dreamer like Sylvie, it represents something magical.

That Sinking Feeling (1979), Forsyth's first feature, owes its origin to the fact that, wanting to get into features but without any previous experience in working with actors, he spent time at the amateur Glasgow Youth Theatre. 'I sat around simply listening to what young people talked about, how they said it, what they laughed at, and what they found interesting.'[42] From this he got the idea for the film. 'Most of the boys there had broken into somewhere or other in the past, so basing the story on an elaborate robbery was fairly obvious. There's a plumber's warehouse round the corner from the community centre, and it all fell into place.'[43] The film, which Forsyth describes as a 'fairy-tale for the unemployed', is an amusing and charming parody of the 'heist' genre, in which a group of unemployed Glasgow youths plan an elaborate robbery of stainless steel sinks from a plumbers' supply depot. It was, in fact, the beginning of Forsyth's interest in taking familiar scenes and situations and giving them a comic edge. For example, during a tense conversation in a boat on a lake, the leader of the gang tells a new recruit of the need for 'skill, courage and determination' while the poor boy is being sea-sick over the side, and paying no attention.

While shooting his films he remains flexible, always looking for different ways to cover the scenes. An imaginative example of such improvisation in *That Sinking Feeling* is the scene in a car when the gang leader, having tried to drown himself in a bowl of cornflakes, tells his two friends that there must be more to life than suicide. As they all get out and walk away, the camera pulls back to reveal the car, without wheels, marooned in a wasteground—a fitting commentary on the boys' sense of helplessness. This idea was a late addition, replacing the scripted scene set on a traffic island in the middle of a busy intersection. It came to Forsyth the night before it was due to be shot, when he was driving home and spotted an abandoned car in a desolate Glasgow cityscape. 'When I'm making a film, all my antennae are out and I'm very sensitive to everything. A week earlier I wouldn't have noticed that car. I used to worry about the level of invention. Now I feel quite relaxed and let things happen.'[44]

Despite *That Sinking Feeling*'s minute budget of £6,000, financing still proved to be an ordeal. After being turned down by the BFI Production Board, money was ingeniously raised from a variety of sources: breweries, whisky distilleries, C&A, various youth organizations,

---

[42] *Sunday Times*, 7 June 1981.
[43] *Guardian*, 13 June 1981.
[44] Interview with the authors.

£3,000 from the Scottish Arts Council, as well as small personal contributions. But the most important factor was that the crew and cast
were prepared to work for nothing more than a stake of the profits. The
film was a great success at the Edinburgh and London Film Festivals. It
hit a chord with the strong sympathy felt at the time for independent
film-makers, and, as a Scottish film, it was so unexpected. 'It came from
nowhere. All of a sudden there was a Scottish film. If I had been a
London film-maker, I wouldn't have got half the attention.'[45] The film
was, astonishingly, the first narrative feature film made by a Scotsman
living in Scotland. Forsyth remembers how strange it was at the first
screening to hear people reacting spontaneously to his film. 'We came in
as scruffy layabouts from Glasgow, and came out as real film-makers,
with people coming up to us and wanting to talk about the film. It was a
very odd feeling.'[46] But despite the originality of the film, and the
enthusiasm of its reception, it was still difficult to get the over-cautious
exhibitors to show it. It was only the later success of *Gregory's Girl*
which secured the film a place on the national exhibition circuit.

Forsyth had been trying to get *Gregory's Girl* (1980) off the ground
for several years. It was an original and engaging adolescent comedy
about a football crazy teenager, Gregory, who is replaced on the school
team by the new star striker, Dorothy, with whom he falls in love.
Forsyth had approached the BFI for finance, but once again, to no avail.

I thought that my pedigree would be irresistible. I was a nice provincial boy with
plenty of experience, working in an area of Scotland deprived of just about
everything. But the BFI managed to resist me and my project for two years
running. I remember one torment of a meeting when I tried to explain that
Gregory's Girl was really a structuralist comedy. I would have jumped through
flames for their £30,000. I suspect my script was too conventional although
nobody actually told me as much.[47]

It was finally the independent producers, Clive Parsons and Davina
Belling, who, impressed by *That Sinking Feeling*, offered to take on
Gregory's Girl. They managed to raise the budget of £200,000 with
quite a revolutionary deal in which Scottish TV agreed to a no-strings
joint investment with the NFFC.

The film deals with 'normal' adolescents, the type who 'still cycle to
school, pick their noses, have pimples and ordinary haircuts'.[48] Much of
the humour derives from reversing sexual stereotypes: it is the girls who
have the mature, realistic attitudes, especially to love, as revealed during
their metalwork class, whereas the boys in their cookery lesson are seen

---

[45] Interview with the authors.
[46] *Movie Maker*, May 1984.
[47] *Sight and Sound*, Autumn 1981.
[48] Forsyth in *The Times*, 5 Jan. 1985.

to be naïve romantics. Watching Dorothy play, one boy deplores her role in the team, but Gregory is thrilled; 'it's modern', he says with an adoring smile. Typically, Forsyth uses comic details to point out everyday absurdities—here those of school life: a penguin nonchalantly walking past an open door, or the boys' toilet which, in break-times, is transformed into an impromptu shop, with doughnuts displayed on the toilet seats. Deservedly, the film achieved national acclaim, winning a BAFTA award for best screenplay, in fact beating Colin Welland's Oscar winning script for *Chariots of Fire*.

Forsyth's next film, the fifty-minute *Andrina* (1981), made for BBC Scotland, is the only work he has directed for television. It is an adaptation of a short story by the Orkney writer, George Mackay Brown: a fable about a retired sea captain living in a remote village and the stories he tells each evening to Andrina, a mysterious girl who visits him to cook him his meals and listen to his tales. 'One of the things that strikes me about George's writing is the sense of how people's lives are ritualised in an enviroment such as the Orkneys.'[49] Forsyth saw television as a perfect medium for the scale of the project:

TV drama leaves me cold when it gets involved with massive concepts. For me, television is about addressing an idea in terms of maybe two or three characters. And I think that has a special value for the audience as well as the film-maker, because it lets you deal with a more intimate side of things, and to make a slightly more personal statement than the cinema allows.[50]

But charming as the film was, he admits that it was not something he would have thought of doing, had he not actually been asked by a colleague in the BBC.[51] 'It was one of those challenges which I couldn't turn down . . . to extract a storyline from the old man's ramblings, and to try and formulate that into a dramatic structure.'[52] However, his experiences while shooting in the Orkneys were to be the inspiration for many of the ideas in his next feature film, *Local Hero*, which he was developing at the same time. He saw the mesmerizing effects such a remote place could have on outsiders, when the wardrobe lady fell in love with the islands and left her job with the BBC to move there. Each night he saw the skies illuminated by the Aurora Borealis: 'From this I had the idea that the head of the American oil company, Happer [Burt Lancaster] should be an amateur astronomer so that as well as sending his negotiator, MacIntyre, to buy the village, he could also send back reports of celestial activity.'[53]

[49] *Radio Times*, 28 Nov.–4 Dec. 1981.
[50] Ibid.
[51] Roderick Graham, Head of Drama for BBC Scotland.
[52] *Radio Times*, 28 Nov.–4 Dec. 1981.
[53] *Time Out*, 11–17 Mar. 1983.

John Gordon-Sinclair and Dee Hepburn in Forsyth's highly acclaimed *Gregory's Girl*, 1980. © Channel 4. Photograph: BFI.

Local Hero (1982) was a testing ground for Forsyth. Puttnam[54] commented at the time, 'Bill knows in his heart of hearts that *Gregory's Girl* was an over-praised film. And that does, I think, present him with a few problems. In a way, he's using *Local Hero* to test himself. Is he as good as they say, or only as good as *Gregory's Girl?*'[55] Ironically Puttnam found it surprisingly difficult to raise the budget of £2 million, despite the huge international success of his previous production, *Chariots of Fire.*[56] Goldcrest, the same backers, put up the development money and agreed to fund half the final cost, but the remainder could not be raised. It was only when Forsyth won the BAFTA award for *Gregory's Girl* that Goldcrest decided to back the project completely.

The film tells of an American oil corporation's attempt to site a massive refinery in and around a Scottish fishing village. Far from opposing the scheme, the wily villagers, sensing a possible fortune, encourage it. At the same time, MacIntyre, the company executive sent over to negotiate the deal, is transformed by his experiences there. The slow pace of life, the natural beauty of the area, and the strong sense of community, all in such contrast to his life in America, finally seduce him. The film is essentially a collage of absurd and beguiling characters. Few are what one would expect: the village fishermen who never fish, the oil company's Scottish agent who speaks an unbelievable number of languages, the hotel owner who is also the village solicitor, publican, and appointed negotiator. But although this borders on farce, these characters are portrayed with such warmth and subtlety that they are always believable. As Alan Parker comments, 'Bill's got this incredible humanity; you even like his villains because he really loves people.'[57] With this fine balance, the audience can laugh at these characters' innocent eccentricities without losing sympathy for them. In one scene, MacIntyre is chatting to several of the men of the village who are standing around a baby in a pram. He innocently asks whose baby it is. The embarrassed men all look at each other silently; although they know who the mother is, any one of them could have been the father, although this is missed by MacIntyre. Such subtlety of humour, directed with great sensitivity, is one of the unique qualities of Forsyth's work. Starring Burt Lancaster, with music by Mark Knopfler of *Dire Straits*, the film was a critical and box office success, confirming that Forsyth's talents were far from ephemeral.

His next film, *Comfort and Joy* (1984), like most of his previous

[54] Puttnam had actually been asked to produce *Gregory's Girl* two years earlier but had turned it down.

[55] *Sight and Sound*, Autumn 1982.

[56] Made in 1981. Produced by Puttnam and directed by Hugh Hudson.

[57] Andrew Yule, *Puttnam: A Personal Biography*.

work, was set in Scotland. In fact, it was all filmed within a ten-mile radius of his Glasgow house. The film explores the effects on 'Dickie' Bird of his longstanding girlfriend leaving him. 'I wanted to explore the idea of someone's life suddenly emptying because someone else walks out on them, and them having to radically alter their lifestyle and live off their own resources, because it was similar to something I had just been through.'[58] Dickie becomes a character searching for a justification for his existence. Fighting his own sorrow, he becomes entangled in a local ice-cream war between rival Italian families for the control of the streets of Glasgow.

Forsyth hoped that with this film he would break out of his reputation as the maker of gentle comedies. Certainly it is much more poignant than anything he had made previously, largely because Dickie's personal crisis is so prominent. The problems he faces are more serious and fundamental by comparison with those of characters in Forsyth's other films. It is much easier to laugh at MacIntyre in *Local Hero*, as he is so out of place in a small fishing village. For Dickie, his predicament is too painful to ridicule, as it is his whole world which is collapsing around him. What makes the film particularly moving is the tenderness with which he is portrayed. He is a helpless victim. But without humour, his despair would be too much, and it is through humour that Forsyth transforms what should be the most devastating scene, when Dickie's girl-friend leaves him, into a comic one. It is a cosy evening before Christmas; Dickie is setting up the fairy lights for the tree, when suddenly Maddie gets up and starts packing her belongings—virtually everything in the flat—into boxes in preparation for leaving. Dickie looks on, astonished. All she can say is that she meant to tell him earlier, but now there was no time to discuss it because the removal men, mutual friends, were due to arrive at any minute. In a state of shock when they do arrive, Dickie actually helps them carry the furniture to the removal van.

Forsyth uses the absurdities of Dickie's daily routine as a disc jockey—such as the news bulletins which are a perpetual stream of disasters, the absurd look-alike competitions, the banality of recording radio commercials—to contrast with the seriousness of his deepening depression. But he also takes time to emphasize that things are rarely as absurd as they appear. When Dickie confronts one of the leaders in the ice-cream dispute and tells him how crazy he thinks it is to argue about ice-cream, the leader replies, 'And what deadly serious occupation are you involved in, Mr "Dickie" Bird!' Through such means, the film has a memorable, bitter-sweet quality.

[58] *Films and Filming*, Aug. 1984.

Despite the greater depth and complexity of the film, it did not receive the unreserved praise of Forsyth's previous features. First, the audience expected something less serious, and secondly, the film itself lacked coherence. It was really an uneasy mixture of two separate stories: the absurdity of the real life ice-cream war underway in Glasgow at the time, and that of an individual's period of crisis. It nevertheless represented a growing confidence in Forsyth as a director, which was to reach its highest point to date with his next film, *Housekeeping*.

Made in 1987, the film marked new departures for Forsyth. Most importantly, it was the first time he had directed outside Scotland (in fact he moved to America for two years to work on the project). It was also the first time, *Andrina* excepted, that the script had been adapted from a book. Set in a small Canadian backwater town in the 1950s, it is the moving and humorous story of the upbringing of two orphaned sisters, first by a succession of elderly relatives and then by their eccentric aunt, Sylvie (beautifully played by Christine Lahti). It focuses on the relationship between the two sisters and how they grow apart; one wanting to be accepted by the other girls in the town, the other searching for her own identity, finally settling for an itinerant life with her aunt. Forsyth was much impressed by the book, which was written by Marilynne Robinson. 'I thought it was more like the kind of films that I wanted to make than my films were . . . It dealt with supernatural things and ground them in reality. I love things that admit magic into our lives.'[59] So great was his admiration for it that he insisted that all he wanted was to make a 'promotional movie of the book. I had no ego as a film-maker as far as this project was concerned.'[60]

Housekeeping is his most absorbing and sensitive film. With its slow pace it strikes a perfect balance between the exploration of the characters and the absurdity of everyday life. As usual the humour is subtle, but here it is particularly wry. An example is at the beginning of the film, which shows how the girls were orphaned. Their mother, close to a nervous breakdown, leaves them with some elderly relatives. Apparently on her way home, she gets her car stuck in a muddy field. After a short wait a group of young boys arrive and help her push it out, only to watch aghast, as she drives over a cliff and into a lake.

The film benefits from being much more focused than Forsyth's previous work. It has a claustrophobic air—attention is so concentrated on the central characters that the viewer is virtually oblivious to life around them. Unfortunately the film did poorly at the box office. One suspects that, as with *Comfort and Joy*, the audience did not respond to Forsyth's apparent change of approach. But the film also

[59] *Listener*, 19 Nov. 1987.
[60] *Sight and Sound*, Winter 1987/8.

Forsyth directing Christine Lahti in *Housekeeping*, 1987, his first film made outside Scotland. © Courtesy of Columbia Pictures Industries Inc. Photograph: BFI.

appears to have suffered from the internal politics of the studio which had backed it, Columbia Pictures. It was given the go-ahead when David Puttnam was the head of the studio,[61] but its release was caught up in the aftermath of Puttnam's resignation, when his successors were unwilling to put sufficient money into advertising films he had commissioned.

Breaking In (1989) was an interim project undertaken during a frustrating period when Forsyth was unsuccessfully trying to raise the finance for *Rebecca's Daughters*.[62] By contrast with his previous projects, in this case Forsyth was approached to direct by a small, independent American production company.[63] The film, based on a script by John Sayles,[64] is the story of an oddball pair of safe-crackers, an ageing pro, Ernie (Burt Reynolds) and an innocent amateur, Mike (Casey Siemaszko) who join forces for what turns out to be a series of misadventures. As Forsyth conceived it, the film is really about 'two lonely people', and clearly it was this aspect which attracted him to the project. Ernie is forced to keep on the move and lead an isolated life to avoid suspicion. Mike is the lonely new boy in town, whose hobby is to break into people's homes, not to steal anything, but just to feel a part of their lives—to eat their food and watch their televisions. But there are, of course, absurdist touches. Ernie lives so close to an airport that massive jets loom over the street, obliterating the sky and shaking the house.

Sayles's script is unusual, clever, and humorous, but it suffers from a lack of development of the relationship between the main characters. Although the film is charming and received a favourable reception in America,[65] it represents something of a step back for Forsyth, a move away from the sophistication of *Housekeeping* and a return to the broader humour of his earliest films.

As Forsyth has gained confidence as a film-maker, his films have

[61] The film was taken on after a deal with Cannon fell through because the lead, Diane Keaton, withdrew six weeks before shooting was due to start. Forsyth comments, 'Basically I forgot to woo her . . . She presumed that she was going to have influence on the project and sent me pages of notes. Not knowing any better, I just read them and put them away. I didn't know I was supposed to re-write the script for her.' (*Films and Filming*, Dec. 1987.) Puttnam agreed to back the project on the condition that Forsyth would direct a more straightforwardly commercial film for him at some unspecified future date.

[62] Based on a script by Dylan Thomas, written while he was under contract at Gainsborough Studios in the 1940s.

[63] Act III Productions. *Breaking In* was the third film they had produced. The two previous films were: *Stand By Me* (1986) and *The Princess Bride* (1987), both directed by Rob Reiner.

[64] Directed and wrote: *Return of the Secaucus Seven* (1979), *The Lianna* (1982), *Brother from Another Planet* (1984), and *Matewan* (1987).

[65] In spite of the way the film was mishandled by the distributors (who were determined to give it a wide release, instead of the usual practice with more marginal films of releasing them in a few cinemas, relying heavily on word of mouth recommendations, and letting the film run for a long time). *Breaking In* opened simultaneously in 400 cinemas; it was playing at the same time in seven theatres in Boston, with the result that the film could not sustain full houses and so exhibitors began to withdraw it.

From right to left, Forsyth directing Burt Reynolds and Casey Siemaszko in *Breaking In*, 1989. © 1989 Breaking In Productions, Inc. and The Samuel Goldwyn Company. Photograph: BFI.

evolved thematically and stylistically, *Housekeeping* representing his most polished work. But this process has resulted in two major problems for him. First, he has been type-cast by the audience on the basis of his earlier films, and secondly, he has had difficulties in finding financial backers for his more recent projects.

Forsyth's early success now curses him. It established his reputation as the maker of light comedies. 'Charming, whimsical and in America, "cute"—those are the words people continually choose to label my films. I don't think they fit at all. I certainly hope not anyway.'[66] What really annoys Forsyth is that the audience sees these aspects as the main focus of his films rather than just a part of them, ignoring the way he has changed. 'I'm not being the normal bitchy artist complaining about critics. It's not a matter of being appreciated, it's just the simple matter of being understood, which is more important to me[67] . . . I'm moving away from comedy . . . I'm much more interested in oddness and eccentricity . . . I'm much more serious than people think I am.'[68] He feels that ironically his films have been misunderstood more in England than America. 'If you ask someone in England what *Comfort and Joy* is about, they'll say it's an Ealing comedy about an ice-cream war. If you ask someone in America, they'll say it's a comedy about depression—which is a lot nearer to what I was intending.'[69] In England he believes that the audience sees the surface comedy, but misses the irony completely, a complaint voiced by the much more caustic satirist and fellow Scotsman, Lindsay Anderson. Forsyth believes that the American audiences are able to appreciate the films because of the directness they share with the Scots, which the English lack.

Although finding financial backers has always proved difficult, it has seriously hampered his recent projects. After his initial prolific output of four feature films in five years, it took nearly two years to finance *Housekeeping*. *Rebecca's Daughters*, after two further fruitless years, has now been abandoned by Forsyth. One suspects that financiers would prefer him to play down the serious side of his films in favour of a more humorous vein. The reason for this is related to the potential audience for his work. As the commercial failure of *Housekeeping* proves, Forsyth's films have become increasingly marginal in terms of the people who are actually interested enough to go and see them. But unlike other marginal British directors, e.g. Jarman, Roeg, Greenaway, who all enjoy an identifiable core of loyal supporters, Forsyth's films still rely on the volatile mass audience. As their viewing habits are unpredictable, the

---

[66] *The Times*, 24 Nov. 1987.
[67] NME, 27 Oct. 1984.
[68] *Western Mail*, 13 Sept. 1984.
[69] NME, 27 Oct. 1984.

most effective means of attracting financiers would seem to be to minimize their risk by keeping budgets as low as possible. Although by Hollywood standards[70] they have always been extremely modest, they have escalated rapidly. *That Sinking Feeling* cost £6,000, *Gregory's Girl* £200,000, *Local Hero* £2 million, *Comfort and Joy* £1.5 million, and with *Housekeeping* and *Breaking In* budgets have stabilized at around £3–4 million. The problem with these larger budgets is not only the initial hesitancy of financiers, but also, once committed, their desire to impose conditions on the director. Making *Breaking In* proved to be a painful experience for Forsyth because of the pressures exerted by the production company and the consequent compromises he had to make. Similarly, an American finance deal for *Rebecca's Daughters* demanded totally unacceptable casting approval. Reducing budgets would arguably offer the prospect of greater independence for Forsyth to make the films he wants, as well as providing alternatives to a dependency on American money, which he feels entails the most compromises.

However, faced with these dilemmas, Forsyth is at present very depressed. He is uncertain of his future direction, even questioning whether he wants to continue to make films at all. One hopes that this is just a transitory phase, for he would be sorely missed.

[70] The average budget of a Studio film is now in the region of $12–15 million.

BILL FORSYTH

You spent many years making sponsored, industrial films before you directed your first feature. Didn't you find that time frustrating?

No, it wasn't really; just being a film-maker was enough for a few years. It didn't matter what you were directing, because inside your head you were a 'film-maker'; the equal of Truffaut[1] or Louis Malle. I suppose that's a measure of how we live inside our own fantasies really. In a way, I still feel the same kind of film-maker as I did then; I don't feel any more successful. Maybe that proves the elusiveness of my success. You see, I feel that film-making is just a job you do. You either enjoy it or you don't, and I certainly don't enjoy shooting the films.

You don't?

The cast and crew may do, but I don't. Maybe it's the old Calvinist Scottish thing, but I don't feel I should enjoy myself when working. At the beginning of each shot I always feel I should be more prepared than I am. It's frightening. The very worst moment of all is the first half an hour of the day's work, when a shot is being set up and everyone is standing around, staring at me, waiting for me to tell them what to do. That's awful.

So why then did you decide to make feature films?

I just felt that it was my entitlement to make movies. There was a myth with my friends in Scotland, and it was half a joke, that maybe De Laurentiis[2] would see one of our movies about ship-building and send us a script to make. I think that everyone half wanted to believe that this was how you got to make feature films—that somewhere at the end of the rainbow this would be the reward for all the struggling. Years and years could pass for some people, just having that dream. I suppose I

---

[1] François Truffaut: French director and a core member of the French 'Nouvelle Vague'. Films include: *The 400 Blows* (1959), *Jules and Jim* (1961), *Stolen Kisses* (1968), and *Day for Night* (1973).
[2] Dino de Laurentiis: international film producer. Produced his first film aged 20. Produced some of the Italian 'Neo-Realist' films; Fellini's *La Strada* (1954), *The Nights of Cabiria* (1956). Later turned to producing expensive, international spectacles, and in the early 1960s set up his own studios near Rome. Other films include: *The Bible* (1966), *Death Wish* (1974), *Three Days of the Condor* (1975), and *Hurricane* (1977).

realized that, in the end, it was never going to happen like that. But it was thirteen years of doing other things before I decided to make a movie myself. And I just fell into doing *Gregory's Girl*, my first script. I asked a couple of writer friends to write it for me and they all said 'no' for various reasons, the last one using the excuse that as I knew what I wanted, I should write it—and so I did. And I found that it was surprisingly easy because there are so few words on a page—it's mostly spaces and dots. It's not the chore that I imagine writing a book is, where you've actually got to fill every part of the page.

What were the major formative influences on your film-making? I sense an affinity with Dylan Thomas, particulary in Local Hero *with the rather eccentric village characters.*

It's difficult to see where the influences come from. I wasn't conscious of any direct influence from Dylan Thomas, but it's often impossible to know from whom you're stealing if you read and think a lot. I read more than I watch movies, so I think I would come more under the influence of writers than film-makers. I remember thinking that that's quite convenient, because the stealing is less obvious; I mean no one would suspect me of stealing from Saul Bellow, although they would from someone like Louis Malle.

I remember that in an interview about Housekeeping, *when asked whether you would recommend someone reading the book or seeing the film, you chose the book.*

Well, I think books are much more important. You have more opportunities to express yourself than you have as a film-maker. Because as a film-maker you're reducing the breadth of the area that you're working in with every step you take. It's a struggle to put across a complex notion in a movie, because you're already working in a limited band of audience reception.

Are there any film-makers whom you particularly respect at the moment?

One film-maker in whose footsteps I would care to follow is Robert Altman.[3] I've never sat down and made a list of my ten favourite films, but if I did *McCabe and Mrs Miller*[4] and *The Long Goodbye*[5] would be on it.

---

[3] Robert Altman: American director who has consistently challenged the traditions of American cinema. Films include: *M*A*S*H* (1970), *McCabe and Mrs Miller* (1971), *The Long Goodbye* (1973), and *Nashville* (1975).

[4] 1971, directed by Altman. It is a non-heroic Western centring on the ill-fated love between a gambler (Warren Beatty), and a whorehouse madam (Julie Christie).

[5] 1973, directed by Altman. An adaptation of Chandler's novel with Elliot Gould as Marlowe.

Do you have much contact with other British directors?

Not really; film-makers are lonely people, but I'm fairly close to Mike Radford[6]—we went to film school together. We show each other our scripts and the rough cuts to our films, but generally we talk more about fishing than film-making. I have passing relationships with quite a few others; I wouldn't say that I'm good friends with Alan Parker, but we've passed through each other's orbits enough times to enjoy each other's company. Pat O'Connor[7] I know quite well and there are a few others I know slightly. Derek Jarman's someone whom I enjoy meeting and talking to, whenever we do come across each other. I think he's a very important guy in Britain. And someone else who's very interesting, although he hasn't made a movie for a while, is Barney Platts Mills.[8] But we never discuss film-making. I might ask a cameraman or an actor what so-and-so is like. 'What does Robert Altman do when he directs and rehearses?' Beyond that you don't actually know how other directors work. The only time that you have social intercourse is either at festivals or award ceremonies, where there's a certain amount of jealousy and uptightness in the air.

Anyway, those award ceremonies are just a farce. The film business, more than any other business in the world, takes itself so seriously. I mean, maybe I don't hear about it, but where do the surgeons give each other awards for the best scalloped kidney—they've got much more right to it? Or where do social workers get together and have the award for the most suicides diverted? The buffoonery and self-congratulation that goes on in the film business is just absurd. I saw in the paper today that a bunch of actors and directors were complaining that the Academy Awards this year were demeaning to their profession because the 'show' lowered the dignity of the proceedings. They're all giving each other gold-plated statues full of sand and they're talking about the dignity of the proceedings. You can't take that seriously.

*Virtually all your sponsored films were made in Scotland, as well as your early features—*That Sinking Feeling, Gregory's Girl, Local Hero *and* Comfort and Joy. *You must feel very attached to the country?*

Not so close as a film-maker. The principle is very simple and has always remained the same—it's just the most comfortable place for me to live

[6] Films include: *Another Time, Another Place* (1983), *Nineteen Eighty-Four* (1984), *White Mischief* (1987).
[7] Irish director. Films include: *Cal* (1984), *A Month in the Country* (1987), *Stars and Bars* (1988), and *The January Man* (1988)—his two most recent films were made in Hollywood.
[8] Iconoclastic British director of very low budget films. Films include *Bronco Bullfrog* (1970)—made for £20,000, *Private Road* (1971)—for which he arranged his own distribution. In the same year, while he was 27, he was appointed a governor of the BFI. But he suddenly stopped film-making and settled in Scotland. *Hero* (1982), his first film for over a decade, was written in Gaelic, and funded by Channel 4.

and have a base. I've taken great pains to say to people that although I live in Scotland, I don't necessarily feel that I'm going to spend the rest of my life working there. I'm sure my relationship with Scotland as a source of material will come and go.

In the past you seemed to be an ardent, even passionate champion of films with provincial British themes rather than anything too internationalized.

I can imagine saying things like that, but the chances are that I was in America when I said it rather than in Britain. You see I've never been a kind of leader in Britain, or taken a stance of offering leadership.

Your films seem to avoid the dramatic structures of traditional Holly-wood films.

Yes, the unseen and the unstated seem to me to be much more important than the dramatic structure of movies, which is generally based on dialogue interchange between characters saying what they think and acting on what they say. This is just banal. But the conventions are so deeply rooted now that its difficult even to step a pace or two aside from this path and still find an audience, unless you're someone like Nicolas Roeg and you're being overtly extreme in the way you want to take an audience. My interest in the minor, quiet things is probably why it's difficult to find a large audience for my films.

Do you think that part of the problem in reaching a wide audience is that the loose dramatic structure lets events flow over the audience, with the result that they are more observers than participants?

Well, you see, if you're trying to get the audience's attention, the more conventional you are, the less of their attention you're really going to get, because they're just going to fall into the trap of following the story and nothing else. So, in order to make anything interesting, you've got to stop the story, or the characters, in their tracks, and reveal something different about them or take them on a different road.

So it is the unexpected which really becomes the story?

Yes, exactly. You see the thing is, because I don't like conventional films, I'm always trying to subvert them. I'm always trying to find ways to sabotage and undermine movies in general, because I don't like them. There might be a situation in which one character is virtually guaranteed to behave in a certain way, but as far as I'm concerned they may be diverted by the smallest of incidents on the side, or motivated by eccentric circumstances. I find it almost instinctive now to make heroes behave unlike heroes. But again it's a problem. Americans have a resistance to characters who are at all marginal—outsiders like Sylvie in *Housekeeping*—unless they are romantic, beguiling, or sentimentalized.

Is this why you also aim at a sensitive, even low key, style of acting for these characters?

Yes, I've always tried to minimalize the way in which a performance is perceived—to bring things down to a naturalistic level. If I made a movie which was dramatically conventional, I would have to direct actors in a very different way. I'd have to ask them to do much more melodramatic things than I've ever asked them to do before, and I don't know whether I'd be able to do that. It bothers me sometimes when I see what people will do up there on the screen. I think, 'Oh God, imagine asking someone to do that—to scream and shout, or cry, or wriggle over a bed in that way. How could you ask another human being to do that?'

Your camera style is also rather quiet; you use a lot of long master shots and avoid any kind of dramatic editing.

That's right. Usually I've just one rule of thumb, which is to try to make the camera serve the action, not the actors *per se*, but the action . . . I think that because of over-emphatic acting styles and over-emphatic camera styles, the audience's options become limited, and they have increasingly been asked to do less and less work. I would hate to be accused of trying to manipulate an audience into a particular emotion at any one time.

What do you look for when casting?

Casting is very important. I spend a lot of time on it, probably because I don't think I'm a very good director. It's one of the safety things I do. I pick very good people, who are really the characters, before I start out. So, only once or twice have I been forced to coax a performance out of someone.

Why do you think you have this urge to concentrate on the abnormal and absurd aspects of people's characters?

I don't actually think I've spent an awful lot of time examining other people. I suppose my interest is actually based on self-scrutiny. I've probably spent too much time examining myself and realizing the ludicrous posturing an individual does *vis-à-vis* what he thinks he is and what the world thinks he is. When I was younger I had quite an active dream life—a virtually schizophrenic life—and I think that's probably quite an influential thing.

Like Walter Mitty?

Yes, exactly. I was another character with a name, a background and

At the height of Forsyth's popular success. From left to right, a beardless Forsyth discusses a scene from *Local Hero* (1982) with Burt Lancaster, Peter Riegert, and others on a remote Scottish beach. Courtesy of Goldcrest Films and Television.

characteristics. And it was me but it wasn't me. For instance, he was ten years older than I was, and unlike me led a glamorous, exciting life. This was probably more instructive for a film-maker than all those years making documentaries, because I was dealing with plot, characters, and incident.

All the main characters in your films are dreamers as well.

Yes, and I suspect that most people are like that to one degree or another.

How do you approach script-writing?

Well, unlike America where you have to keep coming up with ideas for films, I just work on an idea until it happens. I spend a long time, maybe a year thinking about the script before I start writing. On *Comfort and Joy* I mulled over the idea for five years. This is the part I like best, where you still have total control. For about six months I take notes, thinking things through and structuring them on bits of paper, but not actually writing. I think the real secret is not to sit down at the typewriter too early. The result of all this is that I've never been in a position where I've had a script which I haven't made into a movie. Most people, I would imagine, have some dream project which they haven't realized yet. But I've worked from script to script and from film to film, and ticked them off as I go along. But it's probably a stupid way to work.

In what way?

In two ways. Maybe I'm just learning now in the last couple of years that things have to be more fluid; for instance, I can make a movie like *Breaking In* because I've got a year to kill. Maybe I'm just growing up as a film-maker as far as the practicalities of film-making are concerned. Also, I think my films probably all suffer from being under-written in terms of the potential that they have, although I think the dialogue's pretty good. It's a lesson I could learn from people in America—the way they work and produce drafts and redrafts. I've never done any more than rewrite the original script. I've always had a kind of empty pride, feeling that there's something cute about actually doing the writing in three weeks and then making it, as happened with *Comfort and Joy*. It's also such a relief to finish a script that the last thing you want to do is to revise it. But the material deserved more attention than that, and one of the disciplines I'm trying to adopt is to take script-writing more seriously.

But it seems that while shooting you have always been open to modifying the script, taking it as a guide rather than anything more rigid.

Well, I think to be creatively flexible is one of the prime requirements of a film-maker. It's almost a reflex, that when you're shooting a scene, constantly to look for ways of making it different from what's on the page. Most of the enjoyment I get from film-making is that moment when you do change something and it works, and you're seeing it for the first time. There's really nothing quite like the pleasure of that first enjoyment of a new element.

Can you give an example?

Most of the best ones don't end up in the movie because by the time you get back to the cutting room, you've gone off them. I remember one in *Local Hero* when the American and the young Scottish boy are talking on the beach and behind them the church is emptying. I thought this was in itself very funny. But it was actually Chris Menges who said, 'Wouldn't it be interesting if the boy saw the people,' and this was an unusual thing for him to say, because at the time his ideas on story and character were very sparse. So we just did it and that was it, a 'moment'. I was killing myself laughing when we were shooting it. That laughter for me was the best thing of all. I was acting like an audience, enjoying it like a movie, and to my mind I couldn't see how anyone in the world wouldn't double up with laughter watching that. And that's the best moment of creativity when you feel that.

Do you see any problem in not working with your own material as you have done with Housekeeping, Breaking In, *and also your Dylan Thomas project?*

Well, *Housekeeping* was my project entirely. I privately bought a two-year option on the book and I adapted and wrote the screenplay, and only at that stage went around trying to finance it. I put that project together in the same way I would a project I had written from scratch. So it wasn't as if someone said to me, 'Hey, why don't you make a movie out of this book?' With *Breaking In* it was the first time that I'd compromised that way of working by taking a script which had been written by someone else. At the same time, I did have quite a hand in substantially adapting it to suit the way I wanted to tell the story. But it's not something I'd readily do again. As soon as there's a third party in there, another author—the script-writer John Sayles, then every opinion of mine carries less weight. It becomes just one opinion among others. This is certainly something that I would like to avoid in the future and the solution is very simple. Just write your own material.

With *Rebecca's Daughters*,[9] Dylan Thomas had written a fairly good

---

[9] Scripted by Dylan Thomas while he was under contract at Gainsborough Studios in the 1940s. It is a tale about the Tollgate Riots in the 1840s when the Welsh rioters dressed up in their wives' clothes to tear down the toll gates as a protest against oppressive landowners.

story—it was a fun yarn. He treated historical events in such a way that you're completely drawn into the characters and not the spectacle. So it didn't need to be extensively revised. But I didn't really feel very close to the material, and only took it on for pragmatic reasons. It was offered to me right after I'd finished *Housekeeping*, and looked as if it would start immediately. It was originally going to be shot in Scotland, which was very appealing because I'd been working in the States for the previous two years. But it proved harder to finance than the producer [Chris Severnick] originally thought, and with each problem I became less and less interested. Finally, at the end of last year [1989] I pulled out of the film, and was glad to be rid of the cumbersome commitment. But once again, maybe it has been a false step for me, taking me away from original material. I think that I have been sidetracked by other people's projects.

You seem to be becoming increasingly involved in America, and less in Britain; both Housekeeping *and* Breaking In *were made and financed there.*

Well America really *is* the film business, even if you work in Britain. I've been trying to find a way of operating around this fact, but it's very difficult. It's virtually impossible to finance even a modest movie in Britain at the moment, and even to do this you always need some American money. In the finance scenario for *Rebecca's Daughters* 30–50 per cent of the budget was going to be from America. An individual film-maker doesn't really have a choice to decide where he's going to raise his finance anymore. You simply can't ignore America, and therefore you can't ignore the American market—even I, who try to maintain a distance from it.

But the lower the budget, the less important the American market becomes in terms of breaking even?

Yes, but it's still always a factor, and you can't escape it.

How easy has it been to raise American finance for your films?

Well, the problem with the studios is that all that they offer me are big movies. I suppose Alan Parker would tell you the same story, that Orion is quite happy to give him $20 million to make *Mississippi Burning*, but if he wanted $6 million for a script, he would have a hard time. They're quite happy to hire you as a working talent; to have you process their material for them, to pay you a lot of money, and give you a lot of freedom, but then you're generally not doing the film you want to make. There is also a wealth of independent sources of finance springing up; companies like Act III who did *Breaking In*. But I found that quite difficult because there are actually so many marbles lying around, that it's hard to know which one to grab!

Why then did you find it so hard to finance Rebecca's Daughters?

I don't know. It was a great shock to me. Admittedly it was within the top budget range for independent movies in Britain—$4–5 million. This didn't seem a lot of money to me at the time, because I used to think that $5 million was a low budget. But the business has changed so much since then that that's not true anymore. Even for a historical subject such as this, the condition for American money was casting approval—they were suggesting that Michael J. Fox would play a nineteenth-century Welshman! It was because I couldn't get it off the ground that I made *Breaking In*. It was the first movie I've made for reasons other than 'inspiration'. It was very different from *Housekeeping*, my other American movie, which I spent two years trying to finance and which in the end was only incidentally set in America.

So why did you do Breaking In?

When you begin making movies your motives are all very pure. That lasts maybe three or four films; then, not so much 'cynicism' but 'practicality' descends on you. *Breaking In* became an interim thing. It was also a bit of an experiment to see what it was like to make a movie in America, with an American cast and an American story; albeit something which wasn't absolutely mainstream, and something that I felt familiar with. It was, at least, a chance for me to see where I stood *vis-à-vis* the American audience; both how far I had to go to get them, and how far they would be prepared to involve themselves in what I did.

But your films have been comparatively popular in America.

The movies I'd made before had reached an audience there, but always within a very esoteric area. People liked them for patronizing reasons. They were made in Scotland and seemed strange, exotic, foreign, small, and interesting. *Breaking In* was a way of judging how much of an American audience I could really get.

Do you feel then, that the audience hasn't made enough effort in the past to appreciate the serious side of your films?

Yes, that's right. I've come to the painful realization that movies play a very small part in people's lives. They don't use them to be spiritually enriched, but want to be diverted in a very superficial way. Take the ice-cream war in *Comfort and Joy*. I was trying to point out the absurdity of any war by catching people unaware, by making them laugh at an ice-cream war, and then asking them to think why this is less absurd than any other war. But the reviews showed how the film was misunderstood. It was even described as 'a Disney movie for adults'.

But people may respond to parts of your film in the way you would like them to.

But is it worth it for those little bits? The truth is that I'm too proud to be simply a processor of entertainment, which is what most people want from a movie. Just to be rewarded by tiny little successes between the cracks is not enough at the moment for me.

What would be enough?

I would want what I think doesn't exist anymore, lots of people who watch movies in a serious, reflective way. Maybe it's just because when I was younger I watched movies in that way and wanted to learn from them. Maybe people still do this, but I seem to have lost touch with them. And if this doesn't happen then there is no point in making films.

This sounds like the lead character in Comfort and Joy *who needs to find justification for what he does. In the end, the fact that his radio programme comforts an old woman in hospital is enough.*

Yes, but it's awfully hard work.

How far are you ever willing to make compromises with financiers?

Well, I am resistant to most of the easy roads on a movie that financiers and producers would take—in casting, or in simplifying stories in such a way that an audience would accept them. I don't want to package dreams for anyone, because this undermines the kinds of thing that I want to say, which are fairly realistic. So I think I've probably got a reputation for being fairly stubborn, even mischievous in my ambitions for a movie. Standing back and looking at myself, I think that the reason I do all this is that I really dislike movies so much that I'll do everything I can to undermine them, even my own. It's also because I am actually scared to make conventional movies. I'm scared to own up to the sentimentality and banality involved. It would be much easier if I could jump that hurdle—I'd have a much happier time.

Probably with *Breaking In*, I made more compromises than I've made before, and yet I still find myself in a situation where I'm reeling from the effects of defending the film that I wanted to make. And this was with a script which wasn't my original material, and for which I shouldn't feel obsessive ownership, and for a film which I undertook virtually to pass the time. So I think that will tell you how little I would tolerate compromise. I know that I've just got to decide whether to make conventional films or not; it's as simple as that. But there's no point in me trying to make conventional films and then subverting them.

Disk-jockey Alan 'Dickie' Bird (Bill Paterson) is accosted by an unlikely admirer in Forsyth's *Comfort and Joy*, 1984. © Weintraub Screen Entertainment. Photograph: BFI.

But surely it's not so black and white? There's such a diversity of film-makers in Britain who all find an audience. For directors like Derek Jarman their audiences may be relatively small, but at least their work is being seen and appreciated.

I think they enjoy it though. If I enjoyed the film-making process then it would be easier to ignore all the drawbacks. But as you say, it is possible for people to survive and make their own personal movies. I'll just have to see what's possible for me. That'll be interesting.

I sense that your recent experiences in America and the failure to raise the finance for Rebecca's Daughters *have changed your feelings about the practicalities of film-making for a director such as yourself.*

Yes, I feel myself at a crossroads. I'm reassessing whether I really want to continue making movies at the moment. I just don't seem to fit in, and am in a position of frustration right now. I suppose I was a little idealistic when I first started and have now realized that every so often you have to do things that you really don't want to do just to be able to continue a career as a film-maker. I'm beginning to learn all these compromises that go on, and I'm keen to get back home[10] and lick my wounds.

I think the problem is that I've been too close to the production end of things in these last two movies. Before then I was protected by good producers, and also by my lack of interest. When I was working with David Puttnam on *Local Hero*, or with Clive Parsons on *Comfort and Joy*, I just let them do what the producer did. I didn't get involved in any financing, attend any meetings, look at a budget, or get involved in any marketing or distribution debates. So maybe its just having been exposed to all this and the compromises which pour down on a project that has made me a little sour. Maybe I should just try to stay away from that as much as possible and get back to good producers who can protect me from it.

But it seems that it will always be hard for film-makers outside the mainstream.

Yes, I suppose that's all I'm saying really; it's tough, and at the moment the rewards just seem to be very practical ones, but not much else. Maybe I'm just retreating because I've been away from home for around three years, and I've also been away from my home spiritually—in the

---

[10] Interview conducted in Los Angeles after the making of *Breaking In*.

sense of not being close to my own material, not having written the original scripts for my past few films. So maybe this is just a little homesick cry that I'm uttering. I don't know. I'm going to have to spend a year or two writing original material[11] and trying to get a project off the ground, and I'll see if I become inspired doing that. I'm certainly not inspired working as a kind of 'process' film-maker—I've barely done that with *Breaking In*, but I got an inkling of what it was like and I know that I don't enjoy it.

Has your passion for film gone?

Well, right now I couldn't maintain that I had any passion for film-making.

But you did have?

Ten years ago there was passion. You see the passion comes from wanting, before the event, because it is a passion to prove something, to make movies. And after that you realize that a movie is just a movie—it's not going to cause an earthquake; that your best film isn't better than John Huston's or Alan Parker's—it's just a good movie. And so the passion ultimately fizzles out because of the limitations of the goal; because movies are really not that important. At the end of the day you're sitting with an audience of four or five hundred people and all they want to do is to be entertained. They don't want to be moved, to have their hearts race, to cry—well, they want to cry because someone is dying of cancer, not because they are watching the most sublime film they've ever seen. You see we're dealing with a medium which really only wants to involve itself in the superficial manipulation of emotions.

I used to love film, I used to be extremely passionate about it, and it makes me quite sad to realize that I'm not. As soon as you mentioned the word I felt the loss.

What do you feel when you look back on what has been twenty-five years of film-making?

Well, I would like to feel that I've been in love with film for at least twenty years out of that twenty-five. Seriously in love with film. And

---

[11] Forsyth is currently working on a script for his next film, *Being Human*, which he describes as a 'poem'. He feels no need to continue to write about Scottish themes: 'There's too much self-analysis in Scotland. It's the sign of an uninteresting mind which has to look in on itself' (interview with the authors). This film (to be produced by Robert Colesberry, who had previously produced *Housekeeping*), is due to go into production in the spring of 1991; it is Forsyth's most ambitious project to date, and will be shot in Scotland, as well as in several continents.

maybe that love is going through some kind of crisis, but I hope it comes out of it soon. I'm sure it's just the difficulty of making things happen in a commercial environment which is distressing me right now. But also, as a film-maker, I'm fairly naïve and immature. I've just got to settle down into some way of working which I can both tolerate and which also works. God knows what it is!

FILMOGRAPHY

BRIEF BIOGRAPHICAL DETAILS

Born in 1946. Educated in Glasgow. At the age of 17 he started working for a local documentary film company. He made two experimental films which were shown at the Edinburgh Film Festival. In 1971, he was part of the first intake to the National Film School. The same year he set up his own production company *Tree Films* in Glasgow. He made his first feature film in 1979.

FILMS DIRECTED BY BILL FORSYTH

1969   *Language*. 17 mins. PRODUCER AND EDITOR: Bill Forsyth. PHOTOGRAPHY: Martin Singleton.

1970   *Waterloo*. 42 mins. PRODUCER AND EDITOR: Bill Forsyth. PHOTOGRAPHY: Jonathan Schorstein and Martin Singleton.

Still Life With Honesty. Scottish Arts Council.

1972   *Islands of the West*. Highlands and Islands Development Board.

1973   *Shapes in the Water*. Highlands and Islands Development Board.

1974   *Tree Country*. Charlie Gormley/Forestry Commission.

1976   *Connections*. Highlands and Islands Development Board.

1977   *The Legend of Los Tayos*. 45 mins. Tree Films. PRODUCER: Charlie Gormley. PHOTOGRAPHY: David Peat. EDITOR: John Gow.

1979   *That Sinking Feeling*. 92 mins. Bill Forsyth Films. SCRIPT: Bill Forsyth. PRODUCER: Bill Forsyth. PHOTOGRAPHY: Michael Coulter. PRODUCTION DESIGNER: Adrienne Atkinson. EDITOR: John Gow. MUSIC: Colin Tully. CAST: Tom Mannion, Eddie Burt, Richard Demarco, and members of the Glasgow Youth Theatre.

1980   *Gregory's Girl*. 91 mins. Lake Film Productions, in association with the National Film Finance Corporation/Scottish Television. SCRIPT: Bill Forsyth. PRODUCERS: Davina Belling, Clive Parsons. PHOTOGRAPHY: Michael Coulter. PRODUCTION DESIGNER: Adrienne Atkinson. EDITOR: John Gow. MUSIC: Colin Tully. CAST: Gordon John Sinclair, Dee Hepburn, Clare Grogan, Jack D'Arcy.

1981   *Andrina*. 50 mins. BBC Scotland. SCRIPT: Bill Forsyth (from a short story by George Mackay Brown). PRODUCER: Roderick Graham. PRODUCTION DESIGNER: Guthrie Hutton. CAST: Cyril Cusack, Wendy Morgan, Sandra Voe, Jimmy Yuill, Dave Anderson.

1982   *Local Hero.* 111 mins. Enigma. SCRIPT: Bill Forsyth. PRODUCER: David Puttnam. PHOTOGRAPHY: Chris Menges. PRODUCTION DESIGNER: Roger Murray-Leach. EDITOR: Michael Bradsell. MUSIC: Mark Knopfler. CAST: Burt Lancaster, Peter Riegert, Denis Lawson, Peter Capaldi, Fulton Mackay.

1984   *Comfort and Joy.* 106 mins. Lake/EMI/Scottish TV. SCRIPT: Bill Forsyth. PRODUCERS: Davina Belling, Clive Parsons. PHOTOGRAPHY: Chris Menges. PRODUCTION DESIGNER: Adrienne Atkinson. EDITOR: Michael Ellis. MUSIC: Mark Knopfler. CAST: Bill Paterson, Eleanor David, Clare Grogan.

1987   *Housekeeping.* 116 mins. Columbia Pictures. SCRIPT: Bill Forsyth (from the book of the same name by Marilynne Robinson). PRODUCER: Robert Colesberry. PHOTOGRAPHY: Michael Coulter. PRODUCTION DESIGNER: Adrienne Atkinson. EDITOR: Michael Ellis. MUSIC: Michael Gibbs. CAST: Christine Lahti, Sara Walker, Andrea Burchill.

1989   *Breaking In.* 94 mins. Act Three Productions. SCRIPT: John Sayles. PRODUCER: Harry Gitters. PHOTOGRAPHY: Michael Coulter. PRODUCTION DESIGNER: Adrienne Atkinson, John Willett. EDITOR: Michael Ellis. MUSIC: Michael Gibbs. CAST: Burt Reynolds, Casey Siemaszko, Lorraine Toussaint.

FILMS PRODUCED BY FORSYTH

1971   *A Place in the Country.* Films of Scotland Committee.

1975   *The Living Land.* Highlands and Islands Development Board.

1978   *The Odd Man.* South Bank Show (Documentary about Scottish writers).

OTHER WORK

1967   *Chronicle.* BBC. ASSISTANT EDITOR: Bill Forsyth.

        Z Cars. BBC. ASSISTANT EDITOR: Bill Forsyth.

        Play for Today. BBC. ASSISTANT EDITOR: Bill Forsyth.

1990   *Cinema Masterclass.* Channel 4. One programme, examining Bresson's *Au Hazard, Balthazar* (1966), hosted by Forsyth.

BIBLIOGRAPHY

Scotsman, 21 Sept. 1979. Forsyth on film financing.

Screen International, 27 Sept.–4 Oct. 1980. On *That Sinking Feeling* and *Gregory's Girl*.

Glasgow Herald, 22 Nov. 1980. Elizabeth Lyon on Forsyth's documentary career in Scotland and how he made his early films, *That Sinking Feeling* and *Gregory's Girl*.

Sunday Times, 7 June 1981. Philip Oakes on Forsyth's early career in film and the financial problems in getting *That Sinking Feeling* off the ground, and *Gregory's Girl*.

Times Educational Supplement, 12 June 1981. Includes discussion of his early experimental films *Waterloo* and *Language*.

Guardian, 13 June 1981. Derek Malcolm on Forsyth's working methods with reference to *That Sinking Feeling*.

Daily Express, 22 June 1981. Ian Christie on Forsyth's attitude to making films in Scotland.

Films Illustrated, Aug. 1981. Neil Sinyard criticizes *Gregory's Girl* and *That Sinking Feeling* for not offending enough people!

Scotsman, 26 Aug. 1981. BBC to screen *That Sinking Feeling* before the customary three years have elapsed since its theatrical release. Forsyth is worried that this will jeopardize its box office potential.

London Evening Standard, 28 Aug. 1981. Sue Summers on Forsyth's anger at the BBC showing *That Sinking Feeling*.

Guardian, 29 Aug. 1981. GTO distributors, who own the rights to *That Sinking Feeling*, made a deal with the BBC to show the film without gaining the approval of the Cinema Exhibitors' Association. The article notes the speed between the deal being made and the film being broadcast.

Sight and Sound, Autumn 1981. The text of a paper to be delivered by Forsyth at a BFI symposium in which he talks about his background in the industry, how he managed to make *That Sinking Feeling* and *Gregory's Girl*, and what he thinks British film-makers should be aiming to produce.

Broadcast, 7 Oct. 1981. On Forsyth's anger at *That Sinking Feeling* being shown by the BBC.

Daily Mail, 28 Nov. 1981. Robin Eggar on *Andrina*.

Radio Times, 28 Nov.–4 Dec. 1981. On Forsyth's 50-minute TV film, *Andrina*, an adaption of a short story by George Mackay Brown.

Vogue, Nov. 1981.

Blitz, No. 6, 1982. On *Local Hero* and the loss of independence Forsyth faces if he decides to work on larger budgeted productions.

Glasgow Herald, 25 Jan. 1982. On *Comfort and Joy*.

New York Times, 25 May 1982. Lawrence Van Gelder on Forsyth's career to date with emphasis on his pre-feature film years.

Screen International, 10 July 1982. Quentin Falk on location of *Local Hero*, with an interview with the producer, David Puttnam.

New York Times, 16 July 1982. Chris Chase on *Gregory's Girl* opening 'wide' in the US, and a discussion of this film and his first, *That Sinking Feeling*.

Rolling Stone, 30 Sept. 1982. Chris Auty's location report for *Local Hero*.

Sight and Sound, Autumn 1982. Quentin Falk on *Local Hero* consolidating Forsyth's reputation.

Allan Hunter and Mark Astaire, *Local Hero: The Making of the Film* (Polygon Books, 1983).

Observer Magazine, 6 Mar. 1983. On location report for *Local Hero*.

Scotsman Magazine, Mar. 1983. Richard Wilson discusses Forsyth's relationship with Scotland, and his views on Hollywood films and their effects on the audience.

Time Out, 11–17 Mar. 1983. Interesting article by Martyn Auty on Forsyth's career, and *Local Hero*.

Screen International, 12–19 Mar. 1983. Colin Vaines talks to Forsyth about the making of *Local Hero*. Alex Sutherland discusses how the film was marketed in the UK.

Films, July 1983. Article assessing and comparing Forsyth's films.

Sight and Sound, Summer 1983. Interview with Forsyth about his perceived reputation.

Glasgow Herald, 12 Dec. 1983. Forsyth to receive an honorary degree from Glasgow University.

Sunday Times, 22 Jan. 1984. Location report on *Comfort and Joy*.

New York Times, 12 Feb. 1984. On *Comfort and Joy* being a sad film, and marking a departure from Forsyth's previous films.

Screen International, 25 Feb.–3 Mar. 1984. Forsyth talks about *Comfort and Joy*.

Movie Maker, May 1984. Interesting interview, particularly on Forsyth's documentary background and first two films.

Sunday Times Magazine, 12 Aug. 1984. George Perry on Forsyth's career to date, *Comfort and Joy*, and his love of Glasgow.

Films and Filming, Aug. 1984. Interview with Forsyth on his career and *Comfort and Joy*.

Glasgow Herald, 18 Aug. 1984. Andrew Young on *Comfort and Joy*. Forsyth discusses how people have misunderstood his work, especially in England; his views on violence in films; future projects; and his approach to developing ideas and writing scripts.

Western Mail, 13 Sept. 1984.

New Yorker, 1 Oct. 1984. On *Comfort and Joy*, and a general review of Forsyth's career.

Sunday Express Magazine, 7 Oct. 1984. On *Comfort and Joy* and Forsyth's career to date.

New Musical Express, 27 Oct. 1984. Interview on *Comfort and Joy*.

Hollywood Reporter, 29 Oct. 1984. Forsyth plans US film with producers Clive Parsons and Davina Belling.

American Film, Nov. 1984. Interview with Forsyth about his view of his career so far.

Alexander Walker, *National Heroes: British Cinema in the Seventies and Eighties* (Harrap, 1985).

The Times, 5 Jan. 1985. On *Gregory's Girl* and its TV première on Channel 4.

Film Quarterly, Spring 1985. An attempt to fit Forsyth's films into a 'Scottish Cinema'.

Film Directions, Spring 1985. Interview with Forsyth about actors; the themes in his work; and the politics of his films.

Glasgow Herald, 21 July 1986. Andrew Young profiles Forsyth, and looks at *Housekeeping*; before the last minute withdrawal of the lead, Diane Keaton.

Listener, 19 Nov. 1987. Interview about *Housekeeping*, including its traumatic pre-production history.

The Times, 24 Nov. 1987. Review on Forsyth's career to date, and *Housekeeping*.

Films and Filming, Dec. 1987. Forsyth talks about the production history of *Housekeeping*, and about plans for *Rebecca's Daughters*.

Time Out, 2–9 Dec. 1987. Interview on *Housekeeping*.

Sight and Sound, Winter 1987–8. Forsyth outlines the making of *Housekeeping*.

Andrew Yule, *Puttnam: A Personal Biography* (Mainstream, 1988).

Films in Review, Feb. 1988. Interview on *Housekeeping*.

4 STEPHEN FREARS

Stephen Frears at home in Notting Hill, London. © H.P. Productions.

ESSAY ON
STEPHEN FREARS

STEPHEN FREARS is a prolific film-maker, having made over thirty television plays and nine films, but one who until recently has gone largely unnoticed. This can, in part, be attributed to the fact that in television drama, where so much of his work has been concentrated, attention traditionally focuses on the writer, actors, and even producer, rather than on the director himself. But typically, Frears seems not to have minded this lack of credit or attention. He is modest, self-effacing, and self-critical, while his appearance is often on what might be described as the scruffy side of casual. He is, as many of his colleagues have pointed out, extremely easy to work with. The writer Hanif Kureishi[1] described it well, 'Stephen is so good at creating a relaxed atmosphere on the set. We all felt free to say whatever we liked at all times and frequently did.'[2] The cameraman Brian Tufano[3] tells a similar story:

His brain is so alert that if you suggest something, he can immediately see the value of it and how it will fit into the narrative. We built up a working relationship where we were able almost to talk to each other without saying anything. For example for *Daft as a Brush* (1975) I just said to him that visually this film ought to be dry stone walls, the type one finds in the Yorkshire Dales and he said that was exactly it. He knew precisely what I meant and that was the sole preparation we gave to the visual interpretation of the film.[4]

Another cameraman with whom Frears has frequently worked, Chris Menges,[5] agrees:

Stephen is good to work with because he doesn't try to tie you up in knots. He gives you a certain amount of freedom and also ideas from which you can build

[1] Frears's script-writer on *My Beautiful Laundrette* (1985) and *Sammy and Rosie Get Laid* (1988).
[2] *Photoplay*, Dec. 1985.
[3] Cameraman on many of Frears's films including *Sunset across the Bay* (1975), *Daft as a Brush* (1975), *Three Men in a Boat* (1975), and *Play Things* (1976). He also filmed Ken Loach's *The Price of Coal* (1977), Alan Parker's *The Evacuees* (1975), and directed several television plays himself.
[4] *Sight and Sound*, Spring 1978.
[5] Cameraman on *Gumshoe* (1971), *Last Summer* (1976), *Cold Harbour* (1978), *Bloody Kids* (1980), although they did not work together after Menges turned down working on *The Hit* (1984). He has worked extensively with Ken Loach; has filmed *Angel* (1982), and *Local Hero* (1983), and done much documentary work, as well as directing his first feature film, *A World Apart* (1987).

. . . When we started *Cold Harbour* (1978), Stephen told me how he was very concerned to get close to the people in the story. That's about all he said, but it gave me the whole conception of how to photograph it.[6]

The composer George Fenton[7] also praises Frears's collaborative techniques. 'It's a gift Stephen has . . . of keeping the product fluid until the very last moment. Thus, you really feel that a bunch of people have endorsed your contribution, and you, in a way, have endorsed theirs.'[8]

Frears's attitude to what his role and responsibility are as a director has remained largely unchanged over the years. 'I have no sense of myself as an originator or as an *auteur* . . . it embarrasses me to be thought more artistic than I am.'[9] He sees himself more as a facilitator rather than a creator, and encourages contributions from all involved. This has made him an extremely popular director with some of Britain's finest script-writers. He is faithful to their themes and preoccupations, and sets his sights at bringing out the script's maximum potential. Alan Bennett praises him as 'very much a writer's director in the sense that he follows your intentions through'.[10] As such he always likes to have the writers on set while shooting, and relies on them enormously. 'I wouldn't cross the road without one,'[11] he jokes. But when questioned why he does not take a more personal and direct stand on the exact form and content of the scripts, even if he does not write them himself, he responds: 'I have no capacity for originating material. I just lack the courage which writers have; also I'd be no good at it.'[12] Although, while working in television drama, he was much influenced by Ken Loach, who improvises extensively on the set, Frears emphasizes their different attitudes to the script. 'I was unlike Ken because I was much more interested in the writer's contribution. For me, TV was a writer's medium.'[13] Christopher Hampton, who scripted *Dangerous Liaisons*, hints at a greater degree of influence: 'Stephen is a rare director in that he insists I be there, whereas most directors bar the writer from the set. He's very open to suggestions. He's like Brecht, who also asked people around him what they thought. But Stephen knows more about what he wants than he lets on.'[14]

[6] *Sight and Sound*, Spring 1978.
[7] George Fenton worked for Frears on the music for *A Visit from Miss Protheroe* (1978), *Bloody Kids* (1980), *Going Gently* (1981), *Walter* (1982), and *Dangerous Liaisons* (1988). He has also scored *Gandhi* (1982), *The Jewel in the Crown*, *Company of Wolves* (1984), and *Clockwise* (1985).
[8] *Time Out*, 1–7 Dec. 1978.
[9] *Film Comment*, Mar.–Apr. 1987.
[10] *Sight and Sound*, Spring 1978.
[11] *Sunday Times*, 26 Apr. 1987.
[12] *The Times*, 2 Nov. 1982.
[13] *Film Comment*, Mar.–Apr. 1987.
[14] *Sunday Times Magazine*, 12 Feb. 1989.

In numerous subtle ways throughout his work Frears maintains his own individual tone. For a start, it is manifest in the sorts of scripts he decides to make. Moreover, it is, of course, apparent in the particular talents which he contributes when trying to put the script on film: a straight-faced, gentle sense of humour; and an attention to detail, particularly in the subtleties of feeling or the evocative creation of an atmosphere or period setting in which he carefully avoids any sense of nostalgia, awe, or whimsy.

In the scripts he has chosen, certain themes and preoccupations, albeit contrasting, do arise. He likes making films about contemporary Britain, many with a highly political undertone. To him, they are 'films about people living on the underside of society in a secret world'. Certainly his early career was heavily influenced by the 'Free Cinema', and the British Social-Realist movements. He looks back to the sixties, as a particularly vibrant period for British film, but one which was never satisfactorily built upon. He was personal assistant to Karel Reisz on *Morgan: A Suitable Case for Treatment* (1966), and to Lindsay Anderson on *If . . .* (1968), both of whom stressed the importance of personal and even political commitment in one's work. They taught him that 'you have to get the real life right.'[15] Frears feels that 'Karel and Lindsay were my fathers really. And you know, parents are terrible things to have. Because I admired them so, I became a bit inhibited and cautious. Not that someone like Lindsay likes that sort of attitude. I think he'd prefer me to go away and make *The Texas Chain Saw Massacre*.'[16] Another director who influenced Frears's commitment to social and political film-making was Ken Loach. Frears's work in television drama in the 1970s came at a time when Loach, a strong polemicist for the left, was the dominant force. However, Frears has never restricted the range of social/political topics he wants to explore. They remain broad, often determined more specifically by the writer, than by any political crusade Frears may wish to champion. They include racism, attitudes to the mentally handicapped, the problems of retirement, the treatment of political refugees in Britain, redundancy and unemployment, and even the American abandonment of their South Vietnamese allies in the mid-1970s. Moreover, Frears never had the documentary background of much of 'Free Cinema' or the Social-Realist movements: 'I have always liked narrative films—especially the work of John Ford, Orson Welles, Robert Rossen, Howard Hawks.'[17] Not surprisingly, he also has a taste for less polemical drama, which is more consciously 'entertainment'.

He has made several comedies, generally of a gentle, witty nature: one

[15] *Observer Magazine*, 19 Apr. 1987.
[16] *Guardian*, 7 Dec. 1980.
[17] *Cinematograph Weekly*, 26 Dec. 1970.

for Tom Stoppard, and numerous highly amusing plays for Alan Bennett and Peter Prince, although more recently he has moved into satire with films for the 'Comic Strip' team.

He is also attracted to certain genres. 'I love thrillers. I can't get enough of them.'[18] Three of his features, *Gumshoe*, *The Hit*, and *The Grifters* (1990) have been thrillers; *Eighteen Months to Balcombe Street* (1978), was a dramatized documentary about an IRA cell operating in London; *Last Summer* was about adolescent car thieves; and he has even parodied the thriller genre (in particular the television series *The Professionals*) in *The Bullshitters* (1984).

In his more recent films Frears has become fascinated by rather bawdy, sexual affairs, both heterosexual and homosexual. *My Beautiful Laundrette* (1985) marked the start, and his interest continued in *Song of Experience* (1986), *Prick Up Your Ears* (1987), *Sammy and Rosie Get Laid* (1988), and *Dangerous Liaisons* (1988). Consistently he handles these subjects with naturalness and subtlety, never sinking to crudity, nor losing a sense of fun.

All these themes suggest a certain rebellion against the norm, a certain idiosyncrasy. But although Frears's films may be off centre, they cannot, even at their most radical, be said to have actually separated themselves from mainstream cinema. His rebellion is not the anarchic satire of a Lindsay Anderson nor the outrageousness of a Ken Russell. Even the sexual elements in his films lack genuine eroticism. His work is much more controlled, more restrained—one might almost say too British to be totally rebellious. This is, in part, because of the inherent restrictions in working so much in television, rather than cinema. Frears does get frustrated: 'Honestly, I just long for someone to say to me "Let's have a bit more sex and violence". On the whole I think that TV is too bloody respectable.'[19] But his approach is certainly not just a product of the system. It is far more deeply rooted than that; nor are his films for the cinema significantly more radical.

The nature of Frears's non-conformist inclinations is better understood by looking at his upbringing. His background was middle class—his father was a GP—and provincial. He was brought up in post-war Leicester, which he describes as a particularly unimaginative city, lacking any fun. Indeed he describes his whole childhood as 'not a story of deprivation but of lack of imagination'.[20] He was sent away to prep school at 8, and then to a public school, where he reacted against his treatment and surroundings. 'I don't think of myself as having been a rebel then: I just sulked when I didn't like school. I actually *wanted* to

[18] Interview with the authors.
[19] *Guardian*, 7 Dec. 1980.
[20] *Observer Magazine*, 19 Apr. 1987.

conform, and I didn't understand my non-conformity. As I've got older I've become a better non-conformist.'[21] But today's radicalism is one which he himself describes as 'hooliganism accompanied by extreme cowardice'.[22]

Frears made his first short film, a political piece called *The Burning* (1967), after leaving Cambridge University and working for a year at the Royal Court. In a sense it was a fable about the end of colonialist fantasies. It concerned a white household in South Africa going through an annual routine of having lunch with relatives, which this year coincided with the outbreak of revolution. It was shot in Morocco on a £15,000 budget and was financed by the BFI and Memorial Enterprises. Suddenly, he comments, he was a 'film director', and he moved on to do a series of children's dramas for Yorkshire television.

Out of one of these films (in which two youths steal a Jaguar car and act out bits from old movies) came the idea for his first feature—*Gumshoe* (1971).[23] It was written by his actor/writer friend Neville Smith. The film is actually about the fantasies of a Liverpool bingo caller (Albert Finney) who advertises his services as a 'private eye' in the local paper, but soon finds himself in deep water. Initially they set out to make a thriller comparable to *Bullitt*[24] but felt uneasy about the violent world they knew little about and which did not suit the English context; 'If we had tried to make a film like *Bullitt* in this country it would end up being called *Pellet* . . . here policemen wear such silly helmets.'[25] In fact it is more a response to the Bogart/Chandler type of movies than a parody of them. 'It's about fantasies and the way they weave in and out of our lives. Everybody in the film is incompetent—even the professional killer.'[26]

After *Gumshoe* Frears settled into television, where he remained for the next ten years, working for both the BBC and Independent Television. He was accused of retreating into television, and partly this was true. As Frears frankly admits, he was particularly suited to working in television, where one is offered projects rather than having to initiate them oneself.

I can't operate unless there's a big organisation behind me. I rather enjoy the idea of being hired. I know it's a rather unfashionable notion. I think I've always been rather sceptical of independence as a notion. If someone offered me freedom, I wouldn't know what to do with it. Panic, I should think.[27]

[21] Ibid.
[22] *Sunday Times*, 26 Apr. 1987.
[23] The title is American slang for a foot-slogging private detective.
[24] Directed by Peter Yates, 1968.
[25] *Time Out*, 10–16 Nov. 1983.
[26] *The Times*, 11 Dec. 1971.
[27] *The Times*, 2 Nov. 1982.

However, particularly in retrospect, he recognized the advantages of being able to work continuously with a range of top writers; good material—the scripts he was offered were more attractive, and of better quality than those from other sources—and competent actors and crews, without having to fall back on conventional solutions (such as the use of a star, or gratuitous sex and violence) to raise the money. Nor did he have to face the interference of producers, which typically occurs in feature films. As he pointed out at the time, 'I'm not permanently embattled with the TV companies about material—nobody has ever tried to cut a scene, or anything like that.'[28] Consequently, this was a prolific time for Frears, and he made some twenty-six films; far more than would have been possible had he been making feature films for the cinema. Their variety owes its origin largely to the authors that he has worked with regularly, in particular Neville Smith, Peter Prince, and Alan Bennett.

After *Gumshoe*, Frears worked on scripts by Neville Smith on two further occasions—*Match of the Day* (1973), and *Long Distance Information* (1979), in both of which Smith acted. *Match of the Day* is about an Everton supporter, Chance (so called because, many years before, his parents took one in a moment of passion) who, because of his sister's wedding, misses the match. The film hilariously recounts the ways he tries to listen to the day's results, and his attempts to pick up a girl while at the reception. *Long Distance Information* is a portrayal of a Presley-obsessed DJ called Christian on the night that the 'immortal king' died. It was an excellent, light-hearted vehicle for a study of a non-religious, leaderless generation of people in need of a new God to worship.

Frears has worked with Peter Prince on four television plays, a feature film (*The Hit*), and Frears's only stage play (*Television Times*, 1980). These plays all have contemporary British settings and represent some of Frears's more aggressive work. *Play Things* (1976) is a black comedy about a young play leader working in a Notting Hill playground, a born appeaser who is made to pay protection money, first by the white kids, and then by the black kids, and finally even by their parents. He is an anti-hero—although one's initial sympathy for him, based on his dedicated ability to turn the other cheek, slowly alters. Ultimately, he has as little sense of responsibility as those around him. *Early Struggles* (1976) was less successful, although, as a study of the domestic trials of a pop musician deserted by his wife and left to bring up his baby, there are some delightful moments and interesting themes. *Last Summer* (1976) is a play of gallows humour following the progress of an amateur thief from the occasional car to seasoned crime. It is not just dramatic and

---

[28] *Sight and Sound*, Spring 1978.

A pensive Frears with Janice Rule on the set of his first feature, a parody of the American detective genre, *Gumshoe*, 1971. © Courtesy of Columbia Pictures Industries Inc. Photograph: BFI.

entertaining with a graphic atmosphere of criminal sleaziness (shot by Chris Menges with his camera often perched on car bonnets), but also gives serious thought to the way in which the need for urban excitement can develop into professional crime. Its success led to the making of *Cold Harbour* (1978), whose title refers to the dreary, joyless London — the harbour to which a young Chilean political refugee flees. To remain in Britain legally, she is forced to marry, and although she eventually achieves safety she gets little else.

Frears's collaboration with Alan Bennett produced very different works. Bennett sees 'two sides to Stephen's films — a nostalgic, gentle side, and another side which is much more to do with action. The things I write probably cater for the first, while the things he's done with Peter Prince, for instance, seem to be the other side.'[29] Frears's long-lasting and close relationship with Bennett has resulted in some of his best films. Plots are simple, locations often in the North, and the work is characterized by Bennett's highly naturalistic dialogue and the humour he inserts into the humdrum routine of ordinary life. Frears and Bennett, are in some ways very alike. They are both hard-working professionals pretending to be lucky amateurs, and they share a fascination with the English character: 'The films I have made with Stephen are all very English, because he's of a similar cast of mind; they're diverse, and, I hope, humane.'[30]

A Day Out (1972), the first television play for both of them, was shot in monochrome, and is a quiet portrayal of an Edwardian summer outing by the Halifax cycling club. Each of the characters is charmingly sketched as they cycle off in their knickerbockers and boots, and the whole film is given great poignancy by the final reference to their future fate in the First World War. *Sunset across the Bay* (1975) is about the loneliness of a retired Leeds couple in the resort town of Morecambe. It is a typical Bennett/Frears play. The characters are plain but three dimensional; and, while full of humour, it has a sense of indignation at the way years of toil are so often unrewarded. *A Visit from Miss Protheroe* (1978) is also about the way the changing world can so often leave one far behind. Miss Protheroe visits a retired friend ostensibly to cheer him up, but in fact only ruins his tranquillity by maliciously telling him that the thirty years of work he had spent perfecting the office's accounting system is now obsolete because of the new computer system. In 1978/9 Frears offered a series of six Alan Bennett plays to the BBC, who rejected them on financial grounds. His response was to try London Weekend Television, and to produce them himself — his only production experience to date, and one which he did not relish. 'I really got no

[29] *Sight and Sound*, Spring 1978.
[30] *The Times*, 27 Nov. 1978.

pleasure out of it at all.'[31] Frears actually directed four of the six plays—
Me, I'm Afraid of Virginia Woolf, Doris and Doreen, One Fine Day,
and *Afternoon Off*—which, as with most of Bennett's writing, pursue
the themes of lower middle class life.[32]

In the early 1980s Frears made three large-scale television films
Bloody Kids (1980), scripted by Stephen Poliakoff, David Cook's *Loving Walter* (1982), and *Saigon, Year of the Cat* (1983) by David Hare.
The former was so successful that it was given a theatrical release three
years later. It was deliberately anti-Thatcherite—an energetic, forceful
portrayal of two school children in the street culture of a society in
decline.

Loving Walter (divided into two parts *Walter*, and *Walter and June*),
starring Ian McKellen, was the first film shown on Channel 4, and was a
frank appraisal of the complex issues surrounding mental health. It is
one of the most direct pieces Frears has made; 'I wanted it to have plenty
of punch, and I hope audiences find it very demanding and provocative.'[33] Jeremy Isaacs, then Chief Executive of Channel 4, was moved to
comment that Ian McKellen's performance was the most powerful and
moving that he had ever seen in television drama.

Saigon, Year of the Cat was a film about the American withdrawal
from Saigon. It was constructed around a romance—rather like
Casablanca, but it was not particularly successful. 'Saigon is about
defeat. It's about rats turning and running—not such popular subjects.
Also, people were surprised that we were making a film about Vietnam,
but not about fighting, helicopters, and people running about.'[34]

During these later years in television, Frears was becoming anxious to
break out and do a feature film: 'I want to get bigger; bigger actors,
bigger toys,—expand.'[35] He also needed to escape the feeling that he had
to make films about contemporary Britain. 'I've been exercising that
responsibility for ten years . . . You can't escape it, but I guess I've had
enough for the moment. You can't go around forever generating that
kind of anger. Maybe I'm just as defeated as everyone else by Britain.'[36]
He failed to raise money for *Prick Up Your Ears*, a feature project he

[31] *Time Out*, 1–7 Dec. 1978.
[32] *Me, I'm Afraid of Virginia Woolf* is about the hang-ups and insecurities of a lonely
English literature teacher at a polytechnic in Halifax; *Doris and Doreen* focuses on two clerks
suddenly faced with the prospect of redundancy and the end of their cosy office life; *One Fine
Day* is about the life crisis of an estate agent estranged from his family and job; *Afternoon Off*
was, however, more lighthearted, and, with some excellent perceptions on the English way of
life, is the story of the stumbling search of a shy, young, Chinese waiter through a Northern
seaside town to find his blind date.
[33] *The Times*, 2 Nov. 1982.
[34] Interview with the authors.
[35] *Time Out*, 10–16 Nov. 1983.
[36] *City Limits*, 7–13 Sept. 1984.

wanted to make with Bennett. But he did fulfil his long-standing desire to make a serious thriller with Peter Prince's *The Hit* (1984), on a budget of £1.2 million. 'Making *The Hit* was such a relief, a way of getting away from all that . . . all the films I seemed to have made recently in Britain were so dismal . . . I suppose I was slightly burnt out on *Bloody Kids* . . . Not having to shoot in the streets of South London, and in the rain is a great weight off your mind . . . It was such a treat.'[37] Two assassins, strikingly played by John Hurt and Tim Roth, come to take their revenge on Terence Stamp, who by turning crown witness had previously sent his criminal associates to gaol for ten years. He had spent these years in exile in Spain waiting and preparing himself for the moment when they would return. The film concentrates on the journey, after his capture, as he is driven by the two assassins to Paris for summary justice, and the tension and atmosphere is built up with great style and panache. It is a film about death and how to meet it, but disappointingly the actual narrative—specifically how the story is set up and finally resolved, is weak. Although it received some good reviews, it was not a great success, and 1984 was a time of minor crisis for Frears. 'I felt somewhat confused so I went to teach in a film school and that sorted it out.'[38] But it was really the success of *My Beautiful Laundrette* in 1985 which gave him new confidence. Frears recognizes its impact, 'I became more hard-headed, more direct, I think, and more angry with the times.'[39]

The international critical and commercial success of *My Beautiful Laundrette* was a great surprise considering that it was a low budget film initially made for television. Besides, Frears and the writer, Hanif Kureishi, did not think it was an option to make it for the cinema. As Frears points out, 'You couldn't seriously have gone out and said to a financier "I'm going to make a film about a gay Pakistani Laundrette owner".'[40] In fact it is about much more than this—the family, racism, homosexual love, Pakistani immigrant subculture, and commercial opportunism. Frears felt that the script was 'a very accurate and ironic analysis of Britain under Thatcher, a really strong piece of radical writing'.[41] It was described by Paul Taylor[42] as 'a story which empties the contents of three old kitchen sinks into one washing machine'. Despite the meagre budget of £600,000, Frears was able to bring the wonderful script (it got an Oscar nomination) to life. The fact that it was not very stylized visually and had a loose narrative structure was particularly

---

[37] *City Limits*, 7–13 Sept. 1984.
[38] *Radio Times*, 15–21 Feb. 1986.
[39] Ibid.
[40] *Stills*, Nov. 1985.
[41] *Sunday Times*, 26 Apr. 1987.
[42] Al Clark (Editor), *Film Yearbook*, Vol. v (Virgin, 1986).

and Laura Del Sol in *The Hit*, 1983. © Zenith Productions Limited. Photograph: BFI.

appropriate for the subject and mood of the film. Frears's dead-pan sense of comedy made the rather dark subject matter seem full of pathos. One of the finest parts of the film is the homosexual love affair between Omar and his friend Johnny, in particular its final scene, described by David Robinson as 'one of the most delicate and touching love scenes in contemporary cinema'[43] — an ending which gives a sense of hope and tenderness to the chaos. Daniel Day Lewis, who played Johnny, jokingly described how Frears directed it:

Stephen was remarkably tactful. Before the love scene, rather than drawing us aside and having a few quiet words with us, he shouted at the top of his voice, 'who's on top?' It could have been a nightmare if you don't get on with the people that are involved. The most shocking thing about the relationship in *Laundrette* is that it's no different to a love that exists between anyone else — and that's really how it should be.[44]

The success of the film at last enabled Frears and Bennett to get the £1.8 million *Prick Up Your Ears* off the ground. The film is about the life and violent death of the playwright Joe Orton[45] (Gary Oldman), and his relationship with his homosexual lover, mentor, and ultimately his murderer Kenneth Halliwell (Alfred Molina). It is based on the biography by John Lahr. The older Halliwell taught Orton about literature and art, while Orton introduced him to promiscuous sexual adventure. While Orton grew in fame, Halliwell, left in anonymity, felt increasingly abandoned. The script has a careful structure; the film unfolds with a series of flashbacks prompted by scenes with John Lahr himself. This serves several functions. First, Lahr's own troubled marriage echoes Orton's and Halliwell's. Secondly, by having the murder at the beginning of the film, it avoids making it too tragic or sensational, enabling the audience to focus on the psychology rather than the plot. Finally, it meant that Bennett did not have to commit himself to a particular opinion of Orton as a playwright — any opinions being depicted as John Lahr's. Bennett is thereby better able to parody Orton. The film itself becomes Ortonesque, and there are some utterly delightful moments, aided by the brilliant performances of both Alfred Molina and Gary Oldman. But it is not an exploration of Orton's creativity; it is a film about a relationship — the relationship between victor and victim typical of so many 'marriages'. Nor did Frears want it to be sensationalist, 'another merry-go-round about the homosexuality and the murder'.[46] In

---

[43] *Sight and Sound*, Winter 1985/6. David Robinson reviews *My Beautiful Laundrette*.
[44] *Photoplay*, Dec. 1985.
[45] The sixties cult, satirical playwright and author of *Entertaining Mr Sloane*, *Loot*, and *What the Butler Saw*.
[46] *Interview*, Apr. 1987.

Daniel Day Lewis in *My Beautiful Laundrette*, 1985. Despite Frears' two decades in television, it took the success of this film to bring him to the attention of the British public. © Mike Laye/Working Title Films. Photograph: BFI.

these respects it was undoubtedly successful. It stands as one of Frears's finest works and one of the best British films of the eighties.

1987 saw the return of the Kureishi/Frears partnership for the £1.2 million *Sammy and Rosie Get Laid*. In America several publications refused to print the 'Get Laid' part of the title to which Kureishi responded, 'if we'd known we would have made it much worse.'[47] It is about the fluid relationship of six extremely different characters, set, like *My Beautiful Laundrette*, against a background of a changing England, race riots, and inner city decay. Indeed the film opens with the irony of Thatcher's speech about the urgent need to tackle the inner city problem following her third election victory. Frears described it as a 'very, very provocative film, written at a time of political despair and designed to stir up some signs of life in the audience'.[48] Compared with *My Beautiful Laundrette*, it is a more confident, although also more polemical film. It is devoid of any sense of charm and spontaneity, and its political punches lack sophistication, and indeed, any great sense of conviction. It was very much pushed into production on the back of *My Beautiful Laundrette*, without adequate time to develop the script. It would also have benefited from being shot on 16 mm., which would have given it a more gritty feel. Nevertheless, it was an amusing comedy of manners; especially in the way in which the sexual liaisons of the various partners are portrayed in the tradition of Sheridan or Vanbrugh.

After completing *Sammy and Rosie Get Laid*, Frears finally decided to accept an American offer, from Lorrimer Films, to make a large-scale project. The film, *Dangerous Liaisons*, was an adaptation of the eighteenth-century novel by Laclos, with a budget of $14 million—huge compared with Frears's previous budgets. Christopher Hampton,[49] who wrote the script, had also adapted it for the stage, where it was a great success on Broadway and the West End. He described Frears's approach: 'Stephen is allergic to the theatre. He hadn't seen the play or read the book, before he read the screen-play. His reaction was to develop those things that were most cinematic, and so the process of adaptation was away from the play and towards the novel.'[50] The setting is eighteenth-century France, and the subject the machinations of aristocratic sexual politics. Although partly a film about lust, it is far more about ambition and manipulation, where the particular aim is seduction. Christopher Hampton believes that the subject matter holds a particular fascination today: 'Recently, both in England and America,

---

[47] *Films and Filming*, Jan. 1988.
[48] Interview with the authors.
[49] Frears had worked with Christopher Hampton before, on *Able's Will* for the BBC in 1977.
[50] Interview with Christopher Hampton, *South Bank Show*, broadcast on 4 Feb. 1989.

institutionalised selfishness has been encouraged so that the characters' behaviour seems to strike a chord. People recognise the greed—not for money since the characters are unbelievably wealthy—but for power.'[51] Frears made a conscious decision to play down the opulent trappings of the period in order to emphasize the story's contemporary relevance. Milos Forman, who made his own film version of the book, *Valmont* (1989) at roughly the same time, took the opposite view and made an epic period piece. But Frears's subtle use of close-up on the actors, avoiding the wide shot where the settings distract the eye, is more successful. The actors cast in the three main roles, Glenn Close, John Malkovich, and Michelle Pfeiffer, were all Americans, whereas traditionally European actors would have been used for a film of this nature. This was partly at the insistence of the studio, but also because Frears wanted to avoid making 'a museum piece, a respectable film. It seemed that it was *popular* material—love, passion, death, and a tremendous plot—which somehow all fits into the American tradition. Americans, like Brando or whatever, are more equipped and used to doing that.'[52] The emotional power of the film, and the tension given by a plot using a challenge as its central narrative theme (rather like a whodunnit in reverse—a will-he-be-able-to-do-it) gave it an audience, especially in America, which was particularly large for a serious period film.

Frears has continued to work in America, where he has attracted the attention of major American directors, such as Martin Scorsese, who has produced his current film *The Grifters* (1990). The movie is a thriller based on the novel by Jim Thompson, which Frears describes as a cross between 'a "B" movie and Shakespeare'.[53] Set in present-day Los Angeles, it focuses on the relationship between three people living in a world of con-artists, or as they are called in American slang, grifters.

Dangerous Liaisons and *The Grifters* seem to represent a growing confidence in Frears. He has widened his horizons. He is moving away from films for television, from British subjects and politics, and, in the process, appears to be fulfilling his desire to make entertaining but still intelligent films in the model of the great American narrative directors.

[51] Press Pack for *Dangerous Liaisons*.
[52] *Start the Week*, Radio 4, 6 Mar. 1989.
[53] Interview with the authors.

STEPHEN FREARS

What do you think was the origin of your interest in films?

When I was at Cambridge it was a terrific time for cinema. The time of the *Nouvelle Vague*, a time when European films were being seen all over the world—Bergman, Antonioni, and so on. It was at that moment that cinema changed profoundly (although American cinema was not very good at all as I remember it). Cinema was becoming intellectually respectable. There was a masterpiece coming out every week—now you are lucky if there is a good film once a year. This was the point that I started going to see films but the idea of actually working in cinema never crossed my mind. My first interest was in theatre.

How did that develop?

Theatre just seemed a way of getting out of the home life I had led. That sounds rather cruel on my parents as in fact my father was a GP and a rather remarkable man. But when you live in those provincial towns, you just want to get out. The local repertory company was very dazzling, and a great influence on me. I used to see those people on stage and see them at parties and so on. Their lives seemed so imaginative and nice, and their work so attractive. Then at Cambridge the theatre was very active; full of people with ambition and I got involved in all that.

Did you have ambition?

Yes, I think so. I had aspirations to lead a more interesting life, surrounded by brighter people. They were the only group of people I wanted to be with, although I didn't really feel a part of them. I didn't feel much sympathy for the very classical sorts of plays they were doing. I was rebellious. I had met Lindsay Anderson on a holiday, I think in 1959, when I was 17 or 18 and was very influenced by what was going on at the Royal Court. Those were plays that I could relate to. But at Cambridge they were doing Chekhov and so on. I remember doing a musical with Richard Eyre[1] as a way of attacking Trevor Nunn. After Cambridge I applied to the BBC and Granada because I knew that all the bright people were going that way—everyone at Cambridge wanted to

---

[1] Later directed *The Imitation Game* (1980), *The Ploughman's Lunch* (1983), *Loose Connections* (1983), and *Laughterhouse* (1984).

be in 'show biz'. Michael Apted[2] and Mike Newell[3] got places. I was turned down. I spent a year or so sharing a flat with Bill Oddie[4] and got a job at Farnham Rep. as an Arts Council Trainee, and then went on to the Royal Court, which was nominally being run by Anthony Page[5] and Lindsay [Anderson]. It was a very complicated place, like Byzantium — full of intrigue. You attached yourself to a patron and Lindsay was the person I felt sympathetic towards.

Was he a big influence?

Yes, I was very impressed by him. I had never met anyone with that sort of intellectual imagination, range, vitality, and sparkiness. But I actually learned more from Karel [Reisz], whom I met through Lindsay when I was 23. Lindsay said to Karel, 'there's a good lad there. Use him as your assistant on your play.' But the play collapsed and Karel asked me to work on his film [*Morgan: A Suitable Case for Treatment*, 1966] as some sort of vague assistant. I didn't know anything but I suppose he found it good to have someone asking direct questions. Karel was a more fatherly and protective man than Lindsay, who felt that one should just stand on one's own feet. Karel took me into his family, and taught me a lot of things and was much more benign. Lindsay is combative, but he is impressive in his solitariness.

Does your preference for more radical drama — politically and in terms of mood — stem from your time at the Royal Court?

I think that my radicalism is emotional rather than political. When I was young I was called 'anti-social'. I suppose I was always scowling in the corner. After my A level exams, which I took when I was very young, I just stopped working until I was sent by my parents to a teacher who prepared boys for university and who had an incredible influence on me. He turned me around in two months, and helped me get into Cambridge. At Cambridge my rebellion was just as uninformed and showed itself in sulkiness and resentment. I felt an outsider and didn't know how to pull myself together.

Today, I choose to live in this area [Notting Hill, London, an ethnically diverse area] because I feel comfortable here — not because of any political views. I wouldn't flatter myself that my ideas are coherent. It's just someone moaning about authority without quite being able to understand it. This developed out of a training in subservience. I was

---

[2] Directed *Agatha* (1978), *Coal Miner's Daughter* (1980), *Gorky Park* (1983), and *Gorillas in the Mist* (1988).
[3] Directed *The Awakening* (1980), *Dance With a Stranger* (1984), *The Good Father* (1986), and *Soursweet* (1988).
[4] Comedian, later one of the Goodies.
[5] Directed *Absolution* (1978), *The Lady Vanishes* (1979).

brought up afraid of authority and still am. It's a familiar English characteristic. Look at Lindsay. He is very authoritarian and rebellious at the same time. People like us come out of these public schools and universities and expect the right to attack them. It's rather a privileged position. Hooliganism is a better word to describe what I do than radicalism. Same with Lindsay. We're just public school boys being anarchic. I think he sees himself as a surrealist, as a sort of Evelyn Waugh. I don't think that is Lindsay's strength. One should look at independent film-makers, like Derek Jarman. They do it for nothing. I have to be paid before I'll go to work. They are heroic. What they risk is far greater than what I risk.

How did you get such a major film as Gumshoe *off the ground with comparatively little work behind you?*

I had had very little experience at that time but I sat down with a friend of mine [Neville Smith] and he wrote a script which became *Gumshoe*. I got Albert [Finney], with whom I had worked in the theatre, because the script was very good. When he said he would do it, people queued up to finance it. I was completely shielded by Albert's position. I must have driven everyone mad. I could more or less handle the material—I understood that, but I didn't know how to shoot a film and I must have behaved gracelessly. I stumbled through it like a blindman, merely with some instinctive sense of what was good.

How did you come to spend the next decade in television?

It wasn't deliberate. It was partly that nobody offered me a job directing a decent movie. Then Alan Bennett said, 'Why don't you shoot this?'—a script that had been shuffled around the BBC for about eighteen months—and so I got the job [*A Day Out*, 1971].

And you just continued working in television?

Yes, in retrospect it was partly rudderlessness, cowardice, and also some sense of survival. Working freelance, I went from job to job. I was very nervous and insecure, although because I had some instinctive sense of what was good material, I went from one good script to another. It also gave me the opportunity to learn. But my early career was generally dominated by a response to the material; although I was offered things that I was expected to like. I didn't direct anything.

Ken Loach was a dominant figure at the BBC at that time wasn't he?

Yes, Ken Loach was the greatest influence on me at that time. He had an enormous impact on BBC drama. His work was absolutely wonderful. *Kes* and his TV films had an enormous influence just because they were so good. His influence, however, was aesthetic rather than political. Also

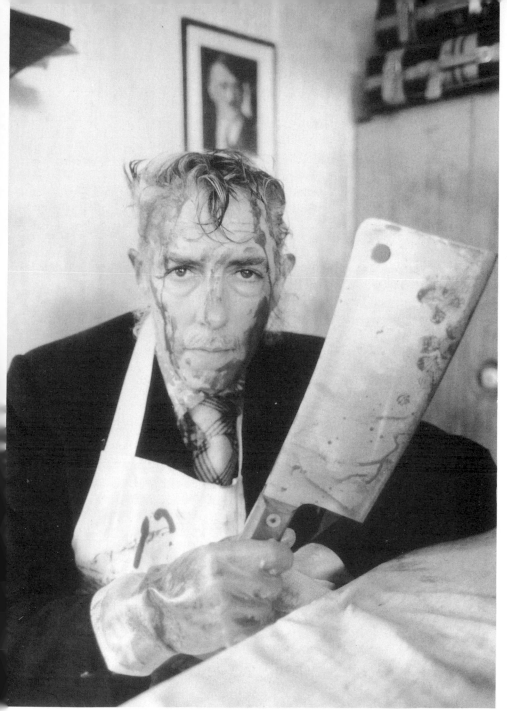

Peter Cook as Mr Jolly in *Mr Jolly Lives Next Door* (1987), one of a series of Comic Strip Productions directed by Frears. © Channel 4. Photograph: BFI.

I remember to an extent, reacting against Loach—having a terrific argument with a cameraman saying, 'Don't do that shot. It looks like a Ken Loach shot.' Really I think that after *The Price of Coal* he ran into an impasse, which I imagine had a lot to do with his personal problems. He became rather rigid and reluctant to change. I always worked with the same cameramen as Ken did. I would say to them, 'Tell him not to shoot in that way, tell him to track,' but Ken would say, 'Listen, it's worked for the last ten years, so why should I change it?'

Was it while working in television that you found yourself forming a personal style?

Yes, the style started to take over for itself. I remember it quite clearly. It was while doing *One Fine Day*—with Dave Allen. I suppose I felt myself more confident with the camera. I find it hard to be specific. I had seen things like *Mean Streets*[6] and *Taxi Driver*[7] from America and they were beginning to influence me, helping me be more specifically cinematic. Before then my style was dominated by the cameramen, especially Chris Menges and Brian Tufano—although *One Fine Day* was actually shot by Charles Stewart. Also, Nat Crosby[8] was an important later influence.

How did these cameramen influence you?

With Brian I really learned what was going on. Initially he taught me to shoot rather classically and straightforwardly. Then we did wonderful long shots and I slowly started to understand it all more and more. Chris was much more . . . 'brilliant'. But he is a very, very dominant, rather 'bullying' chap, and he was trying to find his own way but was rather inarticulate. I did five films with him and love him, and it was a very hard break when I stopped working with him. But he was difficult and terribly dominant. Nat shot *Going Gently* fantastically. He gave emotion and intensity to the work. But we only really worked together after *One Fine Day*. I'm still looking to develop my style but not from the same perspective as before.

Do you feel that there is any difference in the way films should be made for television as opposed to the cinema?

I don't know that I acknowledge a great difference any more, except in so far as certain material is more appropriate for television because it is not economically viable for the cinema. On a deeper level I did think for a long time that genres were central—that cinema was to do with men with guns, and that ended with me making *The Hit* in 1984. Now I don't

---

[6] Directed by Martin Scorsese, 1973.
[7] Directed by Scorsese, 1976.
[8] Among other things Director of Photography on Schlesinger's *An Englishman Abroad* (1983) and *Madame Sousatzka* (1988).

know about that, because I made *My Beautiful Laundrette* for the TV, and it wasn't part of any set categories or prescribed genre, and yet was very successful in the cinema and less so on television. There is a sense, I suppose, in which that very domestic sort of English writing has traditionally not been for the cinema, except that now it is in fact proving rather popular and that's jolly good. When I was making TV films I used to shoot them in a style that was in fact regarded as primarily cinematic. It was very ambitious for television—really terrific. I remember I shot *Walter* (1982) largely on a 'steadicam', and then others did that as well. This was in fact breaking down traditional television story-telling devices in the use of close-ups and all that. I can see that *The Last Emperor*[9] and *2001*[10] are big screen films, but the truth is that I haven't got a clue about what the differences should be.

Why did you leave the BBC and move into film?

Well, one reason was that the BBC offered a privileged system where you didn't have to get all your money back and eventually it got rather unsatisfactory. It was too protected. I started to want an edge to things and so did other people at the BBC. But I don't think of television as a kind of childhood like Alan Parker, who thinks that grown-ups make films. I'd go back to television like a shot if the material is good.

Both as a television director and a film director, how far can you see continuity of themes and style in your work?

This is a question which I find very difficult to answer. I find it very hard to place myself in any graph of creativity. I direct other people's scripts. I do think that I have a specific style although what it is I can't really say. I certainly can see that there is a process where I have absorbed material and regurgitated it in later works. Because other people tell me so, I suppose that there is continuity in terms of themes in what I do. I direct what I like and that was why I was always very happy working with Alan [Bennett], Neville [Smith] and Peter [Prince]. I felt that I could recognize and understand their work *intuitively*. I remember turning down a Barry Hines[11] script at the BBC because I wasn't comfortable with it. It was too polemic and in a way conventional. Also I've never been in a position where I've done things that I haven't wanted to do— although I've done things for television that I'm less enthusiastic about. I do what my guts tell me. You get hooked for something for whatever reasons. So I made *Liaisons* because I thought the story was terrific. All my agents told me not to do it with John Malkovich. But I wanted to do

---

[9] Directed by Bernardo Bertolucci, 1987.
[10] Directed by Stanley Kubrick, 1968.
[11] Many of Loach's films are based on his writings: *Kes* (1969), *The Gamekeeper* (1980), *A Question of Leadership* (1980), *Looks and Smiles* (1981).

it; and with him. In the end I take what these very good writers give me and put it through my own system and the film comes out. That is sufficiently creative for me.

Don't you want your films to be a form of self-expression?

Aspects of my films used to be very private, perhaps because of some neurotic feeling, and I could never understand why something which seemed to me to be wonderful was not apparent to the audience. Now I think that that is completely idiotic. The idea of the enclosed work of art no longer gives me pleasure. The whole joy lies in getting across, in communicating to the audience. I can see that that might not be paramount for other people, but I like to be liked. The idea of audiences all over the world seeing your work is startling. I really want to give the audience a wonderful time, and I view them with absolute respect. So, for instance in *Laundrette*, you put in women who are more beautiful or men who are more attractive, and jokes to make people laugh. It seems to me that this doesn't contradict personal self-expression. I was brought up on John Ford's films, which are entirely expressive of his character. They have sensitivity and intelligence, and are yet enormously popular. But there is no law which says that the audience have to be interested in what you are preoccupied with, and that doesn't mean that you are wrong to make it. It's just bad luck. *Laundrette* was my most popular film, but I'd be very surprised if it was better than a number of other films of mine that I could mention . . . I don't know . . . I suppose that they're the ones that interest me. I think *The Hit* is terrific and although I've never really understood why that film wasn't successful, I suppose the reason I liked it was because of all the peculiar aspects of the characters; the reason the audience didn't like it was because it was not couched in conventional terms, but a combination of the gangster and road movie genres. It's so haphazard whether the audience will like a film or not.

Do you feel that there are any problems in balancing the various respon-sibilities to the crew, the financiers, the audience, and your own views of the material?

No, not at all. That's the joy of it all. That's the fun. But I have no sense of compromise.

How do you conceive your function as a director on a film?

Sydney Pollack,[12] when asked about the director's function, replied 'Well, I turn up on the set in good physical shape.' It seems to me that to

[12] American director of *Three Days of the Condor* (1975), *Tootsie* (1982), and *Out of Africa* (1985).

turn up day after day and to keep going is a big contribution, and in fact I think that *Sammy and Rosie Get Laid* was harder than anything I've ever done in those terms. But I do have responsibility for everything that goes on the screen. Involvement varies from director to director and with the different stages of the film. It really depends on your relationship with your colleagues and whether you feel that they are better left alone. But the director's job is essentially about making everything else work. And responsibility is not synonymous with the 'authorship' of a film. The writers and actors can make such huge contributions. I don't put my name on the posters like Alan Parker—'An Alan Parker Film'. Good for him, but that's not my way.

How do you try and get the best out of your actors?

I tend to cast them well and them I'm just grateful to them. And if they are well cast, I try not to intrude on the journey the actor makes. I would not really know how to help them and would not be very good at that process of rehearsal which actor and director go through in the theatre. It seems to me that the less I have to say the better they will be. Generally speaking, they understand the characters rather better than I do. I just encourage them to relax and invent, and maybe refine what they offer with suggestions. I try and give the actors an enormous amount of breathing space, and confidence. Once they've brought it to life then you can fine tune it, and orchestrate it. But in the end it is the actor who has to bring the character to life. I get enormous pleasure out of what an actor does. It's terrific just watching them.

In casting how do you balance acting ability with the importance of the appearance or 'look' of an actor?

'Looks' are important because you've got to believe in someone. Given that very hard choice between someone who is more accurately cast and someone who is more talented, I suppose I'd lean towards the one who is more talented. But appearance is still important—it certainly was in *Laundrette* and in *Prick Up Your Ears*; for instance Gary [Oldman] was absolutely obvious. In *Sammy and Rosie* casting the singer Roland Gift [Fine Young Cannibals] as Danny was largely because he was so good looking. But I thoroughly auditioned him and he was excellent. I cast Michelle Pfeiffer in *Dangerous Liaisons* because she is very moving. She is also the most beautiful woman in the world, so I don't know where one begins and the other ends. . . . Is she moving because she is so beautiful? . . . Hard to say.

What is your relationship with the author of the script?

I've always had a good reputation with writers—throughout my career people have sent me material which has been terrific—relationships were forged with Neville [Smith] or Peter [Prince] or Hanif [Kureishi] and I've become infatuated with their work. A tradition of having the script-writer actually on the shoot arose by accident, since Alan Bennett, who lives a fairly lonely life, enjoyed the company; but it also makes very practical sense. The script-writer is absolutely central to the film. So for *Sammy and Rosie* you can't take away the fact that it is Hanif's life that I filmed—it would be absurd, posturing, for me to claim more. It's a view from Hanif's window, and I just tried to realize this script as vividly as possible. It's a very, very provocative film written at a time of political despair, designed to stir up some signs of life in the audience.

How involved are you in the writing of the script?

Obviously, it varies. My involvement is pragmatic. I say to the writer, 'What do you mean by that?' They explain it and I say, 'Well, that isn't really what you've done here. Wouldn't it be more interesting to do it this way?' On *Dangerous Liaisons* I am accredited with putting the letters back in the film. *A Day Out* was only a seventeen-page script, and I had to make it more robust. On *The Hit* we never got the ending right. We tried and wrote it three or four times. So it varies, but I've never done any writing myself. I have never been tempted.

How much idea do you have of how the film will look before you make it—in terms of set design, camera movements, colour, and so on?

I don't think I have any sense of how the film will look. I can't pretend that my films are more visual than they are. I don't think of colour with any confidence. I don't really storyboard. The use of close-ups, for instance, in *Dangerous Liaisons* was something that was decided while shooting. The shots were not working. Every line in the script resonated so we just had to come in to see the actors' eyes. I tend to rely on the people around me, they know far more than I do—especially the team on *Dangerous Liaisons*. What the cameraman, set designer, costume designer and so on have to say about their subjects is infinitely more interesting than what I have to say. I concentrate on the bits which I can do—creating conditions in which the actors can work, working on the narrative, or the subtext of the film. I would say, however, that the shape of the scene in my films is largely determined while shooting. I leave fewer decisions for the cutting room than most directors.

Most of your films have had very low budgets. What techniques do you use to keep the budget down without sacrificing quality?

Yes, low budget films are the only experience I've had—films in a tradition which it seems to me owe a great deal to television. Even *Dangerous Liaisons* was on a comparatively low budget for what it was. The way of doing this is through decisiveness on the floor. At a certain point you

From left to right, Glenn Close, John Malkovich, and Uma Thurman in Frears' first American film, the internationally acclaimed *Dangerous Liaisons*, 1988. © 1988 Warner Bros. Inc. Photograph: BFI.

realize that you can shoot something with one shot, that you don't have to do a lot of set-ups, but just get it right. That way it's much faster. I always know where I want the camera. Decisiveness is crucial. In fact, it was actually a bloke called Patrick Cassavetti who really taught me to stay on budget. He was a very good production manager. Once I had learned to come in on budget, 'they' couldn't get me. It's all to do with decisiveness at the point of shooting.

How important is it for you to stay within budget at all costs?

It is part of the job to stay within budget. Also, once you go over budget you actually nail yourself—you lose the film. When you're under budget 'they'—and I am speaking rather paranoically—can't touch you, or get at you.

Wouldn't you like to work with bigger budgets than you have done?

You learn to make money stretch. I actually hope that my work has a sense of economy—although I don't think that it is necessarily a virtue. I'm economic about things by temperament and I'm probably bad at spending a lot of money. Once the money is raised I am pretty responsible. There is a great irony in the fact that I am part of a group of film-makers making films opposing this government, yet in practice, in terms of thrift and industry, I'm the Mrs Thatcher of British film! In fact, on *Sammy and Rosie Get Laid* I made a mistake in how much I asked for— we were under-budgeted and I could have asked for more. It meant that our schedule was too tight and we were working very long hours when it could have been produced in more congenial circumstances. *Prick Up Your Ears* and *Laundrette* were both budgeted correctly, and I can honestly say that it didn't seem worth spending any more on them. Different topics, such as, say, a film on the Napoleonic wars, justify the expenditure of more money.

You have produced six plays for London Weekend Television. How interested are you in the production side?

Not at all—the producers I have worked with, such as Tim Bevan [*Laundrette* and *Sammy and Rosie*], or Jeremy Thomas [*The Hit*] have been very good at raising money and so on—that's their job—I couldn't do it. It's hard enough directing the bloody thing. If one is also the best at producing amongst the team of people one works with—like Francis Coppola in America—one has to do it as well—but I'm not.

How optimistic are you, at the moment, about the British Film Industry?

I think it is terrific that *Rita, Sue and Bob Too*,[13] *Soursweet*,[14] *Wish You Were Here*,[15] *A World Apart*,[16] or *The Singing Detective*[17] have been made and are being seen. I'm very, very fond of Derek Jarman and I think anything in which he is involved is good. They are the film-makers that I can understand and really identify with. At a time when there is a particularly bad government it's ironic that these films are the ones that thrive—a lot of good work has been done in the last few years. But most of these people are highly experienced, and the idea of these middle-aged men causing something which has, in the past, been called a Renaissance is faintly absurd. It's still very hard for young people to get opportunities. Furthermore, the quantity of money involved in all these films is a joke, and you can't run an industry on that. The resources in British film are so limited. The pressure is so enormous. If someone gets a job you don't. You are shoved into competition with each other while at the same time you are trying to develop your 'œuvre' and your style. But I think that many directors have only themselves to blame because of their chasing after American money. The damage was done in the sixties. Now, you only have to get one clown who lays his hands on a few million and its all over . . . Hugh Hudson turns up and blows it all.[18] All this is why I became involved in Greenpoint[19] with Richard Eyre, David Hare, and Christopher Morahan, so that a group of film-makers could get together and support each other. As a group of friends it was very nice to have that psychological security of being able to identify with an office.

What do you think of Channel 4's role?

At the moment there is hardly a film in this country that they don't contribute to. *Walter* was the first film shown on Channel 4—perhaps because it was unafraid or unashamed and they wanted to nail their colours to that sort of mast. In spite of this I was actually rather critical

---

[13] Directed by Alan Clarke, 1986.
[14] Directed by Mike Newell, 1988.
[15] Written and directed by David Leland, 1987.
[16] Directed by Chris Menges, 1987.
[17] Scripted by Dennis Potter, and directed by Jon Amiel, 1986.
[18] *Revolution* (1985), directed by Hugh Hudson, was a financial disaster and brought its British backers, Goldcrest, to the verge of bankruptcy.
[19] Greenpoint Films was set up in 1983 by a group of directors, writers, and producers: Frears, John Mackenzie, Christopher Morahan, Richard Eyre, David Hare, Ann Scott, and Simon Relph. It is a loosely based production company providing creative freedom for the individuals to develop their own projects, as well as acting as a support structure. By providing a pool of creative talent, it gives a sense of continuity to the projects which it undertakes, and in this sense is like a miniature of the old Hollywood studios, which ensured a similar continuity through the people it had under contract. Films made by the company include: *The Ploughman's Lunch* (1983, directed by Richard Eyre and produced by Simon Relph), *Laughterhouse* (1984, directed by Richard Eyre and produced by Ann Scott), and *Wetherby* (1985, written and directed by David Hare, and produced by Simon Relph).

of Channel 4's early films until I was at Cannes and saw their credits on *Paris, Texas*[20] and I thought, 'That's incredible; that's a real achievement; that's progress', and my attitude changed. Now they are, in fact, such a dominant force that I think that there is some truth in the hypothesis that they stifle the independent sector. But they do produce some very good films, and have won a very good reputation abroad. As far as I know, *Laundrette* is almost the only film which they have backed 100 per cent whose cinema sales alone have actually paid for the production. But they are not obliged to look for a return. They fund films primarily for the value of the television screening, and anything they get in foreign sales is jam to them.

*Why do you think that it was your film—*Laundrette*—which was the one to break through financially and what contribution do you think it has made for the industry?*

Well none. By a fluke I happened to be the director of one of these sort of 'television films' being released into cinema. It was inevitable that one would succeed, and it just happened to have been a film I directed rather than somebody else. It's nice to be part of all that—these are films that I admire and support and hope will go on being made. But things change and you have to change with them.

Dangerous Liaisons *sees you making your first film in an American style, for an American company and with a larger budget than before. How much of a break is this for you?*

I think a great deal about this business of going to America. We've all seen people set off for Hollywood and come back with their tails between their legs. But I have no ambition to make Indiana Jones films. I find all that demoralizing. I like my films playing in cinemas, and I love them playing in foreign cinemas. I don't think I see a problem in going out there, shooting a film and coming back. What's the difference between that and shooting in Oldham?

Maybe it will result in a different sort of film?

I'll make a film in an American tradition with the magic and escapism of that genre. I love that tradition. In fact, there are very few British traditions that interest me. *Dangerous Liaisons* is not a European film. Well, it is in some respects—but it's not an English one. It's the stuff I grew up on. I think it's great. In fact, I wish I had made *Dangerous Liaisons* even more American. I thought for a short time of doing it for £1 million as a punk version to attack Milos Foreman.[21] But that wasn't

---

[20] Directed by Wim Wenders, 1984.
[21] Foreman made his own version, *Valmont*, which was released in 1989.

really possible. Lorrimer wanted something of a certain weight and quality with a $14 million budget. For the first time that has meant that I had to think it through. The budget implied certain things.

Did you see that as a problem?

I clearly didn't have any complaints or I wouldn't have done the film. It was a great story. What you have to do is to try and operate those mechanisms of escapism (which I don't pretend to understand) while using intelligent material. It's called 'having your cake and eating it'! We have such an ambiguous position in England with the American influence—so I have become particularly interested in the films of Europeans working in Hollywood. It's rather an eccentric group of people. I'm not thinking of Parker—his solutions will drive him mad. Alan Parker can't resolve the problem of making 'international' films and his belief in Ken Loach. He gets into a real mess. I think he's a wonderful film-maker and he knows his problem. But his solution is absolutely rubbish. Some day he will wake up and realize it. The solutions created by Wilder,[22] Ophüls,[23] or Hitchcock[24] in the forties are much more fascinating.

Where do you think your career will go on from here—do you have any plans?

Whenever I think of a pattern for the future, the next day I will think of the opposite. I'm actually discovering that I'm getting quite old; that the children need time, that you become more cautious and . . . I don't know . . . dozier. But then I don't feel that I've really started yet. I love the surprise involved in the whole question of what one should move onto next. I remember—before *Laundrette*, I was fed up and I had no expectations that anything stimulating would turn up, and then the script just dropped through the letter box. As I started reading I can remember thinking, 'Oh God, this is about race—how boring' and then I read a bit more and my heart lifted. It's great, just great, and incredibly renewing.

---

[22] Billy Wilder: brilliant, powerful, and cynical director. He was born in Austria and fled to France with the Nazi ascendancy to power in 1933. (His mother and many in his family died in the Holocaust.) He braved Hollywood in 1934 not speaking a word of English, where he moved in with the similarly destitute actor, Peter Lorre. He scraped a living as a co-script-writer, before directing his first feature in the mid-1940s. Films of the forties include: *Five Graves to Cairo* (1943), *Double Indemnity* (1944), and *The Lost Weekend* (1945).

[23] Max Ophüls: virtuoso director, pioneering developments in film form—especially *mise-en-scène*, which reflected his strong theatrical background. German-born director who, like Wilder, fled to France in 1933. He went on to Switzerland before leaving for America in 1941. Films of the forties include: *Letters from an Unknown Woman* (1948), *The Reckless Moment* (1949). He later returned to France to make *La Ronde* (1950), *Le Plaisir* (1952), etc.

[24] Alfred Hitchcock: English-born director whose name has become synonymous with the suspense thriller. He moved to America in 1939 where he made some of his most memorable films, among them *Shadow of a Doubt* (1943), *Notorious* (1946), *Rope* (1948).

FILMOGRAPHY

BRIEF BIOGRAPHICAL DETAILS

Born in 1941. Brought up in Leicester. Educated at Gresham's School, Norfolk. Studied law at Trinity College, Cambridge 1960–3. Worked at the Royal Court Theatre 1964–5, assisting on productions of *Waiting for Godot*, and *Inadmissible Evidence*.

FILMS DIRECTED BY STEPHEN FREARS

1967   *The Burning*. 30 mins. BFI Production Board/Memorial Enterprises. SCRIPT: Roland Starke (from his short story 'The Day'). PHOTOGRAPHY: David Muir. EDITOR: Ian Rakoff. MUSIC: Mischa Donat. CAST: Gwen ffrangcon-Davies, Mark Baillie.

1969   *St Ann's* (documentary). Thames TV (Report). PRODUCER: Ian Martin.

       Parkin's Patch. 30 mins. per episode (two episodes *The Deserter* and *The Boys*). Yorkshire TV. PRODUCER: Terence Williams. EXECUTIVE PRODUCER: Tony Essex. SCRIPT EDITOR: Nick McCarty. CAST: John Flanagan.

1970   *Tom Gratton's War*. 30 mins. each episode (five episodes *The Walking Bomb*, *Blind Man's Buff*, *Bridge of Death*, *The Coward*, *Badge of Fear*). Yorkshire TV. PRODUCER: Tony Essex. CAST: Michael Howe.

1971   *Folly Foot* (three episodes *Dora*, *The Charity Horse*, *Know All's Nag*). Yorkshire TV.

       Gumshoe. 85 mins. Columbia/Memorial Enterprises. SCRIPT: Neville Smith. PRODUCER: Michael Medwin. PHOTOGRAPHY: Chris Menges. PRODUCTION DESIGNER: Michael Seymour. EDITOR: Charles Rees and Furgus McDonnell. MUSIC: Andrew Lloyd Webber. CAST: Albert Finney, Billie Whitelaw, Frank Finlay.

1972   *A Day Out*. 50 mins. BBC TV. SCRIPT: Alan Bennett. PRODUCER: Innes Lloyd. PHOTOGRAPHY: Ray Henman. PRODUCTION DESIGNER: Jeremy Bear. EDITOR: Ken Pearce. MUSIC: David Franshaw. CAST: Anthony Andrews, Virginia Bell, Sharon Campbell, Fred Feast, Rosalind Elliott, George Fenton, Brian Glover, Paul Greenwood.

1973   *England Their England* (from a book by A. G. MacDonnet). BBC TV

(Sporting Scenes). PRODUCER: Innes Lloyd. EDITOR: Ken Pearce. CAST: Hugh Lloyd, John Muffat.

Match of the Day. 30 mins. BBC TV (Second City Firsts). SCRIPT: Neville Smith. PRODUCER: Barry Hanson. PHOTOGRAPHY: Michael Williams. PRODUCTION DESIGNER: Charles Bond. EDITOR: Ken Pearce. CAST: Neville Smith, Anne Zelda, Bill Dean.

1974 *The Sisters.* 30 mins. BBC TV. SCRIPT: John McGahern (from a short story by Joyce, from *The Dubliners*). PRODUCERS: Gavin Miller and Melvin Bragg. CAST: Marie Kean, Robert Bernal.

1975 *Sunset across the Bay.* 70 mins. BBC TV (Play for Today). SCRIPT: Alan Bennett. PRODUCER: Innes Lloyd. PHOTOGRAPHY: Brian Tufano. PRODUCTION DESIGNER: Moira Tait. EDITOR: Ken Pearce. CAST: Gabrielle Day, Harry Markham, Bob Peck.

Daft as a Brush. 75 mins. BBC TV. SCRIPT: Adrian Mitchell. PRODUCER: Graeme McDonald. PHOTOGRAPHY: Brian Tufano. PRODUCTION DESIGNER: Osten Spriggs. EDITOR: Ken Pearce. CAST: Lynn Redgrave, Jonathan Pryce.

Three Men in a Boat. 55 mins. BBC TV. SCRIPT: Tom Stoppard (from the book by Jerome K. Jerome). PRODUCER: Rosemary Hill. PHOTOGRAPHY: Brian Tufano. EDITOR: Ken Pearce. MUSIC: David Fanshaw. CAST: Tim Curry, Stephan Moore, Michael Palin.

1976 *Play Things.* 60 mins. BBC TV (Playhouse). SCRIPT: Peter Prince (from his novel). PRODUCER: Innes Lloyd. PHOTOGRAPHY: Brian Tufano. PRODUCTION DESIGNER: Moira Tait. EDITOR: Ken Pearce. CAST: Jonathan Pryce, Nigel Hawthorne, Colin Campbell.

Early Struggles. 60 mins. BBC TV (Play for Today). SCRIPT: Peter Prince. PRODUCER: Graeme MacDonald. PHOTOGRAPHY: Tony Pearce Roberts. EDITOR: Ken Pearce. MUSIC: Brian Gascoigne. CAST: Paul Nicholas, Tom Conti.

Last Summer. 60 mins. Thames TV (ITV Playhouse). SCRIPT: Peter Prince. PRODUCER: Barry Hanson. PHOTOGRAPHY: Chris Menges. EDITOR: Mike Taylor. MUSIC: George Fenton. CAST: Richard Beckinsale, Richard Mottan.

1977 *Black Christmas.* 50 mins. BBC TV. SCRIPT: Michael Abbensetts. PRODUCER: Tara Prem. EDITOR: Andrew Page. CAST: Carmen Munro, Norman Beaton.

Able's Will. 80 mins. (Video). BBC TV (Play of the Week). SCRIPT: Christopher Hampton. PRODUCER: Innes Lloyd. CAST: Elizabeth Spriggs, David Massey, Dominic Guard, Di Trevis.

1978 *18 Months to Balcombe Street.* 60 mins. London Weekend Television. SCRIPT: John Shirley. PRODUCER: Barry Cox. EXECUTIVE PRODUCER: John Birt. MUSIC: George Fenton. CAST: Niall O'Brian.

A Visit from Miss Protheroe. 40 mins. (Video). BBC TV (Play of the

Week). SCRIPT: Alan Bennett. PRODUCER: Innes Lloyd. PRODUCTION DESIGNER: Moira Tait. NARRATION: Alan Bennett. MUSIC: George Fenton. CAST: Hugh Lloyd, Patricia Routledge.

Cold Harbour (a.k.a. *The Innocent*). 60 mins. Thames TV (ITV Playhouse). SCRIPT: Peter Prince. PRODUCER: Barry Hanson. PHOTOGRAPHY: Chris Menges. EDITOR: Andrew Page. CAST: Leticia Garrido, Lindsay Ingram, Billy McColl.

1979   *Series of 6 Alan Bennett plays.* London Weekend Television. EXECUTIVE PRODUCER: Tony Wharmby. Frears produced all six, directed four:

Me, I'm Afraid of Virginia Woolf. 75 mins. (Video). SCRIPT: Alan Bennett. PRODUCER: Stephen Frears. NARRATOR: Alan Bennett. CAST: Neville Smith, Carol MacRealy, Thora Hird.

Doris and Doreen. 80 mins. (Video). SCRIPT: Alan Bennett. PRODUCER: Stephen Frears. CAST: Prunella Scales, Patricia Routledge.

Afternoon Off. 75 mins. SCRIPT: Alan Bennett. PRODUCER: Stephen Frears. PHOTOGRAPHY: Charles Stewart. PRODUCTION DESIGNER: Frank Nerin. CAST: Henry Mann, Peter Butterworth, Thora Hird, Alan Bennett.

One Fine Day. 95 mins. SCRIPT: Alan Bennett. PRODUCER: Stephen Frears. PHOTOGRAPHY: Charles Stewart. EDITOR: Andrew Page. CAST: Dave Allen, Robert Stephens, Dominic Guard.

(The other two plays in the series produced by Frears were *The Old Crowd*, dir. Lindsay Anderson, and *All Day On The Sands*, dir. Giles Foster.)

Long Distance Information. 65 mins. BBC TV (Play For Today). SCRIPT: Neville Smith. PRODUCER: Richard Eyre. PHOTOGRAPHY: Nat Crosby. PRODUCTION DESIGNER: Derek Dodd. EDITOR: Ken Pearce. CAST: Neville Smith, Pauline Collins.

1980   *Bloody Kids.* 91 mins. (Released for cinema in 1983) Black Lion/ Palace—BFI. SCRIPT: Stephen Poliakoff. PRODUCER: Barry Hanson. PHOTOGRAPHY: Chris Menges. PRODUCTION DESIGNER: Martin Johnson. EDITOR: Peter Coulson. MUSIC: George Fenton. CAST: Derrick O'Connor, Gary Holton, Richard Thomas.

1981   *Going Gently.* 70 mins. BBC TV (Playhouse). SCRIPT: Thomas Ellis (from a novel by Robert Downs). PRODUCER: Innes Lloyd. PHOTOGRAPHY: Nat Crosby. PRODUCTION DESIGNER: Derek Dodd. EDITOR: Ken Pearce. MUSIC: George Fenton. CAST: Norman Wisdom, Fulton MacKay, Judi Dench.

1982   *Walter.* 75 mins. Central/Randel Evans Productions for Channel 4 TV. SCRIPT: David Cook (from the novel of the same name).

PRODUCER: Nigel Evans. PHOTOGRAPHY: Chris Menges. PRO-DUCTION DESIGNER: Michael Minas. EDITOR: Mick Audsley. MUSIC: George Fenton. CAST: Ian McKellen, Sarah Miles, Barbara Jefford.

Walter and June. 75 mins. (Sequel to *Walter*; shot together and collectively called *Loving Walter*). Central/Randel Evans Productions for Channel 4 TV. SCRIPT: David Cook (from his novel *White Doves*). PRODUCER: Richard Creasey. ASSOCIATE PRODUCER: Patrick Cassavetti. PHOTOGRAPHY: Chris Menges. PRODUCTION DESIGNER: Michael Minas. EDITOR: Mick Audsley. MUSIC: George Fenton. CAST: Ian McKellen, Sarah Miles, Gordon Sinclair.

1983   *The Last Company Car.* 60 mins. (Video). Central TV. SCRIPT: Mike Scott. PRODUCER: Lynn Horsford. CAST: David Ross, Eileen O'Brien, Jim Broadbent, Kevin Lloyd.

Saigon, Year of the Cat. 105 mins. Thames TV. SCRIPT: David Hare. PRODUCER: Michael Dunlop. EXECUTIVE PRODUCER: Verity Lambert. PHOTOGRAPHY: Jim Howlett. PRODUCTION DESIGNER: David Marshall. EDITOR: Oscar Webb. MUSIC: George Fenton. CAST: Judi Dench, Frederick Forest.

1984   *The Hit.* 98 mins. Palace Pictures/Zenith/Recorded Picture Co./Island Alive. SCRIPT: Peter Prince. PRODUCER: Jeremy Thomas. PHOTOGRAPHY: Mike Molloy. PRODUCTION DESIGNER: Andrew Sanders. EDITOR: Mick Audsley. MUSIC: Paco De Lucia. TITLE MUSIC: Eric Clapton. CAST: John Hurt, Terence Stamp, Tim Roth.

December Flower (a.k.a. *A New Life For Aunt M.*). 75 mins. Granada TV. SCRIPT: Judi Allen (from her novel *December Flower*). PRODUCER: Roy Roberts. PHOTOGRAPHY: Ray Goode. PRODUCTION DESIGNER: Chris Wilkinson. MUSIC: Richard Hartley. CAST: Bryan Forbes, Jean Simmons, Mona Washbourne.

The Bullshitters. 50 mins. (a.k.a. *Roll out the Gunbarrel*). Comic Strip Production/Channel 4 TV. SCRIPT: Peter Richardson, Keith Allen. PRODUCER: Elaine Taylor. EXECUTIVE PRODUCER: Michael White. CAST: Keith Allen, Peter Richardson, Robbie Coltrane, Michael White, Elvis Costello.

1985   *My Beautiful Laundrette.* 97 mins. Channel 4/Working Title/SAF Prods/Mainline/Orion Classics. SCRIPT: Hanif Kureishi. PRODUCER: Sarah Radclyffe, Tim Bevan. PHOTOGRAPHY: Oliver Stapleton. PRODUCTION DESIGNER: Hugo Luczyc Wyhowski. EDITOR: Mick Audsley. MUSIC: Ludus Tonalis. CAST: Daniel Day Lewis, Gordon Warnecke, Saeed Jaffrey, Roshan Seth.

1986   *Song of Experience.* 65 mins. BBC TV (Screen Two) (shot in 1985). SCRIPT: Martin Allen. PRODUCER: Innes Lloyd. PHOTOGRAPHY: Nat Crosby. PRODUCTION DESIGNER: Paul Joel.

EDITOR: Ken Pearce. CAST: Nigel Terry, Rachel Bell, Alan Starkey, Alan Bell, Paul Darlow.

Consuela. 45 mins. Comic Strip Production/Channel 4 (shot in 1984). SCRIPT: Jennifer Saunders, Dawn French. PRODUCER: Elaine Taylor. EXECUTIVE PRODUCER: Michael White. PHOTOGRAPHY: Oliver Stapleton. CAST: Dawn French, Rik Mayall, Jennifer Saunders, Peter Richardson, Adrian Edmondson.

1987   *Prick Up Your Ears*. 111 mins. Civilhand/Zenith Production. SCRIPT: Alan Bennett. PRODUCER: Andrew Brown. PHOTOGRAPHY: Oliver Stapleton. PRODUCTION DESIGNER: Hugo Luczyc Wyhowski. EDITOR: Mike Audsley. MUSIC: Stanley Myers. CAST: Gary Oldman, Alfred Molina, Vanessa Redgrave, Julie Walters.

Mr Jolly Lives Next Door. 60 mins. (Released in a double bill with *Didn't You Kill My Brother*). Comic Strip Production/Channel 4. SCRIPT: Adrian Edmondson, Rik Mayall, Rowland Rivron. PRODUCER: Elaine Taylor. EXECUTIVE PRODUCER: Michael White. PHOTOGRAPHY: Oliver Stapleton. PRODUCTION DESIGNER: Grant Hicks. EDITOR: Rob Wright. CAST: Adrian Edmondson, Rik Mayall, Peter Cook.

1988   *Sammy and Rosie Get Laid*. 100 mins. Working Title/Nelson Entertainment. SCRIPT: Hanif Kureishi. PRODUCER: Tim Bevan, Sarah Radclyffe. PHOTOGRAPHY: Oliver Stapleton. PRODUCTION DESIGNER: Hugo Luczyc Wyhowski. EDITOR: Mick Audsley. MUSIC: Stanley Myers. CAST: Shashi Kapoor, Claire Bloom, Ayub Khan Din, Francis Barber.

Dangerous Liaisons. 120 mins. Warner Brothers/Lorrimer Film Entertainment. SCRIPT: Christopher Hampton (based on the novel by Laclos). PRODUCER: Norma Hayman and Hank Moonjean. CO-PRODUCER: Christopher Hampton. PHOTOGRAPHY: Philip Rousselot. PRODUCTION DESIGNER: Stuart Craig. EDITOR: Mick Audsley. MUSIC: George Fenton. CAST: Glenn Close, John Malkovich, Michelle Pfeiffer, Swoosie Kurtz, Keanu Reeves, Milfred Natwick.

1990   *The Grifters*. Cineplex Odeon. SCRIPT: Donald Westlake (from the novel of the crime author, Jim Thompson). PRODUCER: Martin Scorsese. CAST: John Cusack, Angelica Huston, Annette Bening.

COMMERCIALS INCLUDE

Pony, Knorr Soups, Lucozade, Access, Central Office of Information (called 'Back Seat Child', it won the Gold Arrow Award), Diet Coke.

OTHER FILM WORK

1966   Assistant to Karel Reisz on *Morgan, a Suitable Case for Treatment*.

1967   Assistant to Albert Finney on *Charlie Bubbles*.
1968   Assistant to Lindsay Anderson on *If . . .*

PRODUCER

Six Alan Bennett Plays. London Weekend Television (see above).

THEATRE WORK

1980   Directed *Television Times*, by Peter Prince. RSC. Warehouse.

ACTING

1972   *Incident*. Director: Jonathan Gill.
1978   *Long Shot*. Director: Maurice Hatton.

BIBLIOGRAPHY

Cinematograph Weekly, 26 Dec. 1970. Article on the making of *Gumshoe*.

The Times, 11 Dec. 1971. Short article/interview on *Gumshoe*.

Sight and Sound, Autumn 1972. Comments by the director and Alan Bennett about their BBC-TV film *A Day Out*.

Listener, 24 Dec. 1972. Alan Bennett's diary on the making of *A Day Out*.

Screen International, 1 Nov. 1975. Frears discusses his career and the advantages and disadvantages of working in television rather than cinema.

Sight and Sound, Spring 1978. Frears discusses his work and the current climate of television filming. Alan Bennett, Brian Tufano, and Chris Menges comment on working with Frears.

Film (BFFS), June 1978. Interview about his early career and his work for film and television, and his tendency to work with the same writers.

The Times, 27 Nov. 1978. Interview with Alan Bennett on his six plays for London Weekend Television.

Time Out, 1–7 Dec. 1978. Interview with Frears and Alan Bennett about their new series of six programmes for London Weekend Television.

Daily Mail, 9 Dec. 1978. Interview with Bennett on his six plays for LWT.

Film Dope, Apr. 1979. Biographical note with filmography.

Sight and Sound, Spring 1979. Discussion of the series of six Alan Bennett plays done for LWT; how they fit together and their sequence.

Time Out, 5–11 Oct. 1979. Neville Smith and Frears interviewed on *Long Distance Information*.

Guardian, 7 Dec. 1980. Derek Malcolm on Frears's direction of Peter Prince's stageplay *Television Times*.

Daily Record, 30 May 1981. Short interview with Norman Wisdom on *Going Gently*.

Sunday People, 15 Aug. 1982. Short location report and interview with Judi Dench on the making of *Saigon*.

TV Times Magazine, 30 Oct.–5 Nov. 1982. Ian McKellen on acting in *Walter*.

Observer Magazine, 31 Oct. 1982. Article on *Walter*.

The Times, 2 Nov. 1982. 'Triumph over the bleakest handicap'. John Preston interviews Frears about his career on the occasion of the showing of *Walter* on the opening night of Channel 4.

Jayne Pilling and Kingsley Canham (eds.), *The Screen on the Tube: Filmed Drama* (Cinema City, 1983), 20–2. Interesting discussion of *Bloody Kids*, including interviews with Barry Hanson, Chris Menges, Stephen Poliakoff, and Frears. Pp. 33–6 General discussion of Frears's TV drama.

Time Out, 10–16 Nov. 1983. Frears talks about his career, his work for TV, his desire to make films for the cinema again, and the British film industry generally.

TV Times Magazine, 26 Nov.–2 Dec. 1983. Article on *Saigon* and interviews with Frederick Forest and Judi Dench.

Stills, Nov./Dec. 1983. On the collaboration between David Hare and Frears on *Saigon*.

City Limits, 7–13 Sept. 1984. Frears talks about *The Hit*, working in television, and his views on the state of the British film industry.

Stills, Nov. 1985. Frears talks about his methods of working and the people who have influenced him.

Photoplay, Dec. 1985. Article about the motivation behind the making of *My Beautiful Laundrette*.

Sight and Sound, Winter 1985/6. David Robinson reviews *My Beautiful Laundrette*.

Radio Times, 15–21 Feb. 1986. Frears talks about *Song of Experience*, in the 'Screen Two' series on BBC 2.

Guardian, 6 May 1986. Hanif Kureishi on the success of *My Beautiful Laundrette* in the US.

Film Comment, Mar.–Apr. 1987. Frears discusses his work and comments on the drama traditions of UK TV and cinema.

Interview, Apr. 1987. Career and biographical interview.

Sunday Telegraph Magazine, 12 Apr. 1987. Interview with Alan Bennett on *Prick Up Your Ears*.

City Limits, 16–24 Apr. 1987. Article on *Prick Up Your Ears* with comments from John Lahr and Alan Bennett.

Observer Magazine, 19 Apr. 1987. Interview with Frears on *Prick Up Your Ears*.

Sunday Times, 26 Apr. 1987. Interview with Frears on *Prick Up Your Ears*.

Monthly Film Bulletin, (54), May 1987. Biofilmography and comments from Bennett about writing the script for *Prick Up Your Ears*.

Scotsman (Weekend), 6 June 1987. Interview about *Prick Up Your Ears*.

Films in Review, Oct. 1987. Brief interview with Frears on *Prick Up Your Ears* and *Sammy and Rosie Get Laid*.

Screen International, 17–24 Oct. 1987. Title of *Sammy and Rosie Get Laid* may have to be changed for North American exhibition, as many newspapers are refusing to accept ads for the film.

Films and Filming, Jan. 1988.

Screen International, 14–28 May 1988. Report on the financing of *Sammy and Rosie Get Laid*.

American Film, Dec. 1988. Comments from Frears on the set of *Dangerous Liaisons*.

Time Out, 11–18 Jan. 1989. Discussion of the two film versions of *Les Liaisons dangereuses*, by Frears and Milos Forman.

Sunday Times Magazine, 12 Feb. 1989.

Time Out, 1–8 Mar. 1989. Glenn Close talks about her part in *Dangerous Liaisons*.

5 PETER GREENAWAY

Peter Greenaway on location in Italy while shooting *Belly of an Architect*, 1986. © Oasis Film Production Ltd.

PETER GREENAWAY

IN 1980 Peter Greenaway's film *The Falls* was the first British film to win the BFI award for Best Film for thirty years. But it was in 1982 that he really leapt to the attention of the cinema-going public with the surprise success of *The Draughtsman's Contract*. It was a beautiful looking film and for many it simply represented an exciting seventeenth-century 'who dunnit?'. Yet, as many have started to look deeper into Greenaway's films, the responses have been split. On the one hand, he has been met with incomprehension or even cries of 'pretentious'. Alan Parker once said that he would leave the country if Greenaway made another film here. However, others, particularly on the Continent, have been fascinated, and he is a feast for any semiologist who might like to decode and reconstruct films. He is also one of the few film-makers in this country who really has an audience which devotedly will see his films just because they are made by him, and not because of their subject matter, cast, or reviews. The star in a Peter Greenaway film is Peter Greenaway himself. As such he is one of the few art house film-makers able to make movies with sizeable budgets because his films have such a good chance of getting a decent return.

He is a thin, energetic man who likes talking about his work, which he does articulately, and fast, drawing on a wide range of cultural references. He had a London suburban childhood and an education at a minor public school, where his attention was focused on art and literature from quite an early age. He went to Walthamstow Art School in the early sixties, although it was, as he points out, the sort of art school which 'had never acknowledged the word "Bauhaus".'[1] Painting is an occupation which still fascinates him, and many of his paintings and drawings appear in his films. It was at art school that his interest in cinema started, but he gained a practical knowledge of film-making over the next decade by working his way up through the ranks of the cutting room to being a film editor. It was a crucial period in forming his firm beliefs about art and in particular cinema.

This formative period was marked by a growing fascination with the ideas of the Structuralist Movement. This turned away from cinema as

---

[1] Interview with the authors.

an 'illusionist' or 'emotional' medium and concentrated on structures in the hope of clarifying the production process itself. 'I was influenced by all the post-Brechtian alienation techniques of the late sixties.'[2] Structuralists believe that the only truth that art can convey is the fictional nature of the fiction one is creating. Although Greenaway has moved away aesthetically from the harshness and aridity of these early ideas, philosophically he has maintained his belief that human activity is ultimately unclassifiable, or un-mappable, although he feels a persistent conflict on this issue. He describes his unresolved position:

In all my films there is a contradiction between the romantic and the classical; violent, absurd, bizarre subject matter treated with a severe sense of control. Baroque surface and rich romantic detail regimented into numerical grids and structures—which I would like to think shows a wish—against the odds—to create a rational view of the world out of all its chaotic parts. However the structures and controls are always mocked as being inadequate or ineffectual or destructive. Chris Auty once described my films as 'beautiful butterflies trapped down by drawing pins'.[3]

Despite Chris Auty's suggestion, Greenaway's severe sense of control—what he terms 'the classical'—is still the more dominant within this conflict. 'My way of creating cinema is, both in form and content, a classical one';[4] and he has also said, with irony, 'I'm a clerk. I like organising material.' The interest in systems, cataloguing, and codifying is important for Greenaway. It is more than a device—it never ceases to fascinate him. 'It has been suggested that everything exists to be put into a list.'[5] For Greenaway the lists or classification systems are excellent demonstrations of the vain, absurd attempts to create an objectivity and meaning in the world. He acknowledges that they are necessary for any culture and any society, but believes that we should be aware of just how shallow they are, and that art itself is another example of how we build up and enforce such systems.

Not surprisingly, Greenaway uses simple structural devices to regiment his material—forms of classification all logically drawn from the film's themes or setting. *The Draughtsman's Contract* is divided into twelve parts as the draughtsman is commissioned to make twelve drawings. *A Zed and Two Noughts* (1985), a film about coming to terms with death in a zoo, is divided into eight parts because it uses Darwin's eight stages of evolutionary development. *The Cook, the Thief, his Wife and her Lover* (1989) is divided into both ten and seven in order to parallel the ten-course meal and the seven colour-coded rooms of the

---

[2] Interview with the authors.
[3] Interview with the authors.
[4] Interview with the authors.
[5] Interview with the authors.

restaurant in which the film is set. *The Belly of an Architect* (1986) plays on the number seven after the seven ages of Roman architecture, and *Intervals* (1969), one of his first short films, shot in Venice and using Vivaldi's music, was based on the number thirteen with reference to Vivaldi's musical notation.

Greenaway likes to tell of why the number ninety-two came to play such an important part in the structure of *The Falls*, one of whose many themes is ninety-two ways to end the world. 'In the sixties I was much influenced by the American composer, John Cage. One of his recording ideas[6] had been to arrange a series of short narratives on two sides of a record, each one occupying exactly one minute of recording time. Stories of one sentence consequently had to be read very slowly to fill the time and stories of one page had to be garbled. Both extremes approached the non-narrative. It was another solution—a humorous and ironic solution—to discipline narrative—constrict it by time.' In fact, Greenaway had counted the number of stories incorrectly—there were only ninety on the record and not ninety-two as he had thought. Later, while making *Four American Composers* (1983), he got to meet John Cage. 'He was amused that I had built two years of film-making on an error of arithmetic.'[7] Nevertheless, Greenaway still refers in his work to the number ninety-two. The mistake, he feels, does not affect the point he is making.

The use of games, word-plays, and conundrums is another feature of Greenaway's films. They serve to highlight his structuralist beliefs. 'They sometimes may serve a purpose of not serving a purpose,'[8] so helping the audience consider the issues involved in the rules governing aesthetics. They may also provide a source of humour in the tradition of Lewis Carroll, or Edward Lear. In *A Zed and Two Noughts*, a child is taken to the aquarium where she asks to see a red herring. But Greenaway also believes that games play an important part in people's everyday lives, and that the games in his films reflect this reality.

The English are very good at game-playing—they have probably invented most of them—and we are often criticised for hiding behind them. It is said to be a way of ritualizing emotion. But no game is just a game. A great deal of our life is played in a complex series of games whose rules we could tabulate—we all play the game of employee to boss in the various pecking orders, or the game of interviewers to interviewees, adults to children, men to women in the sex game and so on. It's always noticed that cricket has given the English many words and meanings and attitudes that suggest that only the English could have invented it—a landscape game best played in temperamental weather where light and

---

[6] For the record called 'Indeterminacy'.
[7] Interview with the authors.
[8] Interview with the authors.

humidity, psychology, tradition, conservatism are all important. The symbolism of such games as chess is well-known. My intention of course is ironic; some people do not know how to take the irony because they believe that game-playing is not compatible with serious film-making. Film-making itself is a complex game of illusion and bluff played between the film-maker and his audience.[9]

The Draughtsman's Contract is an example of this illusion and bluff, a game with the audience trying to guess who murdered the master of the house from the clues given. *Drowning by Numbers*, is crammed full of these sorts of games, riddles, paradoxes, ironies, and red herrings. In particular, it is dominated by a number count of one to a hundred which appears relentlessly throughout the film creating a sense of 'fate' and adding a further dimension to the narrative (the number one appears boldly painted on a tree, twenty-two is Catch 22, seventy-six and seventy-seven appear on the rumps of live cows and so on).

Greenaway also likes to raise questions about art head on. Much of his work is concerned 'with questions of representation—looking and seeing are not the same occupation, and it's fascinating to examine the differences'.[10] Many of his films are actually about, or can be seen as metaphors for, artists themselves. He portrays them often as manipulated, helpless figures, victims of fickle patronage, as vulnerable men. *Vertical Features Remake* (1978) is a film about another, fictional, film—*Vertical Features*; *The Draughtsman's Contract*, a draughtsman; *The Belly of an Architect*, an architect; *The Cook, the Thief, his Wife and her Lover*, a superb chef, at the mercy of his crude and brutal patron. *A Walk through H* (1978) is about a map-maker; *A Zed and Two Noughts*, about animal behaviourists; and *Drowning by Numbers*, a coroner. These last three can be seen as metaphors for the way an artist must give meaning to meaningless things, must watch, observe and judge. Greenaway admits 'in a way, all my films are about outsiders.'[11] The way that the outsider artist-figures fail to comprehend what they are observing until too late is highly pessimistic.

But there is one subject which Greenaway does find meaningful. 'The thing I enjoy doing most of all is just to experience a landscape and all that's in it—preferably a natural landscape. It is self-evidently the case that the reality of a landscape is much more exciting and profound than any attempt to transform it into an art form.'[12] The natural world offers the opposite of the 'artificiality' or 'methodicality' of art, and Greenaway's films delight in it. *Windows* (1974), *The Draughtsman's*

[9] Interview with the authors.
[10] Interview with the authors.
[11] Interview with the authors.
[12] Interview with the authors.

Bernard Hill with Kenny Ireland and Michael Percival in Greenaway's semiologists' feast, *Drowning by Numbers*, 1988. © Oasis Film Production Ltd.

Contract, and *Drowning by Numbers* are each in their different ways about death in the English landscape. *A Zed and Two Noughts* is preoccupied by man's maltreatment or abuse of animals. *The Belly of an Architect* self-consciously removes everything natural to create a wholly urban landscape. In order better to consider the function of art autonomous from nature, even the colours blue and green are filtered out because of their reference to the sky and grass. In *Drowning by Numbers* Cissie Colpitts 1 says of Madgett, the coroner and artist-figure, that games help his insecurity; that because he is unable to create a rapport with the environment he must play games with nature. For Greenaway, the natural landscape is regenerative and reliable.

Greenaway often portrays women as the real artists. They have the obvious ability to produce offspring, the immortality that men consciously hope to find in, or through, the creation of art. The mother and daughter in *The Draughtsman's Contract* manipulate the draughtsman in order to get an heir; Kracklite, the architect in *The Belly of an Architect*, is too obsessed with his work to even notice that his wife is pregnant; Cissie Colpitts 3 in *Drowning by Numbers* makes sure that she is pregnant by her boyfriend before she drowns him; and Georgina Spica in *The Cook, the Thief, his Wife and her Lover* has much of the sophistication and some of the culture which her barbaric husband aspires to, and ultimately beats him at his own game. Greenaway calls this the 'male–female creativity axis'.[13]

Greenaway likes to point out that in his films the 'interest is in aesthetics not politics',[14] but, as he concedes, 'you can't not be interested in politics, although I don't think it is so necessary or obligatory to use cinema directly as a soapbox; it's difficult to find cinema changing political opinion, television has a greater potential for doing that.'[15] Still, he has made several Labour Party commercials and would be prepared to do so again, and he does consciously raise issues for consideration besides aesthetics, in particular those concerning gender/sex and death. 'Most Western culture and especially cinema seem to have two overriding subjects; sex and death. They surface in a thousand different ways and invariably can be seen to have wider political implications, though I have never wanted to romanticize or sentimentalise either.'[16]

Many of his films have feminist overtones. 'Given current female emancipation, the male has to reconsider his position as main protagonist in cinema. The old cinema sexual stereotypes can no longer be available for serious use. But I don't necessarily believe that females are the

[13] Interview with the authors.
[14] Interview with the authors.
[15] Interview with the authors.
[16] Interview with the authors.

stronger sex. I want the films to be a platform, amongst many other things, for consideration of every aspect of the issues of gender'.[17] Moreover, almost all Greenaway's films have an erotic element in them, particularly *The Draughtsman's Contract* and *The Cook, the Thief, his Wife and her Lover*—although always done discreetly, more by what is not shown than what is shown.

In 1985 he said, 'Death is the new pornographic frontier. We seem to have shed all of our post-Christian problems in the last couple of decades but we still can't handle the problem of death.'[18] It is a powerful and emotive issue and he tackles it directly. He concedes that cinema often deals in images of death—'but in a very unreal and often "safe" way. What happens if we look at it without hope of consolation?'[19] *A Zed and Two Noughts*, in particular, looks at the process of death and decay in a very explicit way, and *The Belly of an Architect* deliberately traces the last painful months of a man dying of stomach cancer and trying to come to terms with it.

Both these elements hark back to aesthetic preoccupations, in that women are seen as the true creators, and death as the frontier which the artist, in one sense, challenges through his work. But Greenaway denies being morbid or really pessimistic. 'Behind this death, evil, and mediocrity is the natural landscape in *Drowning by Numbers* and *The Draughtsman's Contract*, or the urban landscape in *The Belly of an Architect*. These landscapes are magnificent and optimistic. If the films do not celebrate the lives of individuals, they do celebrate life. Many of them have an ebullient delight in the richness and variety of things and many do end with a birth which persistently argues for another try, another chance, another opportunity.'[20] But ultimately his cold-blooded rationalism denies the audience an optimistic faith in the power of man to overcome, or find understanding. Nor does love, in fact, play much part in his world view.

There are two areas which lie awkwardly within Greenaway's approach as outlined above. First, his films are often packed full with complex ideas—intellectual, philosophical, and aesthetic. Consequently there is inadequate time to develop them all with any great depth, and this has often left him open to charges of intellectual exhibitionism. Secondly, there is the more significant problem of inadequate characterization. The characters are frequently awkward pawns within Greenaway's intellectual game-playing. Until recently they have merely been cyphers inhabiting space. However, since working with Brian Dennehy

---

[17] Interview with the authors.
[18] *NME*, 14 Dec. 1985.
[19] Interview with the authors.
[20] Interview with the authors.

important

Important

on *The Belly of an Architect* in 1986, he has been developing the scope for performance in his films. But it still remains limited. Even in his most recent feature, *The Cook, the Thief, his Wife and her Lover*, the characters remain, in spite of their larger-than-life mannerisms, essentially two-dimensional. The trouble is that for his ideas to have a greater impact on the audience, they need to be associated with identifiable characters, and set in identifiable, even if highly stylized, situations. The absence of this does not, of course, hinder the audiences' understanding of the ideas he is trying to transmit, but it does ensure that their understanding remains on a largely abstract level.

These are some of the dominant themes and ideas running through Greenaway's films—he likes to quote Auguste Renoir who suggested that while most creative people only have one idea and spend their whole life thinking about it, one idea is more than enough for a lifetime. But this is certainly not to suggest that his films are all the same—they show some important changes in attitude over the last twenty years.

He started making his own short movies whilst at art school and then whilst working as a film editor, but is rather unenthusiastic now about the earliest ones. 'Most of them can be seen as editing exercises, demonstrating my delight in putting music and dialogue and image together for the first time.'[21] He regards the densely packed, four minute, *Windows* (1974) as the beginning of his public cinema, and a film which presages everything that came afterwards. 'The film played with cataloguing and the elusive way statistics don't really tell you what you want to know.' Suggested by government statements in South Africa about political prisoners mysteriously falling out of windows, it represents an illustrated list of thirty-seven defenestration casualties, and is set in the beautiful parish of Wardour.[22] The deaths are classified by age, occupation, and cause of death. As the acme of authority, Greenaway's own voice-over, both as narrator and director, is a finely ironic way of presenting the statistics. Even the harpsichordist who has been playing the film's music jumps, in a death-pact with a seamstress, from a window into a plum tree.

Dear Phone (1977) was another early short, and is an intellectual farce about 'the uses and abuses of the telephone' as a metaphor for communication in its widest sense. It represented a visual record of many telephone boxes at every time of day and every type of place. Juxtaposed with each of these is the fantasy written 'text' for the illustration, read by the narrator. Fourteen anecdotes, of progressively improving legibility, all tell of telephone-users with the initial H.C. 'The

---

[21] Interview with the authors.
[22] From which Wardour Street in Soho, London, where many of England's film-making companies cluster, gets its name.

relationship between script, image, and sound is deliberately ur
in the conventional sense, refusing to use the cinema's normal enacu.
of a written text. Every film bar none begins as a piece of writing—I was
asking why take it further? Writing is writing, film is film, yet ninety per
cent of film is slave to a written text—is it necessary for film to be such a
hybrid, bastard art?'[23]

It was largely on the basis of these two films that he came to the notice
of Peter Sainsbury—the head of the production fund at the BFI—who
provided Greenaway with the £7,500 to make *A Walk through H*
(1978). With that forty-minute film the public started to see his work.
Made just after the death of his father who had a passionate interest in
bird-watching, the film's alternative title is *The Reincarnation of an
Ornithologist*. The narrator tells how, minutes before his death, the
maps which he had spent his life collecting were put together and
ordered by his life-long friend and fellow ornithologist Tulse Luper. The
camera shows us, in close up, the paths which the narrator's soul took
across these ninety-two deliberately ambiguous and often misleading
maps.[24] This is interrupted by shots of real birds which are intercut into
the artifice. At the journey's end we return to where we started—the
camera pulls back to show that these are merely drawings in an exhibition and not the world of fantasy we had let ourselves be drawn into. It
is a metaphor for the romantic imagination. *Vertical Features Remake*
was made in the same year for the Arts Council. Tulse Luper once again
plays a part. It is a paranoiac fantasy about a government Institute of
Reclamation and Restoration which attempts to create a 'dynamic
ordered landscape'. But it discovers that its deceased member Tulse
Luper had made a film in protest (*Vertical Features*) which it makes four
attempts to re-edit. 'It began as a structural, land-art examination of
verticals (like posts and trees) in a landscape, but I rapidly decided that
this was not in itself sufficiently interesting. The business of making the
four films was more interesting than the films themselves—so I invented
an elaborate explanation of how and why the films were made, making a
playful statement about the academic filmic establishment who make
their living off the energy and imagination of film-makers.'[25] As such
Greenaway was able to create an allegory for the reformist and rule-
orientated British film culture, along with an essay on the abuse of the
countryside.

The Falls (1980), was his first feature-length film—indeed it is three
and a half hours long. The £35,000 budget was funded by the BFI. It
represents case histories or interviews with ninety-two characters taken

---

[23] Interview with the authors.
[24] Drawn by Greenaway himself.
[25] Interview with the authors.

from the 'Orchard Falla' to 'Anthior Fallwaste' sections of an imaginary directory of the nineteen million survivors of the VUE (Violent Unknown Event), which had caused bizarre experiences and mutations, in some way to do with birds. Greenaway commented on it, 'It is a catalogue movie, made with an enthusiasm for Tristram Shandy, Borges and, most of all, Thornton Wilder's "The Bridge of San Luis Rey". It is full of apparent red herrings, possible cul-de-sacs, game-playing that could be serious, irritating trivia that might be important and much dubious profundity. It progresses cumulatively with a persistent stop-start motion.' He used his own friends for the ninety-two people and it was a useful editing experience for him as he experimented at portraying the most bizarre of events as though they were absolutely true. It was critically successful. The American critic Harlan Kennedy described it as 'the cinema's ne plus ultra of poetic pedantry; a hair splitting hosanna to all things statistical; a paean to pseudoscience; Edward Lear wrapped up in the Encyclopedia Britannica.'[26]

The following year Greenaway made *Acts of God* for Thames Television; a documentary about people struck by lightning. In a sense, it was a parallel development of *The Falls* because the lightning, like the VUE, is an incomprehensible and arbitrary event. Greenaway likewise conveys the absurdist humour in a completely dead-pan style. But it was on the basis of the award winning success of *The Falls* that Greenaway was really able to expand out of underground film-making and the BFI agreed to fund *The Draughtsman's Contract* (1982). It was astonishingly successful and remains his best-known film. To an extent it also represents a certain loss of innocence. When Mamoun Hassan, then managing director of the NFFC, asked him whether he was a private or a public film-maker, his reply was that the answer to that question was in the gap between *The Falls* and *The Draughtsman's Contract*. Before then his work was what *Time Out*'s Chris Auty called 'filing clerk fantasy'. But now dialogue, actors and more conventional narrative were introduced. In part this change was again because of Peter Sainsbury's involvement. He admits that a condition of his funding of what was a very large budget for the BFI was that characters spoke to each other instead of to the camera as before. Still, when the film went into production the scheduled £120,000 escalated to £300,000, and Channel 4 bailed them out. All the same, when it was released people were stunned by the production value of the film. One of the reasons for what is really a very low cost film was the BFI's special relationship with the unions, who (believing quite correctly at that time that BFI films did not make money) let everyone on the film, including actors and techni-

[26] *Film Comment*, Jan.–Feb. 1982.

cians, work for a flat rate of £199 a week. Moreover, the wonderful house and grounds where it was filmed cost the very small sum of £600 a week to shoot in.

Greenaway pointed out 'The antique china in the house—which we were given permission to use alongside our own props—was officially valued at three times the total film budget.'[27] Everyone worked very long hours and shooting was completed in just five weeks (while its preproduction had only taken seven!). However, in editing it looked as if it was going to be four hours long. Peter Sainsbury once more stepped in—'I was not going to be party to the making of a 35 mm four hour movie, which in all its elegant digressions would prove to be an entirely esoteric object for consumption purposes, and I got quite pragmatic and quite cynical with Peter.'[28] So it was finally cut down to 110 minutes. 'Peter accepted that the audience shouldn't be allowed to wander in and out of the film as they had been quite happy to do in *The Falls*, that we had to tie it down and give people reassurance that they knew what they were looking at in the development of ideas, themes and narratives.'[29] In Greenaway's own words:

Unlike underground paintings which invariably become overground paintings, underground movies—with one or two exceptions, like Buñuel's *L'Age d'Or* [1930] or Lynch's *Eraserhead* [1976]—do not become overground movies and a great many of them repeat endlessly what René Clair[30] was doing in the 1920s. I realized that I could go on making obtuse, recherché films for ever and ever that would never be seen beyond the converted. I also had the feeling that I was coming to the end of my subsidized life. So there was a certain pragmatism involved. I certainly wanted to come out of the experimental-movie closet and seek a wider audience; it's difficult to believe the sincerity of any film-maker who says he does not wish to seek a larger audience.[31]

He saw it as perverse to make a virtue of low budget film-making or remaining intentionally obscure. He described the final cut of *The Draughtsman's Contract* as 90% satisfactory.

The plot is in the form of a seventeenth-century enigma. The aristocratic wife of Mr Talman (Janet Suzman) contracts a celebrated but arrogant young landscape artist, Mr Neville, to make twelve drawings of her husband's estate in return for sexual favours while her husband is away. The lady's daughter insists on sharing her mother's

[27] Interview with the authors.

[28] *MFB*, November 1982.

[29] *MFB*, November 1982.

[30] Along with Renoir and Carné, he was one of the great French directors of the 1930s. In 1924 he made the film *Entr'acte* which laid down an experimental-movie vocabulary of slow-motion, eccentric camera-angles, surreal sound-track, use of dance and trick photography that have been the characteristic stock-in-trade for many experimental movies made since.

[31] Interview with the authors.

contractual obligations, although in actual fact she only wants the heir her husband cannot give her. The daughter's haughty German husband looks on as the drawings are completed. Each drawing, however, turns out to have allegorical details such as an unexplained jacket, or a ladder to a second floor window. When Talman's body is discovered these suggest that the draughtsman had foreknowledge of the murder. He doesn't notice this until the daughter points it out, and even then is too arrogant to worry. But when he returns to execute a thirteenth drawing, he himself becomes the victim, is blinded, and then killed. As to who had actually killed Talman, Greenaway enigmatically, even jokingly, responds 'Everybody was responsible, because everybody had reason to gain from his death. So, like the Murder on the Orient Express, everybody is guilty.'[32] The lack of resolution was however annoying for many viewers.

But as might be expected, the film also makes some serious comments about cinema and art. At one time it was set in contemporary England 'but the character would then have had to be a photographer, not a draughtsman—perhaps a David Bailey type—and that would have stopped me doing many of the other things that are so important in the film.'[33] The draughtsman can only draw what he sees. As Mrs Talman states 'a really intelligent man makes an indifferent painter, for painting requires a certain blindness—a partial refusal to be aware of all the options.' The film is full of verbal puns—there is a reference to Truffaut's statement that 'English cinema is a contradiction in terms', when one of the aristocrats sneeringly pronounces that 'English painting is a contradiction in terms'. One of the jokes, which in retrospect he thinks was overplayed, is the use of a naked sprite, a character who poses as a classical statue and watches all that is going on. The film is beautifully shot,[34] with particular attention to the landscape, and it also has a convincing seventeenth-century feel. This was achieved by overplaying certain elements; the wigs are excessive, and the characters' dialogue written with extreme formality and fastidiousness.[35] Some of the film's compositions have consciously been taken from late seventeenth-century painters like de la Tour.

The next film Greenaway hoped to get off the ground was a script of his called *Drowning by Numbers*, but despite his success he could not raise the money. Instead he was able to make *A Zed and Two Noughts* (1985) with a budget of £650,000. It was less tightly scripted or plot-

---

[32] *Cineaste*, Number 2, 1984.
[33] Interview with the authors.
[34] On Super 16 mm. rather than 35 mm. to keep the cost down.
[35] Often through such simple techniques as saying 'the pen of my aunt' rather than 'my aunt's pen'.

The Draughtsman's Contract (1982), with its anachronisms, marked the beginning of Greenaway's public film making. © BFI.

orientated, and in that sense bolder, than *The Draughtsman's Contract*.
It was shot in Holland as a condition of finance, and with a largely non-
British crew. The film starts with a mute swan smashing into the windscreen of a white Mercury, driven by a woman wearing white feathers
called Alba Bewick. Alba loses a leg and her two companions are killed.
Their husbands, Oliver and Oswald, who work in the zoo, try to come
to terms with these deaths through increasingly bizarre experiments and
studies. Oliver and Oswald are the two noughts of the title, and the Z is
the Zoo. The way the film is lit is inspired by the seventeenth-century
Dutch painter, Vermeer (the light source in many scenes is four and a
half feet from the ground, and from the left of the frame as in most of
Vermeer's paintings), and indeed there is much actual reference to
Vermeer in the plot. One of the characters, Van Meegeren, is based on
the famous forger of Vermeer paintings. Greenaway admits that it is a
'strange and difficult film, full of infelicities but it remains a favourite for
the ambition of its themes and its uncomfortable, prickly edge, its
elaborate visual puns and the extraordinarily beautiful photography by
Sacha Vierny.'[36]

His next film *The Belly of an Architect* (1986), however, was his most
conventional, his most humane, and probably, to date, his best. 'I've
realized quite recently, how very personal a film it is.'[37] It tackles the
question of the role of the artist more directly, and with less detachment
than any of his other films. It is about an arrogant American architect
Stourley Kracklite who goes to Rome to organize an exhibition on the
visionary eighteenth-century architect, Boullée—who despite never having realized any of his visionary architectural plans is Kracklite's obsession. The film follows Kracklite's life from the moment his wife
conceives a child on the train crossing the border into Italy, until his
death nine months later. He develops pains in his stomach and has an
obsessive and paranoid belief that his wife is poisoning him. Through his
fixation with the exhibition he forfeits the affection of his wife and his
child to be, and has his responsibilities and eventually the exhibition
itself removed from him. He disintegrates physically and mentally, and
when the doctors diagnose stomach cancer,[38] he decides to commit
suicide with dignity, while his wife is actually opening the exhibition.
One watches this decline with a terrible feeling of melancholy; once
again we see the artist as an outsider, struggling vainly to give his life
meaning, and thereby losing what is important to him, while his wife
produces an heir. 'It's art first, Kracklite second, and everyone else a
poor third.' Fertility parallels are made between Kracklite's stomach and

[36] Interview with the authors.
[37] Interview with the authors.
[38] Both Greenaway's parents died of stomach cancer.

The strength which Brian Dennehy (centre) brought to his character in *Belly of an Architect*, 1986, changed Greenaway's attitude towards his actors. His subsequent films have all demanded powerful central performances. © Oasis Film Production Ltd.

his wife's pregnancy; and significant games are played with the number seven.[39]

A great deal of critical praise went to Brian Dennehy who played Kracklite. He is an actor of the Hollywood tradition,[40] with great physical presence, who felt very strongly about the film—'There is much of myself, and of my own life and emotions up there. On reading the script I was scared to death. I was very moved.' But he goes on jokingly, 'Greenaway is the only really intellectual director I've worked with. We had to talk through translators but I let him be the intellectual, I wanted to be the human being.'[41] This humanizing element was an important and successful development for Greenaway.

After the success of *The Belly of an Architect*, Greenaway was at last able to make *Drowning by Numbers*, and it is probably because it was written some years earlier, in 1981, that it maintained more in common with *The Falls*,—in that it deliberately exposed its aesthetic principles— than with *The Belly of an Architect*. Moreover, it actually centres on a woman from *The Falls*, Cissie Colpitts, who had also been glimpsed in *Vertical Features Remake*, and was the lover of the central character in *A Walk through H*. As in *The Falls*, she is divided into three characters—grandmother, daughter, and grand-daughter (played by Joan Plowright, Juliet Stevenson, and Joely Richardson); in a sense the same woman at different times in her life. Greenaway called the story Billy Goats Gruff in reverse. Each, in turn, drowns her husband (in a tin bath, the sea, and a swimming pool) and has to persuade the coroner Madgett (Bernard Hill), who fancies them all, to conceal the crime. Madgett has a son who is also a sort of *alter ego* at a different age. None of these are real characters, nor are we really supposed to believe in them. The plot is a vehicle for the consideration of many issues, in particular game-playing and male impotency.

His next film (the first of a new three-picture deal he has with the Dutch production company, Allarts) *The Cook, the Thief, his Wife and her Lover* (1989) uses ideas inherent in Jacobean melodrama, but with a contemporary setting. It considers the principle that everything (including ourselves) is edible—consumable—and 'it is, amongst other things, a critique of current selfishness. It is about the contracts and connections between greed, power, sex and violence and about the repetitive cycle— both literal and metaphorical—of everything passing from mouth to

[39] Seven Hills of Rome, Ages of Man, Days of the Week, Colours of the Rainbow, Major Planets, Dwarfs, stages of Classical Architecture, and so on. 'It is said that seven is the highest number of objects that man can recognize numerically without actually counting them—the difference between looking and seeing—an analogy for Kracklite's predicament.' [Interview with the authors.]

[40] He starred in films like *First Blood* (1982), *Cocoon* (1985), and *Best Seller* (1987).

[41] *Observer Magazine*, 6 Sept. 1987.

Helen Mirren and Michael Gambon (centre standing) in Greenaway's disturbing and internationally acclaimed *The Cook, the Thief, his Wife and her Lover*, 1989. © Oasis Film Production Ltd.

anus.'[42] Unlike most of his previous films, the narrative is very straight-forward. It is encapsulated in the title. The thief, Albert Spica (Michael Gambon), is a vulgar and violent gang-leader, who, in an attempt to assume a higher social status, frequents the cook's (Richard Bohringer) elegant restaurant, with his band of hoods[43] and his persecuted but sophisticated wife, Georgina (Helen Mirren). She eyes a quiet, bookish man in the dining room, and begins to have an affair with him. Each evening they make passionate love somewhere in the restaurant—in the toilets and various parts of the kitchen. Ultimately her husband finds out, and brutally murders the lover. But he is beaten at his own violent game by his wife, who, with the help of the chef, has her lover cooked, and forces Albert to eat him before being executed.

It is a disturbing film to watch, reminiscent of Pasolini's *120 Days of Sodom* and Lynch's *Blue Velvet*,[44] in its view of pleasure and sophistica-tion being founded on incredible baseness. Throughout, there is a con-stant air of menace with the ever-present threat, or actual eruption, of violence and debasement. This feeling is heightened by a strong sense of claustrophobia; every scene takes place in and around the restaurant, and the few exteriors shown are all at night. Contrasting with this is the film's visual beauty. It marks Greenaway's first foray into a widescreen format, and he exploits it well through the use of long, lateral tracking shots. The set design is also magnificent: the dungeon-like eighteenth-century kitchen, and the flamboyant nineteenth-century dining room.

Like *The Belly of an Architect*, the performances are strong, particu-larly those of Helen Mirren and Michael Gambon. Greenaway recog-nizes the film to be 'an attempt to learn a lot from my experiences with Brian Dennehy.'[45] Despite this, what was absent from the script was the necessary depth of characterization. The audience is largely an observer, shocked rather than moved by what it sees; and this is emphasized by the coldness of Greenaway's rationalist approach as well as the distanc-ing effect of the wide shots and the controlled use of close-ups. It nevertheless shows his willingness to handle material aimed at a wider audience, and with it he won the 1990 London Evening Standard Award for best film-maker.

One of Greenaway's longest ongoing projects has been the television version of Dante's *Inferno*, *A TV Dante*. This is a collaboration with the painter Tom Phillips,[46] made for Channel Four and starring Bob Peck as Dante and Sir John Gielgud as Virgil. Dante, in a sense, is an ideal source

[42] Interview with the authors.
[43] Their costume and appearance is mirrored by the group of over-dressed soldiers in a large Dutch painting by Frans Hals that hangs on the wall of the restaurant dining-room.
[44] A film which Greenaway greatly admires.
[45] Interview with the authors.
[46] He also translated *The Inferno* for the programme.

for Greenaway, in that his work shows complex and rigorous structure, and has numerous metaphors, images, and word-plays to work from. In 1984 Greenaway and Phillips made an experimental template for Canto V; an instructive experience for both of them. What has been produced shows the imaginative, almost surrealist ways that the latest video technology can be used to construct a highly complex, artificial form on the screen. To make it a more personal work Greenaway has picked up all the references in Dante to birds, insects, and games; to the old questions of art itself and of creativity in relation to death and immortality. Although closely related to the original text, visual references are made to the contemporary world using extensive archive footage; familiar images of hell are conjured up from the post-nuclear, post-Nazi-holocaust world. For example, when Dante talks of his beloved Florence wracked by civil war, we are shown pictures not only of Florence, but also of modern divided cities like Belfast; or—to capture an image of the vastness of hell—groups of photos, as if from police files, of modern figures (some even from the film world) who might inhabit the inferno are flashed up on the screen.

Being made specifically for television, it is designed to be recorded on video and then watched by the audience at their own speed, stopping and starting at will like reading a book.[47] This presumed method of viewing is crucial, given the way visual footnotes are used[48] and the way so many picture boxes are imposed on top of, around, or beside each other to create a layering of image; it is more dense than any of Greenaway's earlier works. Even in some of his feature films it is not ideal to watch Greenaway's work without a video recorder, because watching in conventional fashion and at normal speed orientates one towards following a story which may be subsidiary to what the film itself is about. Greenaway concedes that film is not like a seventeenth-century Dutch painting which one can view in a museum in one's own time, or a Borges fiction in a book, which can be taken at a reader's own pace. Film moves at twenty-four frames a second. But it is a problem hard for Greenaway to come to terms with. 'Sometimes I think I'm not really a film-maker but a painter working in cinema, or a writer working in cinema.'[49] Making films to be seen through a video recorder, and taking on board real characters and narrative techniques, are two ways in which Greenaway is better fulfilling his aims.

[47] Reminiscent of *Four American Composers* (1983), where titles even appear on the screen telling the audience to turn on their video recorders, and at the end to rewind and rewatch.
[48] Interview segments with Dante experts commenting on obscurer parts of the text are placed in a small box on the screen; a clever device but one which tends to hold up the drama.
[49] Interview with the authors.

PETER GREENAWAY

To what extent was your style of film-making influenced by your education?

I had one of those educations offered by a minor English public school, arbitrary discipline, traditions to be preserved, much bullying, High Anglican—where the cadet corps is run by the Divinity Master. Later I found Anderson's *If . . .* painfully accurate. I hated the regimentation and the institutionalism. But I did benefit in two ways. I left crammed with English literature and a passion for history. I was encouraged to draw, and take an interest in European painting. I decided then that I wanted in some way to become associated with painting, but I had no idea how.

Does your interest in structure and number-games stem from any interest in mathematics at school?

I was lousy at maths. My interest in formal games, numerology, and cataloguing cannot be found there—I failed O level maths five times.

When did you actually start to want to make films?

Going to Art School was a breath of fresh air—the novelty value lasted for years, and there I tried to make some sense of an accidental discovery—Bergman's *The Seventh Seal* [1957]. That film really changed everything. Almost all down to one film. I was familiar enough with British and American movies but I had seen them without ever feeling that film-making would be important for me. I saw *The Seventh Seal* nearly two performances a day for five days. I realized that here was something I was used to from painting and the novel. It had metaphor and symbolism and literal meaning, preposterous notions of Death playing chess, of history, mythology. It was a costume piece like none I'd ever seen—and all with a strong drama you didn't want to end. For the next two years I went through a crash course in European cinema—seeing everything I could. My likes and dislikes were indiscriminate but, looking back, probably focused most on Antonioni, Pasolini, Godard, and Resnais.[1] I saw their films eagerly as they arrived in England, —with Godard that was regularly.

That was what I wanted—to make films regularly. I bought my own camera, a clockwork 16 mm. Bolex, and I applied to the Royal College of Art film-school. Not surprisingly I was not one of the twelve out of four hundred applicants who got a place that year.

How did you get into the industry then?

Well, I spent eighteen months after leaving art-school trying to arrange successful painting exhibitions and, thanks to the encouragement of the BFI educational department, keeping my interest in cinema alive by writing totally unreadable articles like 'The Relationship between Chirico and Alan Resnais'—making connections between two thousand years of European painting and eighty years of cinema. I have tried to destroy every article. For a time I wanted to be a critic like Raymond Durgnat of *Films and Filming*. I didn't know how to get into the industry. It's a problem that never changes. Totally by chance I had got a temporary job as a doorkeeper at the BFI, thanks to an ex-Bluebell girl receptionist on the BFI front-door. From there I got work in the distribution department and so the opportunity to see hundreds of short films from the archive—including many European underground movies since the twenties. It was a second course in European cinema and maybe more valuable than the first for these were experimental, rarely seen, personal, and cheaply-made movies. It was encouraging—if movies like this could be made, then there might be a chance for me. From there I started as a 'third-assistant-editor-on-trial' in the broom-cupboard cutting-rooms of Soho . . . But, after-hours, I had access to film-editing equipment and I started buying, borrowing, and acquiring raw film stock. Very slowly I started to make very modest 16 mm. films, developing theories of my own about non-narrative cinema—largely determined by my absurdly meagre means of production. I then got a proper job at the COI (Central Office of Information)—nine to five, paid at the end of the week, union-supervised film-editing . . . and I slowly rose up the ranks from sweeping up the trims to becoming an editor.

---

[1] Michelangelo Antonioni: Italian director whose films are characterized by a concentration on psychology and coincidence rather than narrative. Films include *L'Avventura* (1959), *The Red Desert* (1964), and the British film *Blow-Up* (1960).

Pier Paolo Pasolini: controversial Marxist Italian director whose films include *The Gospel According to St Matthew* (1964), *Oedipus Rex* (1967), and *Theorem* (1969).

Jean-Luc Godard: French writer/director of the European New Wave. His films often jettison narrative threads, and have semi-surreal elements. Films include: *Breathless* (1959), *Vivre Sa Vie* (1962), and *Alphaville* (1965).

Alain Resnais: controversial French director, who like Greenaway started as an editor. Films include: *Hiroshima Mon Amour* (1959) and *Last Year at Marienbad* (1961).

Your films reflect a strong fascination with natural history and the landscape. Was that also an early interest?

Yes. My father was a business-man in Shoreditch but at every available opportunity he was an amateur natural historian. His passion was ornithology, but of a self-taught and quirky variety. He made me sit down to listen to Percy Edwards[2] on the radio on Sunday afternoons. We did not particularly get on and the relationship got worse—much worse—as I got older. I am sure I picked up my interests in landscape from him but through a process of osmosis rather than being a good pupil. When I was nine or ten I started to collect insects. Trying to identify them and sticking them neatly in boxes was, I think, as important to me as finding and catching them. Nonetheless I kept it up until my early twenties. The listing and cataloguing probably started here . . . and in a curious way came again into play when I worked as an editor at the COI which is, or was, the 'information' arm of the Foreign Office. It would tell the rest of the world about the British way of life—how many sheep-dogs there were in North Wales, how many Japanese restaurants in Ipswich. This classifying and listing—in a hundred different ways— has become part of my film language. But it is used in a subversive or anarchic, rather than a merely informational, way.

How important do you think it is for film directors, making films for an audience, to have a specific, personal style, be an 'auteur', or draw on personal experience and preoccupations, as you do?

For me cinema is an art form. In Britain, a lot of people wouldn't agree. In England, traditionally the theatre is the art-form for drama—cinema is for 'entertainment'. Stoppard can argue and discuss difficult and arcane material. Ayckbourn is encouraged to experiment with unconventional structures and staging. But if these activities are considered in cinema you are liable to attract the label 'pretentious'. Pretending to what I wonder? Pretending to suggest that cinema is a very sophisticated medium perfectly capable of handling any and every dramatic and narrative and visual possibility. On the Continent, there is, or was, a tradition of encouraging personal film-making. If one believes in the widest possible range of film-making activity, no film-maker should be castigated for not conforming to the dominant mode. There is no one way to make a film. The word 'auteur' has acquired—perhaps from those who are frightened of it—dangerous overtones of indulgence. After all, it means no more than 'author'. You can—without fear or argument—use the word to describe the maker of a novel. I have a set of characteristics that welded together, I suppose, can be recognized as a

---

[2] Percy Edwards was a famous bird and animal sound impersonator.

specific, personal style—though they are not manipulated to be 'stylish'. They arise naturally out of personal interests—both in content and form. I would say that it is very difficult to put raw, neat autobiography onto the screen. By its nature it's messy, unstructured, self-obsessed and I don't think anybody's really interested. But to iron away all the idiosyncrasies, interests, and obsessions of a personal voice is probably certain to make a bland product.

Are you at all optimistic about personal film-making in this country?

I wish there was a more developed climate in Britain for personal film-making, but since there is scarcely a climate for any film-making at all, that's a great deal to ask. I am sure we get the cinema we deserve . . . which seems to be stretched between a desire to please an American audience and an obligation to service a social conscience. A social conscience—if you really are serious—can be more effectively put to use by working in television, by becoming a politician or—best of all—a primary-school teacher. For me, up until three years ago, the BFI and then Channel Four had supported the most interesting personal film-making—though primarily in terms of content and rarely in structure.

That now seems to be over. The luminaries have passed on—or away. Neil Jordan[3] seems to have relinquished his auteur status and moved into a middle commercial ground. Mike Radford[4] moved to a safe neutral space in a British context but now seems silent. Bill Forsyth is working abroad. Nic Roeg has consistently been his own man—long may he go on being so—though I thought *The Witches* was a sad, if calculated, move.

Bill Douglas[5] is quiet and Terence Davies[6] is struggling for funds. Derek Jarman remains almost alone as a working and uncompromised personal voice though maybe to less exposure than before. I have admired Terry Jones[7] and Terry Gilliam.[8] This is not because I would make movies like they do but because they pursued anarchic and irreverent

[3] A director whose films include *Angel* (1982), *Company of Wolves* (1984), *Mona Lisa* (1985), and two American movies, *High Spirits* (1988) and *We're No Angels* (1989).

[4] His films have developed from *Another Time, Another Place* (1983) to *1984* (1984) and to *White Mischief* (1987).

[5] Films include: *Comrades* (1986) and the trilogy *My Childhood* (1972), *My Ain Folk* (1973), and *My Way Home* (1978).

[6] *Distant Voices, Still Lives* (1988), *The Terence Davies Trilogy* (1974, 1980, 1983).

[7] One of the Monty Python Team, and director of their *The Holy Grail* (1974), *Life of Brian* (1979), and *The Meaning of Life* (1983). Also made *Personal Services* (1987) and *Erik the Viking* (1989).

[8] American animator and director who also started with Monty Python. *Jabberwocky* (1976), *Time Bandits* (1981), *Brazil* (1985), and *The Adventures of Baron Munchausen* (1988).

tangents away from the English obsessions of realism. Ridley Scott[9] is another example—I personally would wish for more substance in his work but his earlier movies, for me, are self-aware. I enjoyed their energy and the sharp referential use of art-forms outside of cinema.

A lot of names—but it's not easy to be optimistic. In this country, at least, perhaps they have more Past than Present. My favourite film-maker west of the English Channel is not English—but to me doesn't seem American either—David Lynch[10]—a curious American-European film-maker. He has—against odds—achieved what we want to achieve here. He takes great risks with a strong personal voice and adequate funds and space to exercise it. I thought *Blue Velvet* was a masterpiece.

How English a film-maker would you consider yourself to be, then?

The French have tried to stress Celtic origins because I was born in Wales. That is really not very significant since I left when I was three. I suppose the characteristics of irony, black humour, a pronounced interest in words and landscape and game-playing are English. My films perhaps relate to an English literary tradition that also includes Edward Lear and Lewis Carroll, and to English landscape painting. I do have an appreciation of what it is to be English and I cannot imagine working anywhere else. But my sort of cinema does not tap into the British film traditions of Grierson,[11] Cavalcanti[12] or Jennings[13]—traditions of realism.

I suppose my films work in a tradition of art cinema which is not English. But even on the Continent there is not much of that around any more. In the sixties you could find some twenty or thirty new foreign and non-American films every week in London. Now how many?

So you think that this is a problem reflected in audience attitudes as well?

Happily the audiences in this country are growing again. My best audiences have been the French. More recently German. In France it seems there is an audience that goes to the cinema prepared to find a

---

[9] Commercials director who moved into features with great success. *The Duellists* (1978), *Alien* (1979), *Blade Runner* (1981), *Legend* (1985), *Someone to Watch Over Me* (1987), *Black Rain* (1989).

[10] American director of *Eraserhead* (1976), *The Elephant Man* (1980), *Dune* (1985), *Blue Velvet* (1986), and *Wild at Heart* (1990).

[11] John Grierson: distinguished British documentary film-maker. Films include: *Song of Ceylon* (1934), *Night Mail* (1936).

[12] Alberto Cavalcanti: Brazilian film-maker who worked with Grierson in the 1930s. Films include: *Went the Day Well* (1942).

[13] Humphrey Jennings: another distinguished British documentary film-maker, whose work includes *Listen to Britain* (1941) and *Fires Were Started* (1943). Cf. Lindsay Anderson's view that Jennings was a 'poetic' not a 'realist' film-maker.

cinema of ideas which provoke and speculate—not only about the world but about cinema itself. They have long been interested in cinema as an art-form—after all, they invented it. You can buy more than a dozen serious weekly film magazines there and cinema is often discussed on French television in a different way than we can expect here. England has the patrician *Sight and Sound*[14] and the unadventurous film-previewing of the BBC's *Film 89*[15] and not much in between or on either side.

Intellectual cinema seems to be very important to you?

I am often accused of being an intellectual exhibitionist—the New York critic, Pauline Kael (and I acknowledge the wit) said of *The Cook and the Thief*: 'Greenaway is a cultural omnivore who eats with his mouth open.' It is curious that any attempt to use material away from the orthodox—to engage in any speculative discussion not normally used in cinema, to use historical reference, and to be prepared to discuss aesthetics—should arouse such hostility. The cinema community makes itself look very conventional—their model for cinema seems narrow. I like making films that speculate. 'What happens if you put this sort of idea with this circumstance? What do you get when you combine this view of things with these sets of pictures? If this is considered a conventional moral position, what if you exaggerate it? What happens when you turn this phenomenon on its head?' As much of the cinema language as I can muster is put to service these speculations. These procedures are commonplace in literature and in painting—why deny them in cinema? There is an anti-intellectualism in England that brings out its worst prejudices when challenged by cinema. I would like audiences to think about movies as well as to feel the need to emotionally identify with their content or characters. It is true that I want to provoke. I want to proselytize for a cinema of ideas without necessarily arguing for any particular idea. I'm not overtly a political film-maker but I suppose there is a strong political motive in this position, for I am arguing for cinema for its own sake, and for its ability to hold thought and ideas without necessarily having to use conventional illustrative drama, without necessarily demanding that an audience should be battered into suspending disbelief or that such a thing is cinema's sole function. I do not deny the validity of 'emotional' cinema but there are other ways to make movies than *Kramer versus Kramer*.[16] I wonder sometimes, if you wear your heart so much on your sleeve, it must ultimately stay there.

[14] *Sight and Sound*: established, quarterly BFI publication on film.
[15] *Film 88*, *Film 89*, *Film 90*, etc.: popular BBC television film review programme presented by Barry Norman.
[16] Written and directed by Robert Benton with Dustin Hoffman and Meryl Streep (1979).

How do you balance this personal film-making with your desire to communicate to a large audience?

I have often thought it was very arrogant to suppose you could make a film for anybody but yourself. Every member of an audience wants something different and probably wants something different on different nights of the week. To consciously fix the equation to satisfy your own personal investigation and make it communicable to vast numbers of people is difficult, and also manipulative. For me it doesn't work like that. I have an idea or a series of ideas and without any thought for a hypothetical audience, try to make the film. I can also take no credit for the balancing—that is my producer's doing. It is he, Kes Kassander, who makes the calculations that say how much personal film-making can be paid for by what audience. The audiences I can expect, of course, are negligible by the standards of successful Hollywood cinema. However they have so far been large enough to support the projects. The films are made very cheaply. *The Cook and the Thief* by September 1990 had earnt $8 million in America and about the same in the rest of the world. When the film was made I doubted that it would ever be seen by an American audience at all. It has been suggested that being able to continually make films is a measure of the best sort of success.

Does this apply to all great artists? What in your mind are some of the characteristics of great art?

I suspect that all the great cultural landmarks of the last two centuries have in some way been provocative and probably been speculative—to be one may mean to be the other. The first works of Impressionism and Cubism, for example, were not commissioned or painted with a specific audience in mind and their newness and strangeness was certainly regarded as a provocation. Of course, to paint a painting—any painting—is considerably cheaper than to make a film. However, if any work of art is worth consideration, it must be planned and organized—some artists may be able to dash off a brilliant work rapidly but you may be sure that they've served a long apprenticeship. It must try and offer something new, although seeking novelty in itself can end up as mere gimmickry. I think any work of art which is worthwhile gives up its meaning slowly and you have to work at it—an activity that can make the satisfaction greater. I also believe that all great works of art acknowledge their own existence in some way. They see themselves from the outside. It is said that *Hamlet* is a play about play-acting and Rembrandt's *Nightwatch* is a painting about painting. I believe there is only one truly great film-maker who made cultural landmarks that stand comparison without embarrassment in this company and that is Eisenstein.[17] All his films have this self-knowledge. From his very first film *Strike*, it was there—an extraordinary piece, full of hundreds of images all working on different levels and saying, 'look, this is how film works'.

But it is always a problem to understand what role art plays and especially what role it plays in terms of the cinema. If we believe that art is the only thing that endures and is capable of being rediscovered with undoubted value every generation—unlike most other aspects of civilization—what price cinema?—a very perishable medium with little history, needing special equipment to make it visible. I think audiences should be made aware of the artifice in the cinema. There is really nothing there except a projection-beam and a white screen.

Is the desire to highlight the artificiality of film the reason why you do not place dramatic narrative very highly?

I would have thought the narrative of *The Cook and the Thief* was a dramatic narrative—even a melodramatic narrative. When I first started to make films there was a shared feeling among underground film-makers to make non-narrative cinema. It was a reaction—never of course shared by overground cinema—that the psycho-drama approach, working with the dominance of actors, was making films that essentially were illustrated novels. There must be alternate ways of doing it. Literature had experimented with different ways before the dominance of the novel. However if you throw narrative out, you have to find other ways of structuring. Some people, taking to heart Godard's 'cinema is truth twenty-four frames a second'—organized their work by counting frames or time intervals, or other sorts of formulae of a public or private sort. I suppose the most well-known manifestations of this were the early films of Andy Warhol—they were tedious, but nonetheless it was important that the issues were discussed.[18] I used number and alphabet counts—universal systems as alternatives to the anecdotes of narration. It could produce arid results. However, I began to feel I was denying myself what I really wanted to do—which was to tell stories. I began to introduce simple narratives and the process has continued—always keeping alive a concern for those early disciplines—never relying entirely on the anecdotes of narrative. I have always found it easy—too easy—to

---

[17] Sergei Eisenstein, one of the great Russian directors. His first film *Strike* (1925) was made when he was twenty-six. His silent films were experiments to find appropriate new cultural forms, untainted by a 'bourgeois' past, which would instil a revolutionary consciousness in the masses. Other classic films made by him include *Battleship Potemkin* (1925) and *October* (1927).

[18] In the 1960s Andy Warhol worked as a film-maker in partnership with Paul Morrissey on American underground films which were intentionally 'passive' and 'empty'. *Empire* (1964) was an eight-hour stationary view of the Empire State Building.

write stories. *The Falls* has ninety-two in three hours. *Dear Phone* has thirteen in twenty minutes. *Death in the Seine* has twenty-three in forty minutes. In *Drowning by Numbers* perhaps the situation reaches its most complex. The film is propelled by both a narrative and a number count which are often interchangeable—the narrative indicating the characters' so-called 'free choice' and the numbers representing some sense of destiny. At the end narrative and number count complement one another, the action has been completed and the film is over.

The use of these structures and devices attracted me to the New York composers of the early sixties: Terry Riley, Philip Glass, Steve Reich. The composer Michael Nyman was an enthusiast and a musician for Reich for a time, and has written most of the music for my films. I think, whilst using a lot of these systems, my films are also critical of them and mock them—demonstrating that they are artifices. But our lives are governed by such artifices. As well as the demands for a strong narrative content I am sure a good movie needs a sound sense of structure and I also believe a film needs an overall sense of metaphor. Content and structure and metaphor. I think without all three it is difficult to find a really good movie. I would add a strong sense of surface, but that is a painterly concern perhaps and a personal one. I believe that much cinema—much American cinema—is too content heavy. Much European cinema may be too structure conscious. I can also think of filmic trends that might be seen as too metaphor heavy.

Most of the audience are not aware of the full levels of metaphorical meaning in your work. Does this matter?

I am sure it doesn't. But by audience reaction I can see that many are intrigued. Many French audiences are aware of the many levels and are entertained. I am sure everyone agrees that our language, our actions, the way we use the media, are always multi-layered in meaning. We are used to sending out many messages simultaneously—though not all of them are necessarily picked up. Every word has a variety of meanings and nuances which are received differently in different contexts. The cinema is very capable of holding multiple meanings. I deliberately want to create a rich film-fabric acknowledging the diversity of multiple meanings—literal, symbolic and metaphorical. Our lives are constantly full of this interaction—why can't we permit it in film?

Do you mind wild interpretations being made of some of your films?

If you make something for public consumption—however small the audience—it must be theirs to interpret as they wish. You cannot control it. Borges said, 'It's much more difficult to be a reader than a writer.' In some ways the viewer has to work harder because the material is not his.

To comprehend it, he or she has to refashion it in some way—see it more personally, relate it to his or her experience, sympathies, education, cultural interests, systems of values, even ethics and morals. However I am sometimes amused by the interpretations. The French organized a semiotic conference on *The Draughtsman's Contract* that, I am told, failed to progress much beyond the title. The film has been seen as an essay on fruit symbolism, the political history of Northern Ireland, pro-Thatcherism, anti-Thatcherism, pro-feminism, anti-feminism, and many other things. *Time Out* published an ironic article that argued that the film was an exposé of English League Football. All films are open to different interpretations—*E.T.*[19] was seen by some as a movie on the resurrection myth with the alien as Christ.

What about all those who miss most of this multi-layered symbolism?

The Draughtsman's Contract can be seen as an Agatha Christie tale and there is much in the film to enjoy besides the use of conceit and allegory. It's absurd for people to say, 'Christ, I'm not clever enough to catch all these references'.

Which are the aspects and areas of film-making which you enjoy most?

I think there are three areas involved. The first is conceiving the idea—which is the most enjoyable stage of all because everything is possible—and the writing of the script. This you do entirely on your own, in a garret, in isolation, and it can be quite lonely. Then there's the middle period when you film the thing on location or in a studio like a general of an army. Most of the time you are not being a film-maker but a chaperon or a nurse or a clerk or a bully or an amateur psychologist looking after people's egos and trying to preserve your own. The third part is coming back with everything you have shot into the cutting room and you sit down to look at the thousands of images which you have trawled in.

Of these three parts, I prefer the writing first, the editing second, and filming third. But things are changing. I am beginning to enjoy collaborating far more than I used to. I used to imagine a situation where for eternity you were forced to film three minutes of usable film a day. I am starting to understand why some people might think that that was the very opposite of a punishment. I have been accused of wanting too much control—of having too much obsession for every detail, though not by the people I collaborate with. If you spend a minimum of nine months of your life on a film that has been yours from the start, you'll want to look after it very carefully and make sure that what the public eventually sees is the film you wanted to make.

[19] Directed by Steven Spielberg and written by Melissa Mathison (1982).

Could you describe how your ideas coalesce into a script? How the script develops into this complex structure with its many layers of meaning?

The films often start as a string of speculative ideas, not as a ready-made story, or a plot or a set of characters or an event. Taking *A Zed and Two Noughts*, it began with a fascination with twinship. Being a twin is perhaps the nearest you can come to meeting yourself. Many world mythologies share a theory that we are born as twins and in most cases the second twin dies in the womb leaving us incomplete, always looking for the lost half. We compensate by trying hard all our lives to pair with a stranger. Our bodies constantly remind us of this duality—this two-ness—nostrils, feet, breasts, testicles, ovaries. I travelled a great deal with *The Draughtsman's Contract* when it was being shown around the world. It was an opportunity to visit the zoos of every capital city. I have always been fascinated by zoos—three-dimensional encyclopaedias, living dictionaries of animals, yet a continuation of man's reprehensible relationship with animals. The first animal prison was the Ark where animals went in two by two. Berlin Zoo—an animal prison inside a human prison—gave me the idea of putting these and other speculations into a film. The main characters would be twins who work in a zoo as animal behaviourists—a field popularized by David Attenborough whose commentaries have become authoritative in the public imagination.

Approach Darwin through Attenborough. How do animal behaviourists think about their subject—how do they relate their anxieties with their studies? The greatest loss I could imagine would be the death of my wife. So kill the twins' wives in a car-crash—the most possible and yet gratuitous of events. Grief-stricken, the twins try to use what they know best—natural history—to comprehend the event. To complete the circle, the crash is caused by an animal—a swan. We now have the beginnings of a plot to explore many things: the absence of meaning in gratuitous death; is death pre-determined; how do religion and science deal with the problem; is Genesis or Darwin the most likely myth; what other myth-systems try and answer the question? The classical humanist myths have all the stereotypes of every story ever told. To fit all this together the zoo is staffed by the equivalents of the characters from Mount Olympus. Venus de Milo is the zoo prostitute, the gatekeeper is a lightly disguised Mercury who wears the zoo-colours, a silver hat with wings—the symbol and colour of Mercury. Pluto, God of the Underworld, is the Keeper of Reptiles who makes an alternative animal collection of black-and-white animals because there is no colour in Hell . . . and so it goes on, involving Vermeer and Vermeer's faker Van

Greenaway on location of *Drowning by Numbers*, 1988. © Oasis Film Production Ltd.

Meegeren, mythological animals, the conception of the centaur, besti-
ality. I was creating a rich and detailed world full of literal and metaphorical and referential meaning that is totally artificial but mirrors our own.

*You said that you were starting to enjoy shooting the film and working
with actors more now?*

I am learning a great deal with each film and enlarging the vocabulary all
the time. In *The Belly of an Architect*, Brian Dennehy taught me much. I
have tended to use English actors with a theatrical background to
sustain the long scenes without cutaways, to use stylized language and
be very familiar with artifice and multiple-layered language. I was
frustrated that some of those things had to be jettisoned in working with
Brian, but was much rewarded by many things he demonstrated that I
had been denying. He demonstrated the self-evident truth that complex
ideas needed as much help as possible from the actors. People have
rightly congratulated his performance but have said that the film is less
layered than previous or later films. I suggest that the film is just as
layered but, by his approach, he made it seem less consciously so. I still
haven't yet discovered how to have both the artifice and the performance—both the self-consciousness and the suspension of disbelief. They
deliberately alternate in *The Cook and the Thief* and that film is
certainly an attempt in some ways to use what I had learnt from working
with Brian.

Since working with Sasha Vierny on ZOO *you have used him as your
cameraman on all your films?*

Yes. He is a brilliant cameraman. I had been enthusiastically watching
his photography for years in Resnais's films—from *Hiroshima Mon
Amour* to *Stavisky*. He worked with Buñuel on *Belle de Jour* and for
Duras and Ruiz[20] and many others. The more he is encouraged to be
experimental, the more excited he becomes. There is a deliberate catalogue of the various ways to light a set in ZOO—by daylight, dawn-
light, twilight, candle-light, fire-light, moonlight, starlight, search-light,
by cathode-tube, arc-lamp, neon-lamp, projector-beam, car headlamps,
fog-lamps, and many others. He made a rainbow in that film that
competed with the sun. And all with great economy of means and very
fast. I'm sure if you gave him a lit candle and two sheets of newspaper he
would come up with a solution to any film-lighting problem. He insists
on two sheets—one for him to use and the other for you to read while
you wait.

---

[20] French directors. Marguerite Duras' films include *India Song* (1974); Ráúl Ruiz has made
City of Pirates (1983), *Hypothesis of the Stolen Painting* (1978), *Of Great Events and Ordinary People* (1978), and *Three Crowns of the Sailor* (1983).

You are doing an increasing amount of work for television. Is this the direction that you think your career is moving?

I prefer to make films for the cinema. The scale, the commitment of the audience, the repeatability, the in-the-dark exclusiveness, the way the product is distributed still make it preferable. A bigger than you are, noisier than you are, experience. However, the cinema—after only ninety years, about the length of the Golden Age of Dutch painting—is dying: it does not have the same pull as it had in the forties and fifties. I used to think that television had a much more reduced vocabulary than cinema—but I have come to realize that the language can be as rich. But it is a different language with different rules and characteristics. I think we should explore those characteristics to the utmost, though there is an equation that seems to suggest that as the aesthetic possibilities increase enormously so the uses to which television is put become increasingly smaller and more banal. *A TV Dante*—part of which was completed recently for Channel Four—is an ambitious project to see how the new television vocabulary can be put to service an old venerable text. I want to be able to use both media and currently am working on a version of *The Tempest* called *Prospero's Books* to see how the cinema and television vocabularies can be put together.

Do you actually shoot things significantly differently for the smaller, as opposed to the big screen?

Because of the scale, television obviously does not do great favours to wide shots which I use a great deal in the cinema. It is poor on complex and detailed compositions which have lots of contrast, and is not very sympathetic to the very dark and the very light. Television sound-tracks, compared to the cinema, are poor. It is perhaps more effective with moving than static material. It also has a very limiting 1 to 1.33 screen aspect ratio. However, it is more efficiently edited, cheaper to make, and has a very cheap shooting-ratio. For me, the greatest advantage is its post-production possibilities which have left cinema far behind.

You have a reputation for working extremely hard, and being very prolific.

Maybe. At the end of *The Belly of an Architect*, Kracklite is asked several questions by the police which are also in context the final questions that anyone can ask, the barest significant details. 'What is your name? What's your profession? Do you have any children?' Kracklite replies, 'Is that all?'. Working hard and being prolific in cinema—to good or bad or indifferent ends—must be merely decorative compared to those details. Being a film-maker for me is very enjoyable indeed and I would like to continue. But, in the sum of things, it is not truly a very significant or important thing to be. I am sure that the questioning, the doubting, and the arguments and alternatives and ironic tone of my films suggest the same.

FILMOGRAPHY

BRIEF BIOGRAPHICAL DETAILS

Born in 1942 in Wales, and brought up in Wanstead. Studied at Walthamstow College of Art, where he trained as a painter. Between 1964 and 1965 he worked at the BFI's distribution department. He then worked as a film editor, and spent eleven years cutting films, including numerous documentaries for the Central Office of Information. From the mid-sixties to the mid-seventies he also wrote novels and short stories, and some illustrated books (all unpublished).

PAINTING EXHIBITIONS

He first showed his work at a 1964 exhibition called 'Eisenstein' at the Winter Palace, at the Lord's Gallery, and has held exhibitions of his paintings at Arts 38, the Curwen Gallery in 1976, and the Riverside Studios in 1978. One-man shows in 1989 at the Arcade, Carcassonne, and the Palais de Tokyo, Paris; in 1990 at the Nicole Klagsbrun Gallery, New York; the Australian Centre of Contemporary Art, Melbourne; the Ivan Dougherty Gallery, University of New South Wales, Paddington; the Cirque Divers, Liège; Shingawa Space T33, Tokyo; Altium, Fukuoka; the Danny Keller Gallery, Munich; the Video Galleriet, Copenhagen; Xavier Huskens, Brussels.

FILMS AND TELEVISION WORK DIRECTED BY PETER GREENAWAY

1966   *Train.* 5 mins.

       Tree. 16 mins.

1967   *Revolution.* 8 mins.

       Five Postcards from Capital Cities. 35 mins.

1969   *Intervals.* 6½ mins.

1971   *Erosion.* 27 mins.

1973   *H is for House.* 9 mins. (re-edited 1978). NARRATOR: Colin Cantlie, Peter Greenaway, and his family. MUSIC: From Vivaldi's *Four Seasons.*

1974   *Windows.* 3½ mins. NARRATOR: Peter Greenaway. MUSIC: Rameau's 'The Hen'. CALLIGRAPHY: Kenneth Breese.

Water. 5 mins. MUSIC: Max Eastley.

1975   *Water Wrackets.* 12 mins. NARRATOR: Colin Cantlie. MUSIC: Max Eastley. CALLIGRAPHY: Kenneth Breese.

1976   *Goole by Numbers.* 40 mins.

1977   *Dear Phone.* 17 mins. CALLIGRAPHY: Kenneth Breese.

1978   *1–100.* 4 mins. MUSIC: Michael Nyman.

A Walk through H. 41 mins. BFI. PHOTOGRAPHY: Bert Walker and John Rosenberg. MUSIC: Michael Nyman. NARRATOR: Colin Cantlie. CAST: Jean Williams. CALLIGRAPHY: Kenneth Breese.

Vertical Features Remake. 45 mins. Arts Council of Great Britain. PHOTOGRAPHY: Bert Walker. MUSIC: Michael Nyman (theme music Brian Eno). NARRATOR: Colin Cantlie.

1979   *Zandra Rhodes.* 15 mins. COI.

1980   *The Falls.* 185 mins. BFI. SCRIPT: Peter Greenaway. HEAD OF PRODUCTION: Peter Sainsbury. PHOTOGRAPHY: Mike Coles, John Rosenberg. EDITOR: Peter Greenaway. MUSIC: Michael Nyman (additional music by Brian Eno, John Hyde, Keith Pendlebury). With the voices of Colin Cantlie, Hilarie Thompson, Sheila Canfield, Adam Leys, Serena Macbeth, Martin Burrows.

1981   *Act of God.* 28 mins. Thames TV. SCRIPT: Peter Greenaway. PRODUCER: Udi Eicler. PHOTOGRAPHY: Peter George. EDITOR: Andy Watmore.

1982   *The Draughtsman's Contract.* 108 mins. (a.k.a. *Death in the English Garden*). BFI in association with Channel 4 Television. SCRIPT: Peter Greenaway. HEAD OF PRODUCTION: Peter Sainsbury. PRODUCER: David Payne. PHOTOGRAPHY: Curtis Clark. EDITOR: John Wilson. PRODUCTION DESIGNER: Bob Ringwood. MUSIC: Michael Nyman. CAST: Anthony Higgins, Janet Suzman, Anne Louise Lambert, Hugh Fraser, Neil Cunningham.

1983   *Four American Composers.* 55 mins. each episode. Channel 4 Television/Transatlantic Films. Four television documentaries on John Cage, Robert Ashley, Philip Glass, Meredith Monk (based on the Almeida Concerts of September 1982). PRODUCER: Revel Guest. PHOTOGRAPHY: Curtis Clark. EDITOR: John Wilson.

1984   *Making a Splash.* 25 mins. Channel 4/Media Software. SCRIPT: Peter Greenaway. PRODUCER: Pat Marshall. MUSIC: Michael Nyman.

A TV Dante—Canto 5. Channel 4 Television (the first part in a series of 33 Cantos to be made in collaboration with Tom Phillips). SCRIPT: Peter Greenaway and Tom Phillips. PRODUCER: Sophie Balhetchet. PHOTOGRAPHY: Mike Coles, Simon Fone. EDITOR (FILM): John Wilson. EDITOR (VIDEO): Bill Saint. CAST: Suzan Crowley, John Mattocks, Donald Copper.

1985   *Inside Rooms—26 Bathrooms.* 25 mins. Channel 4/Artifax

Productions. SCRIPT: Peter Greenaway. PRODUCER: Sophie Balhetchet. PHOTOGRAPHY: Mike Coles. EDITOR: John Wilson. MUSIC: Michael Nyman.

A Zed and Two Noughts. 115 mins. BFI Production/Allarts Enterprises/Artificial Eye Productions/Film Four International. SCRIPT: Peter Greenaway. PRODUCERS: Peter Sainsbury, Kees Kasander. PHOTOGRAPHY: Sacha Vierny. PRODUCTION DESIGNER: Ben Van Os, Jan Roelfs. EDITOR: John Wilson. MUSIC: Michael Nyman. CAST: Andrea Ferreol, Brian Deacon, Eric Deacon, Frances Barber, Joss Ackland.

1986   *Belly of an Architect.* 118 mins. Callendar Company/Film Four International/British Screen Hemdale/Sacis. SCRIPT: Peter Greenaway. PRODUCERS: Colin Callender, Walter Donohue. PHOTOGRAPHY: Sacha Vierny. PRODUCTION DESIGNER: Luciana Vedovelli. EDITOR: John Wilson. MUSIC: Wim Mertens. CAST: Brian Dennehy, Chloe Webb, Lambert Wilson, Stefania Casini.

1987   *Fear of Drowning.* 30 mins. Allarts (for Channel 4). SCRIPT: Peter Greenaway. PRODUCER: Paul Trybits. CO-DIRECTOR: Vanni Corbellini. CAST: Peter Greenaway (Presenter), and the actors and crew of *Drowning by Numbers.*

1988   *Drowning by Numbers.* 119 mins. Film Four International/Elsevier Vendex Film. SCRIPT: Peter Greenaway. PRODUCER: Kees Kasander, Denis Wigman. PHOTOGRAPHY: Sacha Vierny. PRODUCTION DESIGNER: Ben Van Os, Jan Roelfs. EDITOR: John Wilson. MUSIC: Michael Nyman. CAST: Bernard Hill, Joan Plowright, Juliet Stevenson, Joely Richardson, Jason Edwards.

A TV Dante—Cantos 1–8. KGP Production in association with Channel 4/Elsevier Vendex/VPRO. Made in collaboration with painter Tom Phillips. CAST: John Gielgud, Bob Peck, Joanne Whalley.

1989   *The Cook, the Thief, his Wife and her Lover.* 120 mins. Allarts/Erato Films/Films Inc. SCRIPT: Peter Greenaway. PRODUCER: Kees Kasander. PHOTOGRAPHY: Sacha Vierny. PRODUCTION DESIGNERS: Ben Van Os, Jan Roelfs. EDITOR: John Wilson. MUSIC: Michael Nyman. CAST: Michael Gambon, Helen Mirren, Richard Bohringer, Alan Howard.

Death in the Seine. 40 mins. Erato Films/Allarts TV Productions/Mikros Image/La Sept. CAST: Jim Van Der Woude, Jean-Michel Dagory.

BIBLIOGRAPHY

BFI Production Catalogue, 1977–8. Articles by Greenaway and Sainsbury on *A Walk through H.*

Monthly Film Bulletin, Sept. 1978. Tim Pulleine on *A Walk through H.*

Time Out, 1–7 Dec. 1978. A study of Greenaway's early work especially *A Walk through H.*

Sight and Sound, Spring 1979. Greenaway's background and aims in his early films.

BFI Production Catalogue, 1979–80. Articles by Greenaway and Simon Fields.

Time Out, 11–17 Apr. 1980. On the screening of his films at the ICA.

Film, Apr. 1980.

BFI News, Jan. 1981. Profile on Greenaway.

Monthly Film Bulletin, Jan. 1981. John Pym on *Vertical Features Remake.*

Sight and Sound, Spring 1981. Article by Chris Auty on *The Falls.*

City Limits, 30 Oct.–5 Nov. 1981. Interview with Greenaway on *The Draughtsman's Contract.*

Screen International, 19–26 Dec. 1981. Interview on *The Draughtsman's Contract.*

Sight and Sound, Winter 1981–2. Harlan Kennedy on his early work and *The Draughtsman's Contract.*

Film Comment, Jan.–Feb. 1982. On his work, especially *The Falls.*

BFI Special Collection, 1982, on *The Draughtsman's Contract.*

Monthly Film Bulletin, Apr. 1982. Articles by John Pym and Robert Brown on *Dear Phones, H is for House, Intervals, Water Wrackets,* and *Windows.*

Sight and Sound, Autumn 1982. Article by Jill Forbes on *The Draughtsman's Contract.*

Monthly Film Bulletin, Nov. 1982. Greenaway and Peter Sainsbury talk about *The Draughtsman's Contract.*

Guardian, 12 Nov. 1982. 'Illusion at the Point of a Pin', by Chris Auty.

Time Out, 12–18 Nov. 1982. Interview on his work.

Time Out, 17–30 Dec. 1982. Article by Greenaway on the enclosure of wild animals in zoos.

International Herald Tribune, 31 Dec. 1982. 'A Public Triumph for a Private Movie', by Mary Blume.

Stills, May–June 1983. On Greenaway's career, and especially *The Draughtsman's Contract.*

New York Times, 19 June 1983. 'Very English, Very Eccentric', by Stephen Harvey.

Film News (Australia), July 1983. On Greenaway's aims, and his films to date.

Sunday Times, 14 Aug. 1983. 'The Film-Maker's Contracts', by Adam Mars Jones.

American Cinematographer, Sept. 1983. Profile of Greenaway.

Time Out, 3–9 Nov. 1983. On *Four American Composers*.

Art Forum, Nov. 1983. 'Breaking the Contract'; interview by Stuart Morgan.

Time Out, 12–18 Dec. 1984. On *Four American Composers*.

Cineaste, No. 2, 1984. Interview on *The Draughtsman's Contract*.

Guardian, 27 June 1985. 'Rotterdammerung', by James Park, on *A Zed and Two Noughts*.

Guardian, 4 Dec. 1985. 'Catcher of the Wry', by Quentin Falk, on *A Zed and Two Noughts*.

Time Out, 5–11 Dec. 1985. A fun 'A–Z' guide to Greenaway and his films, especially *A Zed and Two Noughts*.

City Limits, 6–12 Dec. 1985. Greenaway on *A Zed and Two Noughts*.

New Musical Express, 14 Dec. 1985. 'Telling Fibs', by Jonathan Romney.

Monthly Film Bulletin, Dec. 1985. Article by Chris Auty, with an interview by Tony Rains.

Screen International, 4–11 Jan. 1986. Greenaway talks about his films and his approach to film-making.

Broadcast, 9 May 1986. Short interview on his career.

Sight and Sound, Summer 1986. Greenaway writes about the film by the Quay brothers, *Street of Crocodiles*.

Film (BFFS), June–July 1986. On the working relationship between Greenaway and Michael Nyman.

Sight and Sound, Summer 1987. Interview on *Belly of an Architect*.

Observer Magazine, 6 Sept. 1987. 'Alimentary Lesson', by Lesley Thornton, including an interview with Brian Dennehy on working in *Belly of an Architect*.

Video Week, 28 Sept. 1987. Interview with Greenaway on *Belly of an Architect*.

Independent, 10 Oct. 1987. 'Bricks and Martyr', by Kevin Jackson.

Films and Filming, Oct. 1987. Interview on his career, especially *Belly of an Architect*.

The Times, 9 Nov. 1987. 'The Director's Contract', by Chris Peachment, with an interview on *Belly of an Architect*.

Screen International, 13–20 Feb. 1988. On some of Greenaway's current and planned projects.

Hollywood Reporter, 31 May 1988. Details of a contract with the Dutch producer, Kees Kasander of Allarts Ltd., to produce three films.

Listener, 4 Aug. 1988. 'Safety in Numbers', by Bill Pannifer, on *Drowning by Numbers*.

City Limits, 1–8 Sept. 1988. Interview with Greenaway about his work, especially *Drowning by Numbers*.

Monthly Film Bulletin, Oct. 1988. Article on Greenaway, especially *Drowning by Numbers*.

6 DEREK JARMAN

ESSAY ON

DEREK JARMAN

DEREK JARMAN has been called 'the *enfant terrible* of the British film industry', 'the English Andy Warhol', and David Bowie has even referred to him as 'a Black Magician'. He is certainly a highly original and visionary, independent English film-maker, whose daring and resourcefulness are inspiring. He is probably wilfully marginal, largely because, unlike so many artists in similar situations, he has always resisted the constant temptation to yield to commercial pressures. Peter Sainsbury while head of the BFI production board commented, 'Derek Jarman is an artist of note, for what he has achieved is outstanding while what he has had to contend with is grossly symptomatic of our cultural malaise.'[1]

Charming and open, Jarman is prepared to talk to anyone interested—although his enthusiasm and passion concerning the many issues on which he has thought-provoking opinions sometimes spill over into rhetorical or contradictory remarks. He lives in what has been compared to a monastic cell; an austere one-room apartment off Charing Cross Road with a large bed taking up most of the space, and with his paintings decorating the walls. In the same vein, he also has a small wooden house on the Dungeness beach which he uses as a retreat when he is not working. To Ken Russell, Jarman is 'Probably the last true bohemian'.[2]

Jarman stresses that his early formative years are crucial to understanding the nature of his films—'all creativity has its roots in one's childhood'.[3] In particular there was a nascent tension between radicalism and conservatism which was to be a powerful force throughout his life. Brought up in a middle class, military family, he had what he describes as 'the most conventional English education, public schools, and so on'.[4] It was while at school that he became interested in art. He was strongly influenced by one of his teachers, Robin Noscoe ('an absolutely wonderful man'[5]), who introduced him to William Morris and the philosophy that art and creative pursuits are, in fact, a whole

[1] *Three Sixty*, May 1985.
[2] Ken Russell, *A British Picture: An Autobiography* (Heinemann, 1989).
[3] Interview with the authors.
[4] *Observer Magazine*, 22 Feb. 1987.
[5] Interview with the authors.

Derek Jarman on the set of *Caravaggio*, 1986. © Mike Laye.

way of living. His father agreed to put him through the Slade Art School if he first went to King's College, London to study English, Art History, and History. It was while studying Fine Art at the Slade that Jarman came to terms with his homosexuality. He was determined to 'mug up the courage to be honest, to break the taboos—for oneself, one's self respect, and other people who might be helped'.[6] His involvement there in the department of theatre design was, in part, because of the tradition that specialization had in attracting gay students.

After leaving the Slade, Jarman designed several opera and ballet sets, including one for the English National Opera and one for the Royal Ballet, but had almost given this up and gone back to painting when, through a chance encounter with a friend of Ken Russell's, he was asked to be set designer for Russell's film *The Devils* (1971), and was involved in a significant way with the script. The huge project 'was a baptism of fire—a year in Pinewood on a Super-Production'.[7] He designed the sets as large as possible, sacrificing detail for scale, so as to be 'as forceful as the sets from an old silent movie'.[8] During the course of the project, Jarman became heavily influenced by Russell's ideas of film. 'There was no better director than Ken to learn from—he would take the adventurous path even when that interfered with the coherence of the film.'[9] It was after working on *The Devils* that Jarman became involved in making Super 8 movies, and five and a half years later he made his first feature film.

The two driving forces that emerged from these years are aesthetic and sexual, engendering an approach which is at once rebelling against anything which can be construed as authoritative, or even just established, and which also yearns for certain traditional values. Such feelings are expressed in his films, particularly his features, with great vehemence. They are, as one critic commented, 'the aggressive expression of something which is quite private'.[10]

Jarman's aesthetic impulse centres on the value of art, not for the viewer but for the creator. 'The only impact which I hope I've had with my films is for people to do the same sort of thing in the same spirit. The end product is not really important, it is merely a witness to the process of creativity.'[11] He treats film in the same spirit as Orson Welles's famous statement that directing a film is like the greatest train set a boy could ever want. Jarman loves the collaborative nature of film-making, and works both with and for his friends—especially those of the gay

[6] *Observer Magazine*, 1 Jan. 1989.
[7] *What's On*, 24 Apr. 1986.
[8] *Movie Maker*, July 1984.
[9] Ibid.
[10] Michael O'Pray in *MFB*, June 1984.
[11] Interview with the authors.

community. Film is seen as a process rather than a product: 'Provided the cast approve of the finished result, find the experience of film-making a joy, I'm happy and could wish for no more.'[12] The process of making commercial films he describes as 'Horrible. Everybody is in their compartment, and it's not creative, and it's not fun. I really feel that work should be a joy, and as a director one is responsible for making it enjoyable.'[13] Sarah Radclyffe, the producer of both Jarman's *The Tempest* (1979) and *Caravaggio* (1986), commented with amusement on some of the problems this kind of an approach can lead to:

The Tempest was a hippy production if ever there was one. Inevitably we all got kicked out of our hotel and had to move into the location, Stoneleigh Abbey, where we all got snowed in. The whole cast helped build the set. We never had call sheets. There was no production manager . . . How it got made, I shudder to think.

As part of this process of creativity much of his work has been highly personal, technically experimental, and also startlingly original by tradi-tional British cinematic standards. He abandons the more accepted ideas of pace and rhythm in editing, and either has a camera which is distinc-tive in its very static nature or one which never stops moving, often a Super 8 mm.[14] His style is, in large part, doggedly anti-naturalistic, abandoning narrative or psychological exploration, and replacing it with an unusual vocabulary of image, symbolism, colour, and sound. The films are often dream-like. Jung's *Alchemical Studies* and *Seven Sermons to the Dead*, he says, 'gave me the confidence to allow my dream images to drift and collide at random',[15] and this element is compounded by the degree of improvisation and collaboration that characterizes his films. Sometimes he dubs evocative sounds, like run-ning water or the sounds of birds, over the images. With his strong background both in painting and set design, he frequently uses imaginat-ive *mise-en-scène* and tableaux, many of them quoting from paintings, and likes to suit the action to the stills. Although many of these scenes are brilliant and exciting, they sometimes fail, just looking crude or awkwardly and obviously 'set up'.

Jarman has been a real pioneer in his exploration of the potential of the Super 8 camera and, more recently, of the domestic VHS video camera. Part of the reason for his use of these very cheap mediums is a conscious rebellion against the unnecessary levels of film technology which he believes to be too institutionalized in the industry. Although he

---

[12] *Observer Magazine*, 22 Feb. 1987.
[13] Ibid.
[14] This refers to the width of the film stock. Most feature films use 35 mm. film. Due to the smaller frame size, Super 8 mm. film tends to become very grainy when enlarged.
[15] Derek Jarman, *Dancing Ledge* (Quartet, 1984).

sees important advantages in technological developments, he fears the way it can 'reinforce the grip of centralised control, and can emasculate opposition',[16] because of the high resulting costs. *Caravaggio* and *War Requiem* (1988) are the only two of his films actually shot on 35 mm.

One of the most important technical areas which Jarman has explored is manifested in two of his features and in particular in his short films. He shoots at three to six frames per second (rather than the standard twenty-four frames per second), a speed that is above the threshold that the eye can catch. He then projects it onto a screen, again at this low frame rate, and refilms it on video at normal speed. The resulting images appear to glide in slow motion, as there are groups of identical frames one after the other, creating a rhythm or visual beat. The fact that there is a degrading of the picture quality when it is copied gives a very painterly texture in colour and grain. The overall result, especially when combined with superimposition, is very dream-like and capable of expressing such diverse moods as the romantic *The Angelic Conversation* (1985) or the aggressive *Last of England* (1987).

In *Art Monthly* Michael O'Pray concluded, 'Jarman's films are the work of a painter in their refusal to tidy up the edges, or to cover the canvas, so to speak.'[17] When asked to justify the validity of this personal painter's approach against that of a more 'realistic' one, he replied, 'I have never believed in reality. The news for me is fiction. People say to me "You make fantastic films," and I say "No, I make documentaries".'[18] Elaborating on this idea he states, 'Why is a Ken Loach film more real than a Ken Russell film? Why is Italian Neo-Realism more real than Michael Powell? For *me* the answer is that they aren't.'[19] When he says that he is making documentaries what Jarman means is that he remains as faithful as he can to the reality of the life he lives, and tries to make films which reflect this. He describes them as 'a *private* analysis of my own feelings and position in the culture, and then a presentation of them'.[20] He finds it quite hard even to consider doing it any other way, unless the subject is a great passion like *The Tempest*. 'Unless one can put one's own voice into a film, then there's an element of dishonesty in it.'[21] 'The most sound foundation for making a film is reflecting one's own life. It's the opposite of politics or media which is never true.'[22]

It is because of his total belief in the efficacy of these values that much of his work—in particular *Jubilee* (1978), *Imagining October* (1984),

[16] Interview with the authors.
[17] *Art Monthly*, June 1988.
[18] *Observer Magazine*, 1 Jan. 1989.
[19] *Afterimage*, No. 12 Autumn 1985.
[20] Interview with the authors.
[21] *NME*, 25 May 1985.
[22] Interview with the authors.

and more recently *Last of England*—mourns or rages against the loss of these values, and the consequent loss of the identity of contemporary British culture and society in the face of encroaching commercialism. 'There's no room in the modern world for art and culture . . . values are subverted by money.'[23] After a trip to Russia in 1984 he commented, 'The structures of censorship in this country are not definitive in the way they are in the Soviet Union. Here it's censorship by money; they ask "Is it commercial?" The idea of freedom within these structures is a myth.'[24] *Caravaggio* was more positive in that it was about the potential which the creative process can have, and about the nature of the accessibility of art. Many of his other films are just statements about the validity of other approaches, in particular the small, personal approach—'films which purposefully disrupt the pattern and say that there should be films made like these as well—that these other ways are important'.[25]

Jarman has, moreover, been especially vocal at meetings and in print and even in some of his paintings over the particular state of the British film industry—the American dominance, the superficiality, and the commercialism. 'Everyone talks about the film *industry*. I was never an industrialist, and I wasn't brought up to be like one.'[26] When talking of the difficulties serious film-makers have in raising money he remarked, 'If Fassbinder was English, he'd still be making Super 8 movies at the Film Co-Op.'[27] Jarman was infuriated about the way British Film Year (1985) presented current British cinema and celebrated new trends in Britain towards 'international' film-making ('In fact an unhealthy pandering to the United States'[28]). 'I saw the cinema of "product" taking over everything.'[29] He criticized Puttnam's work as 'Cinema of the new right, shallow, opportunist, ugly, and a betrayal of care for the weak.'[30] When asked why he seemed to be one of the few protesting voices who refused to accept any compromise of values, he replied, 'It's because I'm gay. It's quite simple. Because I don't have any responsibility to family. All those other people are built in with mortgages, and have to find so many pounds a week to keep the kids clothed, and I don't. I don't actually "own" anything, and I've run the whole thing in debt.'[31]

His homosexuality can, in fact, be seen as the other crucial element in his life. He relishes an open stand over sexual encounters and an abandonment of the responsibilities and inhibitions of sexuality.

[23] *London Evening Standard*, 29 Oct. 1982.
[24] *Scotsman*, 8 May 1986.
[25] Interview with the authors.
[26] *London Evening Standard*, 29 Oct. 1987.
[27] *Saturday Review*, BBC 2, 22 Nov. 1985.
[28] *Time Out*, 3–9 Jan. 1985.
[29] *NME*, 25 May 1985.
[30] *ICA Seminar*, Dec. 1984.
[31] *NME*, 25 May 1985.

'Subservience to family and state remains the pattern from here to China ... sexual encounters can lead to knowledge ... homosexuality can cut across this sad world.'[32] More provocatively he has also said, 'The family is the great law enforcer. It's the thing that keeps the banks in place ... I'm all for promiscuity.'[33] The homosexual he sees as generally more liberated than the heterosexual, and deeply resents the inability of society to come to terms with it properly, especially since the AIDS crisis. He has done a lot to fight Clause 28,[34] to help AIDS charities,[35] and in many ways has become a spokesman for those who are HIV positive.

These sexual elements strongly influence the content of his films, sometimes dominating them totally. *Sebastiane* (1976), his first feature film, was openly homo-erotic, and represented an almost classical study of the male nude, attempting to give fresh images for gay people. Its utopian overtones, he believes, have been made much less relevant by the onset of the threat of AIDS. 'I would *never* make *Sebastiane* again. I wouldn't even consider it. I made different films for different moments in time.'[36] *Jubilee*, set in the seventies against the background of the abandonment of sixties sexual ideas, replaced the traditional images of male violence, with female violence. *The Tempest* gave gay undertones to the relationship between Ariel and Prospero, and Caliban is presented as a camp figure. *Caravaggio* concentrates on the artist's sexual ambivalence. In *The Angelic Conversation*, he wanted to restore the feminine element sometimes lacking in gay films, and so made a beautifully poetic film conjuring up the romantic feelings between two young men on a summer afternoon. It was even accused of 'coming close to a homosexual version of heterosexual kitsch'[37]—an unfair comment more appropriate for *Sebastiane*.

Jarman, like others in his position, feels that he has been 'condemned to being eccentric, peripheral, independent, underground, you name it'.[38] *Sebastiane*, and *Jubilee* were condemned by Mary Whitehouse[39] as 'corrupting, pernicious filth'. 'I've been turned into a renegade by the media.'[40] In fact he considers himself to be of the most traditional and conservative of British film-makers. He turns to the traditional strands

[32] Derek Jarman, *Dancing Ledge*.
[33] *NME*, 24 Apr. 1986.
[34] Part of the government White Paper on education forbidding public money being spent to promote an understanding of homosexuality.
[35] Jarman is actually against charity but felt that in a situation where the government was abandoning its responsibilities, he had no choice but to get involved.
[36] Interview with the authors.
[37] *Afterimage*, No. 12, Autumn 1985.
[38] *NME*, 25 May 1985.
[39] Self-appointed, ultra-reactionary voice of the public conscience.
[40] *London Evening Standard*, 29 Oct. 1987.

of culture—Shakespeare, Benjamin Britten, and Wilfred Owen, Renaissance painting, and the classical male nude—for inspiration, and looks back to the fifties and early sixties before cynicism and the rat-race took over. 'One always has a nostalgia for one's childhood, and of course, mine was in the fairly austere 1950s, before American consumerism completely invaded this country, before the Supermarket if you like.'[41] There is another part of him, however, which finds the controversy a welcome companion and an important element in his creativity. He has a distrust of the successful ('I hate people who win'[42]), and has a natural sympathy for the oppressed and doubtful:

I always knew I had a way of looking at things. It set me apart . . . It's more important to destabilise this situation than to give people any comfort. I feel I was on the rack and should reflect that. I worship doubt. I do not have any solutions, and I'm deeply suspicious of those who do—because the names of God are unknown.[43]

Jarman is brilliant in opposition, but politically one cannot see him as having any other role than that—in a sense, he is the archetypal revolutionary traditionalist, and he calmly admits that his work will have no more effect than 'bearing witness to what is happening . . . Am I tilting at windmills?'[44]

Jarman's first films were shot on Super 8 mm., a format he has continued to use throughout his career. These early works, made in the seventies—which he affectionately calls 'my home movies'—were all silent and were ancillary to his painting. Initially made to show his friends, they soon became increasingly important to him. They are extremely varied in their content. Some are just records of events or people such as *Duggie Fields* (1974),[45] *Ula's Fête* (1974),[46] *Picnic at Ray's* (1975),[47] or *Pontormo and Punks at Santa Croce* (1982).[48] Others are more experimental and imaginative, often using quite literary symbols, especially those centring on magic, fire, and the sun. He wrote in his diary in 1971, 'The pleasure of Super 8 is the pleasure of seeing language put through the magic lantern.'[49] In all of them the camera is constantly moving, always vibrant with the grainy texture and the visual 'beat' created by the speed of filming. Typically the result is one of

[41] *Afterimage*, No. 12, Autumn 1985.
[42] *Observer Magazine*, 1 Jan. 1989.
[43] Interview with the authors.
[44] Interview with the authors.
[45] Duggie Fields is an English pop artist.
[46] Record of a party given to raise money for Ula, who had been fined by a court for trying to take a chandelier from Harrods.
[47] Ray Mouse was a fashion designer in seventies London.
[48] A film of a group of Italian punks playing around by a fountain at Santa Croce, Florence.
[49] Derek Jarman, *Dancing Ledge*.

shifting images with little detail, smudged colour, and slowed down movement giving the whole film a very painterly and often lyrical quality. The way in which the picture lingers on the screen encourages an appreciation of just these qualities rather than those of narrative. The films are unified by music, especially the electronic music of the related groups *Throbbing Gristle*, *Coil*, and *Psychic TV* (who try to transpose into sound William Burroughs's ideas of literary cut-up), but also classical and some old pop music.

Gerald's Film (1976), for instance, is typical of the personal nature of this work. It was a film taken on a trip down to a small lake in Essex where Jarman and his friend Gerald Incandela[50] found a decaying, roofless Victorian boathouse. It is a softly amorous study of the exploration of the building with the light falling through the rafters onto the patterned walls. *In the Shadow of the Sun* (1980) used various bits of film which Jarman had shot in the 1970s, in particular a piece he shot while on a trip to the stone circle at Avebury. The film is full of ritual events and objects of mystical significance, with shimmering images of fire and dancing superimposed onto the picture.

Imagining October (1984), made while on a trip to Russia, was more political. With a budget of £3,800 it was also more ambitious. Its narrative thread is built around the creation of a painting of half-stripped, embracing British soldiers carrying a red flag, from the initial sketches for the work, to its completion, and the celebratory dinner which follows. Poetic titles in the form of slogans are used on the screen to give some control to the images. The film is typical of Jarman's style in that it takes its form from the act of painting and posing for painting, and also in that it is prompted by the documentation of events occurring in Jarman's life—in this case his visit to Russia.

The Angelic Conversation (1985), shot at Dancing Ledge in Dorset, once more started organically—'filming people I like in places and spaces that I like'. Then the BFI gave £50,000 to make it into a full length film. The result is truly one of Jarman's finest pieces. It is an unabashed reverie of love and the male body and it has a glorious, dreamy golden yellow feel to it. On the sound track, Judi Dench reads Shakespeare's sonnets (especially those addressed to a young man rather than the Dark Lady) in sweet echoing tones, and her voice cleverly and decisively prevents any simple reduction of the film to homo-eroticism. Jarman also uses evocative 'sound images' such as the noises of seagulls, the ticking of a clock, or water splashing. The film is full of ritualistic and moody images, many of them prompted by the sonnets themselves: a

---

[50] Gerald Incandela has appeared in and been stills photographer for several of Jarman's films.

Visual experimentation has been one of the hallmarks of Jarman's films. Here in *The Angelic Conversation* (1985) the image has been degraded by being filmed off a television screen. © BFI.

man gazing out of a window; two men quietly kissing, or walking along a rocky coastline; or a man bathing in a river. Unfortunately, the prolonged section showing a boy's body being washed, even though it may have greater erotic appeal, loses the poetic quality which the rest of the film has—with its more elusive, blurred, grainy, semi-abstract shots, many in slow motion, which are so appropriate to the whole mood of the film.

Jarman's Super 8 mm. films are striking in reflecting a very spontaneous, intimate, and private world, something in contrast to many of his features. To Jarman,

There is an uncomfortableness in my film career between that sort of very private world of the Super 8s with light bulbs and the camera (just drifting around, just focusing in on people and things), and the more constructed films like *Jubilee*. I see that now very clearly. Ten years ago I wouldn't have been able to.[51]

The genesis of *Sebastiane* (1976), Jarman's first feature, is indicative of the very informal way in which he stumbled into making feature films. A friend, James Whaley, who had just left the London Film School, suggested the idea of making a feature, and the subject of Saint Sebastian was chosen in large part because of the homosexual overtones of so many of the High Renaissance paintings which depict his martyrdom. The film tells the story of Sebastian, the Roman soldier and ex-favourite of the emperor Diocletian, who is posted to the back of beyond with a platoon of bored soldiers. Resisting the lustful advances of his commanding officer, and incurring the contempt of his fellow soldiers for his piety, he finds ecstasy in his punishments, and almost welcomes his death. The film was conceived of as a collective project for all involved. Paul Humfress, who had worked at the BBC and had professional experience, co-directed it with Jarman and also edited it, but most of the cast were amateurs. Extraordinarily, the script was translated into Latin (it had English subtitles)—the only film for which this has ever been done. There were several reasons for taking this unusual step. Its novelty value gave the film much publicity, and it also allowed an international cast to work together without any language problems. Primarily, however, it carefully avoided the banalities of the usual language in costume drama, and enabled the portrayal of military boredom turning inwards into vulgarity, and homosexuality, to be more convincing. But there were tensions in the making of the film. Jarman commented in his diary, 'James [Whaley] wanted an oil and vanilla film full of Steve Reeves muscle men working out in locker rooms. Paul

---

[51] *Afterimage*, No. 12, Autumn 1985.

Humfress wants a very serious art film, slow and ponderous. I want a poetic film full of mystery.'[52] £15,000 for the 3½-week location shoot in Sardinia was raised from a variety of private sources, including a sale of Jarman's paintings. Post-production was mostly done on credit, until David Hockney, the Marquis of Dufferin, and Lord Kenilworth stepped in to help—even then the total budget for the film was only £35,000. The finished film shows the influence of Ken Russell, especially in some of its more humorous moments, and also of Pasolini's *The Gospel According to St Matthew* (1964). The traditionally sacred subject of a Roman saint is used as a vehicle for sensuously studying the male nude, and the static camera contributes to making the film very graceful. There are also visual references, typical of Jarman, to various works of art—a love scene recalling Rodin's *Kiss*; Sebastian's martyrdom reminiscent of Guido Reni's painting; or the references to Degas's painting of *The Young Spartans Exercising*. In America the film was certified as a sex film, but in Italy and Spain it was a great critical success. In retrospect Jarman comments, 'we had serious aims but looking back at it you can't take our seriousness seriously. We were just having a good time—we were just a bunch of beach boys really.'[53]

Jubilee (1978) developed in much the same sort of haphazard way and on a similarly low budget—£70,000: 'The film was cast from among, and made by friends. It was a reckless analysis of the world which surrounded us—the shooting script was a mass of xeroxes, quick notes on scraps of paper, torn photos, and messages, and the resulting film has something of the same quality.'[54] It was what Jarman describes as a 'collage film' about punk, and in particular a criticism of the superficiality of many of the ideas behind their all encompassing rebellion, and the inevitability of their eventual sell-out to commercial pressures. The film opens four hundred years before in 1578 with Queen Elizabeth (Jenny Runacre) transported by her magician into the contemporary devastation of 1970s Punk London, with its bright garish images of burning prams, police out on a vigilante spree, punks dancing round a girl as round a maypole, but with barbed wire for streamers, and a punk version of 'Rule Britannia' as England's entry into the Eurovision Song Contest. But the general reaction by many of the punks and the more conservative critics was not positive.

For an audience who expected a punk film full of 'anarchy' and laughs at the end of King's Road, it was difficult to swallow. They wanted action, not analysis; and most of the music lay on the cutting room floor . . . the critics

---

[52] *Dancing Ledge.*
[53] Interview with the authors.
[54] *Movie Maker*, July 1984.

dismissed it as 'Chelsea on ice' believing reality is an artful black and white film set in some northern industrial town,[55]

Some critics were, however, extremely enthusiastic. Alexander Walker described it as 'a collective nightmare, assembled from dole queue despair, a mirage of anarchy in the youth wasteland, a whirligig of disturbed teenagers from some contemporary bedlam,'[56] and *Variety* hailed it as 'one of the most original, bold, and exciting features to come out of Britain this decade'. Part of its real problem is that although it is undoubtedly original, and although it prophesized the punk 'sell-out' several years later, it has lost its immediate relevance in the way that some of the 'Swinging Sixties' films have—though these were also, in their own way, just as exciting and new in their time.

The Tempest (1979), costing £150,000, mostly funded by Don Boyd,[57] was a much greater success. It dealt with a very traditional, timeless strand in British culture—a Shakespearian play. The project began in 1969 as something Jarman wanted to design, but in 1975 he did his first screenplay. The play was an obsession for him and he gave it a very personal and visual interpretation. But in the weird, wonderful, and magical portrayal, he was not desecrating Shakespeare as some critics have suggested, but fashioning it like a dream. Set in a huge empty country house with endless corridors, it is lit so as to leave huge areas of shadows and darkness to allow the audience freedom for imagination. Jarman uses contemporary idiom, such as the flamboyant homosexual themes, and the famous finale of the film has Elizabeth Welch singing *Stormy Weather*, surrounded by a troupe of dancing sailors. It is not surprising that he has expressed an interest in filming *A Midsummer's Night's Dream* as well. But *The Tempest* met with mixed reactions: while many people admired it and it was a commercial success, teachers' organizations warned schools against it, and Vincent Canby in the *New York Times* rather viciously wrote, '*The Tempest* would be funny if it weren't very nearly unbearable. It's a fingernail scratching along a black-board, sand in spinach, like driving a car whose windscreen is shattered . . . there are no poetry, no ideas, no characterizations, no narrative, no fun.'

Caravaggio (1986), about the great painter of the Italian Baroque whose use of dramatic lighting and shadows (chiaroscuro) was so influential, has been the greatest single struggle and passion of Jarman's film career. He was fascinated by Caravaggio's life and art: 'He was the most inspired religious painter of the Middle Ages, and was also a

---

[55] *Dancing Ledge.*

[56] Alexander Walker, *National Heroes* (Harrap, 1985).

[57] A major independent British producer and occasional director, who has produced or co-produced many of Jarman's films.

Jarman (centre) discussing the finale of *The Tempest*, 1979, with a troupe of sailors. © BFI.

murderer. Imagine if Shakespeare had been a murderer—it would completely alter the way we see his plays . . . also he was a gay painter . . . he was hacking his way out of his situation with a knife. Actually I feel quite a close parallel with my own life.'[58] Jarman wrote the first draft of the script after *Jubilee* but it went through seventeen revisions while he was desperately trying to raise finance. At one stage it was going to have a £5 million budget—huge by Jarman's standards—with Italian money and creative input. Eventually seven years later the BFI stepped in with £475,000. It was a first for Jarman in several ways—his first film made on 35 mm., the first with seriously developed acting parts and a thought-out narrative, 'a necessity imposed by the situation and a reflection of Caravaggio's style'.[59] It is beautifully shot and lit in a manner faithful to the painter. The paintings themselves are shown unfinished, in the process of being painted, but a feeling for the completed work is shown through the tableaux created by the models posing motionlessly for Caravaggio. Although this concentration on the paintings made the film's structure episodic and created problems of continuity, Jarman was trying to capture on film the very process of the creation of a painting in the silence of the artist's studio. The set design is also highly imaginative, belying the lack of money to do any exterior or location shots or even to do any of the complex interiors. One example is Cardinal Del Monte's sumptuous art collection, which they could not afford to replicate, but which is hinted at through the numerous dust sheets which cover recognizable forms. Another example is the way the opulence and foreboding of the Vatican is suggested merely by showing a long corridor of black velvet lit through muslin.

Typically, Jarman introduced into the film many interesting and unusual anachronisms. In the background there is the hum of street sounds, of mopeds or of steam trains; a banker works out interest on a pocket calculator; another character cleans his motorbike; and a rival artist, Boglioni, cooks up gossip on a typewriter while in a tin bath with a towel around his head (also shot as a visual reference to David's *Death of Marat*). Jarman was able 'to pare away many of the mannerisms of my early films without sacrificing spontaneity. Life was breathed into the characters who were no longer ciphers like the characters in *Jubilee*.'[60] Nigel Terry expresses the intensity of the painter with great sensitivity. Overall this is probably Jarman's most significant film to date, and has a sense of convincing authenticity which validates his approach in comparison with any conventional film on the same subjects.

The Last of England (1987), named after a painting by Ford Maddox

[58] *NME*, 24 Apr. 1986.
[59] Derek Jarman, *Caravaggio* (Thames and Hudson, 1986).
[60] Derek Jarman, *Caravaggio*.

Jarman's dark and mesmerizing portrait of contemporary Britain, *The Last of England* (1987), shot entirely on Super 8mm. © Mike Laye.

Brown, is, in the sense that it is dominated by an equally bleak and disturbing view of modern urban society, a sequel to *Jubilee*. Through images and symbolism, it attacks with ferocity the moral and spiritual devaluation of Britain and the people who run it. But unlike *Jubilee*, Jarman breaks altogether from any conventional narrative or even dialogue, with the music and sound effects dubbed on later, and this approach makes it an infinitely better film. As a result, it has much in common with his more private Super 8 work of the seventies—the film being, as he says, 'nearer to poetry than to prose . . . a dream allegory'.[61] It includes several different elements shot in Super 8 and 16 mm. It opens with a flashback structure similar to that used in *Jubilee*—Jarman is shown writing the script at his desk. The other elements are Jarman's father's and grandfather's old family home movies, which contrast with both the scenes of urban decay and with a series of 'imagined sections of a feature film' which are the fantasies, generally sexual, in which the characters find escapism. The film includes images of soldiers masked in balaclavas roaming the streets to finish off the dregs of urban youth; a young man has sex with one of them on a Union Jack; the sounds of bombs and a Hitler speech are heard; a baby is swaddled in copies of the *Sun* newspaper with its famous Falklands headlines; Tilda Swinton is shown ripping and devouring her wedding dress—the shattering and shredding of illusions. Some of these are brilliant; others less evocative, but the camera dances about endlessly, portraying contemporary Britain with a gnarled, grainy, grey texture. To Jarman, 'The film is an assault on the senses; it's made on the attack. I think it's the most adventurous feature film made in this country for years.'[62] He also commented, 'If it doesn't wake up a few people I'm going to retire at the age of forty five and go and paint.'[63]

War Requiem (1989), a film version of Benjamin Britten's oratorio using Wilfred Owen's life as a narrative thread, was made after Jarman had heard that he was HIV positive, and it colours the whole nature of the film. In some ways it is not a typical film for him to have made, being less personal, less intense, and more conventional than his other works. He was suddenly invited to direct it by Don Boyd, after he had suggested the idea in an off-hand way some time before. Jarman's attitude had become much more *laissez-faire* since he had been diagnosed, and he decided to accept the £675,000, 35 mm. project despite only having a mere five weeks for pre-production, a three-week shoot, and five weeks to edit. He saw his role more as an interpreter than a film-maker, and only really felt a commitment to Wilfred Owen rather than Benjamin

[61] Interview with the authors.
[62] *London Evening Standard*, 29 Oct. 1987.
[63] Interview with the authors.

Britten. However, much to Jarman's surprise, the film received more acclaim than any of his other works. Nicolas Roeg commented, 'I think Derek Jarman is a unique and extraordinary film-maker and *War Requiem* is a unique and extraordinary film.' The *Daily Telegraph* commented, 'Never can a musical piece of such magnitude have been translated so powerfully into another medium.'

Part of the reason for this acceptance was of course the subject matter, the music, and the involvement of Laurence Olivier, who was tempted out of retirement for a day to play an old war veteran being wheeled around by a nurse (Tilda Swinton), who read Owen's 'Strange Meeting'. Perhaps it is only in this traditional setting that Jarman's images and approach can be appreciated by the wider British public; the reaction to *Last of England* was more mixed—in a controversial article in the *Sunday Times*[64] Norman Stone[65] wrote, 'No identifiable story, being essentially ninety minutes worth of impressions—rain, "urban decay", etc. . . . but I do not really know what the film is about and fear that the director is all dressed up with nowhere to go.' This is essentially the problem at the root of Jarman's controversial career. The traditional expectations of the audience are overturned, both in terms of his subject matter and approach. By adopting such a style, he makes it generally difficult for a wide audience to appreciate his films, a dilemma he nonchalantly brushes aside. Such viewers can either accept his work for what it is; as do his devotees, or be as concerned for him as he is for them.

[64] *Sunday Times*, 10 Jan. 1988. 'Through a Lens Darkly', by Norman Stone.
[65] Professor of Modern History at Oxford University.

DISCUSSION WITH

DEREK JARMAN

It must have been very exciting being involved in the world of art in the sixties before you moved on to film?

I was secretly very ambitious as an art student, although I never would have admitted it at the time. But while all my friends had become so famous—all that generation were doing so well and making so much money—I was the odd one out. It was really difficult to be going out with David Hockney and the others in our circle—always to be the one for whom they paid. I was much too conventional—middle class, military background. Everyone had to be an individual—it was almost a form of conformity. I was all wrong for the sixties. But it was good fun to be with them. I really enjoyed it, and being gay, I no longer felt so isolated. Here was another world. Until I was 22, I had found it really gruelling, even suicidal. Hockney and his friends completely altered the way I thought, and brought a breath of fresh air to the first years of the sixties. David with his red jacket dancing with Peter Doughety, Patrick Procktor and myself scandalizing everyone. It was fun and exciting and glamorous. I also knew Ozzie Clarke, the most famous designer of the sixties, and then of course there were all the people they knew—the Stones, Marianne Faithfull, and so on. It was terribly glamorous.

What was the genesis of your interest in films?

It was really completely haphazard. Until the 1960s it centred on the cinema at the end of the road which mostly showed musicals and war films. The first film I got really excited about was *Wizard of Oz*[1]—it filled my childhood with dreams and nightmares. But, the first 'grown-up' film I was taken to was *La Dolce Vita*,[2] when I was 17 or 18. It seemed so very extraordinary that I mark that as the moment when I realized that film could really interest me; that it could be more than just a way of filling in a rainy afternoon. At university everyone was interested in Italian films, especially those of Fellini or Antonioni. The underground films came in when I moved to the Slade. Kenneth Anger's *Scorpio Rising* made a huge impression—particularly the look of his films—and then from 1965 all the Warhol films arrived. Here was a type

[1] Directed by Victor Flemming, 1939.
[2] Directed by Federico Fellini, 1960.

of film-making which, it seemed to me, was close to something one could actually do oneself. He just picked up the camera and filmed his life, even out of focus. I loved that. It was an exciting and refreshing attitude, especially as I was so afraid of cameras with their complex light meters and things. But with his films I realized that it didn't matter if you didn't adhere to all the technicalities and rules. I think his films brilliantly reflected the vacuity of everyday life.

How did you get involved in actually making films?

The first moment I actually collided with the feature film was with Antonioni's *Blow Up* (1966). Antonioni came to one of my friends at the Slade to find extras for one of the nightclub scenes in the film. I nearly did it, but his costume lady refused to allow me to because of my very 'with it' clothes, and I refused to wear the Yardbird's tee-shirt she tried to foist on me. I wasn't going to have somebody tell me what to wear—I was much too vain and concerned with 'the look'. Until 1970 I was involved in theatre design and painting, and then I met a friend of Ken Russell's on a train from Paris. She told him about me, and Ken came round to my studio—possibly the first New York style loft in London—and I think that really impressed him. He impulsively gave me the job of designing sets for *The Devils* (1971) and that's how it all began. I learnt a lot on *The Devils*, it was an incredible experience. After that I got involved in Super 8 films, and I was able to do it all myself. You just put the film in the camera, press the button, and a film comes out. Eureka! They became part and parcel of recording my world along with my painting. Now I think that it was fortunate that I was not actually trained in cinema. Many painters like Francis Bacon or Van Gogh were never trained. There is no correct path into film, there is no set way. I never thought of myself as a film-maker, rather than a painter, until after *The Tempest*; and the transition only really came about because no one bought my paintings, and everyone watched the films. I still quite like the idea that people should think of me as a painter who dabbled in another art form, namely cinema, and made several features.

How did your art training influence your future attitudes to film-making?

In school I was taught art as a whole way of life, as a way of exploration, and as something very moral. It was a very 'William Morris', very Victorian education. I believe that to put the world on an even keel again it is necessary to encourage everyone to engage in a creative approach to life; I think it makes you more human somehow. So I make films to create—like painting. The fact that I was brought up as a painter says everything. I believe that film should be just as personal. My first criterion for judging a film I watch, irrespective of whether I like it or not, is: do I think the author had a deep need to make this film? Can you feel the commitment as you watch it? So what I attempt in my films, in my art, is to uncover things about myself, and the world around me—things I might be only dimly aware of. That's what interested me in *Caravaggio*, a film about the creative process; Caravaggio himself worked in the same way as I do, using people he knew in his paintings, so that his art is part and parcel of his life.

Aren't there practical, especially financial, problems to this approach?

Yes, normally film is second hand! The director or producer buys a script, but he is being no more than the equivalent of an illustrator to a book, rather than its creator. Even if a script is written by the film-maker, financiers decide whether it should or should not be made. But can you imagine Francis Bacon having to write out what he might paint next year? This is all impossible for a painter, and I can't cope with it either. It has always been my problem, in the context of British cinema, and I've watched others like Frears who can job it. But I just can't do it unless its directly related to my own life. But then why should I have to be a director (in the ordinary sense of the word)? I'm not. Why should I have to have these mass audiences? The audience should melt into the night. Your primary responsibility is to yourself, and the people you work with.

The co-operative nature of film-making seems to be very important?

You should try and create an environment where people can be creative with people coming up with ideas. The chance for people to come together to make something is wonderful. There are certain shots that are impossible in my films because I've never shouted through a megaphone, or used a walky-talky. I hate the feeling that I'm pushing people around. I always talk quietly to people. I always try and get to know the names of the extras, and have a chat to them. This my approach. I have an extreme aversion to authority—it's a reaction to my military background.

Do you find any tension between trying to be an artist producing his own individual work and working with people in this collaborative way? The works you make are all highly individual . . .

I don't think there is any tension. They just come out that way—although some films like *Last of England* have more improvisation than others like *Caravaggio*, which is classically made in 35 mm.

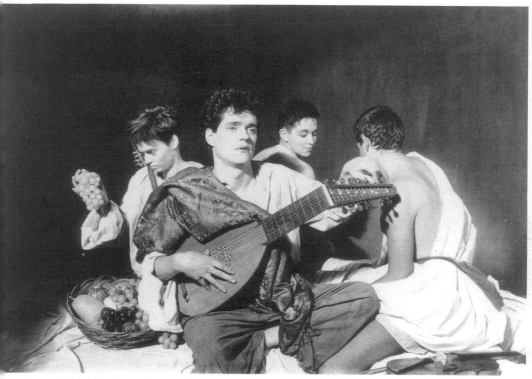

Jarman's classicism. A set piece from *Caravaggio*, 1986. © Mike Laye.

What do you feel about politically or socially committed film-makers who feel that their responsibility is to the audience, and changing society for the better?

I feel a little suspect of those who adopt good causes. The road to hell is, after all, paved with good intentions. That's my answer to those who deliver a message based on external themes which they want to impose on the audience. Even Ken Loach living in Bath, and then making films about northern mining villages, is external to the situations he films. But some like Terence Davies make backyard movies—his own world, with sensitivity and love.

If these directors look at these issues from a distance doesn't it, in fact, make it easier for them to comment?

Maybe . . . maybe that's true . . . it's possible. But I feel that it should be a personal passion. Couldn't Loach for instance have made a political film about Bath and its surroundings. It's like Gauguin going to Tahiti. I've always wondered about Gauguin—a bit suspect. No! Maybe you're right. Perhaps it's Loach's dream of Paradise.

You feel that it's more honest to make these personal films?

Well, it's not so much to do with 'honesty' or 'sincerity'. I think 'sensible' is the right word. Straightforwardly sensible. I'm obsessed with this because the personal film is so rare. No one ever really raises this issue at all in film. I think that if the whole film world were like that, I would be against it. I say this because it's never discussed in film, but it's always discussed in painting. I don't see why there should be a difference. The problem in film is that there is no tradition of this here. Most film-makers wouldn't know where to begin—they're illiterate in this sense. If you said to Alan Parker, who is actually a really sweet man, 'What are you going to do about your life which is of interest to the viewer?', he would look at you non-plussed and say 'The cinema has nothing to do with that' or 'I don't have anything to base this on.' Neither, of course, is the audience educated to watch all this. But I do believe that lurking at the back of all film-makers' minds is a film of their life, as opposed to their public films—Cocteau's magnificent failure is *The Testament of Orpheus* (1959) or Pasolini's tragic *120 Days of Sodom* (1975).

All your films except Caravaggio *and more recently* Last of England *and* War Requiem *have been funded privately and all have been extremely inexpensive. This must have been a severe constraint?*

I'm the film-maker who has gone the furthest with the least. I made all my films for about £1 million, when the present cost for a low budget feature film is a million and a half. This puts me in a true perspective and this is in part why my films look the way they do. From the beginning I had to write my own scripts, partly at least because no one else could understand how to make a film for so little. Generally people who write films are not film-makers, so they have no idea how to be economic. They tend to write 'so and so walks across the road and walks into this . . .'—already a million pounds has been spent on the first ten minutes. I can't handle the narrative approach because it is too expensive! I have really made a virtue of necessity in avoiding this. Working on *The Devils* I actually learnt to put together the *mise-en-scène* for nothing. So, for *Caravaggio* we had a budget for the whole of the art department of £40,000—including wages. That is nothing—and out of this we had to build all the sets from seventy-two 12 ft by 8 ft units in a warehouse out in the Isle of Dogs that wasn't even sound-proofed. Still, the design of *Caravaggio* was as near faultless as could be. Moreover, I had to make great personal financial sacrifices while other people made all the money. *Caravaggio* was the first film for which I was paid—until then I was up to my ears in debt. I was paid £25,000 because it was financed by the BFI—but on *Sebastiane* I only got £1,000, on *Jubilee* £8,000, and *Tempest* £6,000, and nothing directly on *War Requiem*. You make the films as a director; the films make millions; but the financiers own them and none of it comes back to you. It has been so hard getting money together—seven years for *Caravaggio*. What we did on *Last of England* we ran up £200,000 of debts with the idea that someone would either put up the cash, or put us in prison. Fortunately we were bailed out by British Screen, ZDF—German television—and Channel 4 which for the first time funded one of my films directly.

If only I had had decent budgets . . . even a little would have helped incredibly! Perhaps less for lavish sets and more for just longer shoots. On *Caravaggio* it was exhausting—the shoot was only six weeks in total. I think the financial censorship in this country is deeply disgusting. But I never dreamt of big budget films, because I knew that it was impossible if I was to make the sort of films that I really wanted to . . . Listen, I don't regret a thing! I think my films are all right given the constraints under which I operated. That's just the way it was. I would do everything the same way a second time. You know, I sometimes think that it's terribly easy to get a large audience into the cinema, and I'm always surprised at how commercial cinema so often fails. I bet I could make the most wild commercial successes if I put my mind to it—lots of nudity and sex. But being a commercial success is not something to be proud of, as some think it is.

Your relationship with television has been discordant hasn't it?

I was very much marginalized by Channel 4, along with film-makers like Ron Peck[3] and Julien Temple[4]—disenfranchised by a channel for the slightly adventurous commuter. It was impossible to work because they took over the area called 'independent film' which I as much as anyone had created. They were TV people with TV values and thought that films like *Jubilee*, *Sebastiane*, and *The Tempest* were peripheral. Channel 4 was the worst thing that happened to *our* type of independent cinema (although I wouldn't say to cinema in general because, of course, some people worked who wouldn't otherwise have done so). There are sharks behind that cool veneer. They bought my films for nothing—£6,000 for *Jubilee* and £12,000 for *The Tempest*. It was tantamount to stealing them when the normal cost of making one hour of television drama is £250,000.[5] But there is a monopoly—who else in Britain could we sell them to? They are passionate and do feel responsibilities. But they will not do anything to endanger the bedrock of the company—the advertising and the government. They have played my films once, but won't show them time and again like *The Draughtsman's Contract*. It's not because of audience ratings, because mine were quite high. Channel 4 is not the altruistic body people consider it to be.

Whitehouse and people got up and made a stink when my films were shown on TV. My films were shown because the channel bought them, not because I wanted them shown. I never wanted my films on TV in the seventies—*ever*. I never wanted your children to see my films *ever*. I've made my films for art-house cinemas with no real ambition beyond that. I hate television. The television is an end to dreams and reverie. It has replaced the hearth in family houses and is generally destructive to people's lives.

But you have made pop-videos for television.

Well, everyone is co-opted. The eight or nine videos I've done have kept me alive—they are very well paid. Moreover, they put me in touch with the latest technology. They are advertisements but involve people struggling to put across ideas. I've never done any commercials, although I worked for Ken Russell for one coffee advertisement—I was totally broke at the time. I think that some sorts of advertisements are all right. It depends on what message is being put across.

---

[3] Director of *Nighthawks* (1978) and *Empire State* (1987).
[4] Directed *The Great Rock and Roll Swindle* (1979), *The Secret Policeman's Other Ball* (1982), and *Absolute Beginners* (1986).
[5] Despite the high costs of production British television companies will usually only pay £15,000–20,000 for the rights to broadcast independent films which they have not commissioned themselves.

Do you consider yourself a rebel?

I am not an outsider. The one thing I really regret about my career was that I was put into the position of being anything but the most traditional film-maker of my generation. I hope this has not disappointed you, but this is what I really wanted and that's why I did *The Tempest*, which was a passion for me and which deals with and upholds traditional strands of the culture. Yet I was made into a fake revolutionary. The older I get, the more I believe in tradition. The tradition of hedgerows and fields with flowers—in opposition to commercialization or the destruction and rape of the countryside and cities.

Yet the gay side to your films is very progressive, isn't it?

The whole gay thing is crucial to my films. I have found it very hard to come to terms with the strictures of society over homosexuality. Although there has been some improvement since the sixties, it's small. It fills me with fury to think that I can't kiss or hold hands, things that heterosexuals unthinkingly do when they walk down the street together. I find it suffocating, even phobic. It puts you outside society—even the Left in this country still hold old-fashioned views of sex. I sometimes wonder when we can expect to see an openly gay chairman of Channel 4, or an openly gay Prime Minister. I try to keep a balanced view, but when people ask, 'why are your films so gloomy?', it is because of this. In a certain sense I buried all this in the 1970s because things were getting better, but now in the 1980s with AIDS it has opened my eyes and I see the most appalling things happening. But it's all anaesthetized by the media, so no one really notices. This year I've watched ten of my friends die—all the brilliant ones who breezed around; all the boring stay-at-homes survive of course. I've written a very sad poem although it doesn't scan:

> I walk in this garden holding the hands of dead friends.
> Old age came quickly for my frosted generation.
> Cold, Cold, they died so silently.

> Did the forgotten generation scream
> Or go full of resignation, quietly protesting their innocence.
> Cold, Cold, they died.

> We linked hands at four a.m. deep under the city.
> You slept on; never heard the sweet flesh song.
> I have no words.

> My shaking hands can not express this fury.
> Sadness is all I have.
> No words.

Matthew fucked Mark, fucked Luke, fucked John,
Who lay in the bed that I lie on.
Touch fingers again as you sing this song.

Cold, Cold, they died.
Sweet garden of vanished pleasures
Come back next year.

It's like being in a war. I love William Burroughs's idea of picking up the guns, and of going out and creating the gay state, and shooting 'them' down so that we can live in a society untrammelled by all this prejudice. What would you think if I raised the age of consent to 21, imprisoned and beat up heterosexuals for kissing in the street, outlawed the promotion of heterosexuality, illegalized marriage, and had the Church of England preach celibacy? You'd soon wake up to the intolerable world. The anger in my films is totally justified. My films try and be as open as possible, but I resent having to go into an attack position. I wanted to be a traditional film-maker, and have a quiet time. But I know what it's like to live under a totalitarian regime.

Are you politically active? What are your views on British politics?

In a way I'm the most political film-maker around because of the traditions I support. But my films have no 'political' messages. I've never been a member of a political party. Well, I was involved with the Workers' Revolutionary Party for a short while in the early eighties; as an artist, I wanted to find out what they were about—they had an astute analysis of the situation but no answers. Anyway, it isn't possible to believe in democratic government any longer. There isn't such a thing. It's all rigged—at least behind the scenes—so that no one cares whether Neil Kinnock wins, because the corporations will always be able to bring him to heel.

How would you place your work in the context of British films?

It has always been difficult to make films in Britain. Look at what Alexander Korda was saying in the 1940s[6]—and then this country had a feeling of independence! In the sixties people, particularly of my generation, felt that to succeed one had to make American films, and they started to go over to the US where everything was happening. Things went really downhill from 1970. A few people like Lindsay Anderson held out against this trend with Free Cinema, people like Ken Loach moved into TV, and the BFI spluttered on. When the pound collapsed in

---

[6] See the Introduction for details of Korda's production company, London Films. As producer/director, Korda spent the war years in Hollywood but returned to Britain to make such films as *Perfect Strangers* in 1945.

After the violent anger of *The Last of England* Jarman returned to a more subdued classicism in *War Requiem*, 1989. © Anglo International Films.

1974, it became even more difficult to make films here, because there was no funding and film directors like Nicolas Roeg or Tony Richardson went to the US to make their films with US money. When films started to have budgets as huge as *Star Wars*, it was hard for anyone to contemplate making any film here at all . . . In this vacuum a few directors like myself came along and made low budget films for art-house cinemas. The independent film was a new phenomenon and its one or two films a year represented British cinema in the seventies. This is a culture which is so rotten it's deadwood. I used to be in love with it, but now the night is coming in. Sometimes I think that I don't really belong in England . . . Italy, probably, is my second home.

What about British film in the 1980s?

In the eighties of Margaret Thatcher has come the 'British Film Renaissance'. But it was a fake. A group of advertising men who had gone into film—Alan Parker, Ridley Scott, Hugh Hudson, and their chum David Puttnam ran this PR exercise, declaring themselves to be 'British Cinema', using their knowledge of how to manipulate the media—it was very astute. But in this publicity campaign, they had hardly any truly British films. Ken Russell, Nicolas Roeg, and Lindsay Anderson were largely ignored, while they included American films like *Revolution*[7] or *The Killing Fields*[8] with some unbelievably tentative British connection—like an English producer or something. Bill Forsyth was perhaps the only independent director represented in that world. It was terribly upsetting to see. I remember they had an official book—*Learning to Dream*[9] by Mr Park about this 'New British Cinema', particularly the idea of dream cinema, but *The Tempest* wasn't even mentioned. Then Mr Puttnam went over to Columbia Pictures where he said on American television how he had always dreamt of Hollywood as a boy. How can he argue that he wanted to help British cinema—he ony wanted to help himself. Meanwhile Hugh Hudson spent millions on *Revolution* and actually wrecked any chances that anyone had, because after that the City would not put any money into British films. Imagine how many small half-million-pound films could have been made instead! What a fantastic thing that would have been for British cinema! British cinema is dead—it's a dodo—it doesn't exist any longer, though Stephen Frears may roll out a sequel to *Laundrette* once a year.

But surely this publicity made it easier for all British film-makers to raise money and sell their films?

---

[7] 1985, directed by Hugh Hudson.

[8] 1984, directed by Roland Joffé and produced by David Puttnam.

[9] James Park, *Learning to Dream: The New British Cinema* (Faber and Faber, 1984).

I would not disagree with that. But this is a Westland crisis. I happen to feel that the British film should be kept for us; that there should be subsidies. The idea that my views are anarchistic is absolutely wrong. There is essentially a rift between old-fashioned conservatives and new ones—I'm old-fashioned—they're the new conservatives. The film-makers I would have liked to see being able to work more were people like Terence Davies,[10] Ken Loach, Julien Temple, Sally Potter,[11] or Lindsay Anderson. I look back on the British cinema of the last twenty years and I find very few films I would want to keep—Terence Davies' *Trilogy*, Sally Potter's *The Gold Diggers* (1984), possibly Bill Douglas's trilogy;[12] Anderson's *Britannia Hospital* (1982), and glorious Terry Gilliam's *Brazil* (1985).

How do you look back on your career in films?

I see myself at this stage of my life as essentially having failed. Only now at the very end of my career, I'm getting some recognition and acceptance—it's maybe too late now—isn't that strange? As a film-maker I had a huge amount of promise which was never realized in any way whatsoever. It has been such a struggle. *The Tempest* had the best reviews of any British film of the seventies, but I didn't receive a single offer after it. In a way I am telling a tragic and warning tale. The experience of getting *Caravaggio* together over seven years was exhausting—the minuscule amount of money; people actually wanting it to go under; the fact that it was almost impossible to shoot an Italian movie in the Isle of Dogs . . . It was almost too hard, but I wouldn't give in. We sort of brought it off—but it was like climbing Everest. I was really a happy-go-lucky person until the beginning of the eighties. With *Caravaggio* it was all over, I started to look at pieces of white paper and think of what I wanted to do next. I decided that no film was worth so many years of one's life. I just picked up my home movie camera and started to make *Last of England*—a huge tapestry film . . . I don't know why I did *War Requiem*. I'm sort of muddled about it . . . I'm not certain any of my films are any good at all, or of any consequence, except *The Angelic Conversation*. I look at them sometimes, but the feeling is only like taking a pill to reinvigorate you—an illusion. I've no real idea of what my films are like. I find it very complicated to watch other people's films—I can't find my way into most of them, and can only see false acting or another hired prop. I'm sure people feel the same about my films. It's like the way that you look in the mirror everyday but don't know what you look like . . . I don't know. I actually don't really like the

---

[10] Directed *The Terence Davies Trilogy* (1974–83), and *Distant Voices, Still Lives* (1988).
[11] Directed *The Gold Diggers* (1983).
[12] *My Childhood* (1971), *My Ain Folk* (1972), *My Way Home* (1979).

cinema very much. I'm not convinced by it at all. I still think that I should have painted. Maybe I was diverted. The only real thing I like about my films is that it is possible to see my dead and dying friends in all the nooks and crannies, and I like that. It's wonderful.

FILMOGRAPHY

BRIEF BIOGRAPHICAL DETAILS

Born 1942 in Northwood, Middlesex. Went to Cranford School. 1960–3, studied English, Art History, and History at King's College, London University. 1963–7, studied Fine Art at the Slade, London University.

FILMS DIRECTED BY DEREK JARMAN

1970   *Studio Bankside.* 6 mins. (Shot in Super 8 mm.).

1971   *Miss Gaby.* 5 mins. (Shot in Super 8 mm.).

        A Journey to Avebury. 10 mins. (Shot in Super 8 mm.).

1972   *Garden of Luxor* (a.k.a. *Burning The Pyramids*), 6 mins. (Shot in Super 8 mm.).

        Andrew Logan Kisses the Glitteratti. 8 mins. (Shot in Super 8 mm.).

        Tarot (a.k.a. *The Magician*). 10 mins. (Shot in Super 8 mm.). Made with Christopher Hobbs.

1973   *The Art of Mirrors* (a.k.a. *Sulphur*). 10 mins. (Shot in Super 8 mm.). CAST: Kevin Whitney, Luciana Martinez, Gerald Incandela.

1974   *The Devils at the Elgin* (a.k.a. *Reworking The Devils*). 15 mins. (Shot in Super 8 mm.).

        Ula's Fête (a.k.a. *Ula's Chandelier*). 10 mins. (Shot in Super 8 mm.).

        Fire Island. 5 mins. (Shot in Super 8 mm.).

        Duggie Fields. 10 mins. (Shot in Super 8 mm.).

1975   *Picnic at Ray's.* 10 mins. (Shot in Super 8 mm.).

        Sebastiane Wrap. 6 mins. (Shot in Super 8 mm.). MUSIC: Simon Turner.

1976   *Gerald's Film.* 5 mins. (Shot in Super 8 mm.). CAST: Gerald Incandela.

        Sloane Square (a.k.a. *Removal Party, A Room of one's Own*). 12 mins. (Shot in Super 8 mm.). Dark Pictures. PRODUCER: James Mackay. MUSIC: Simon Turner. Made with Guy Ford.

        Sebastiane. 85 mins. (Shot in 16 mm. in Latin with subtitles). Megalovision. CO-DIRECTOR: Paul Humfress. SCRIPT: James Whaley and Derek Jarman. PRODUCERS: James Whaley and

Howard Malin. PHOTOGRAPHY: Peter Middleton. PRODUCTION DESIGNER: Derek Jarman. EDITOR: Paul Humfress. MUSIC: Brian Eno. CAST: Leonardo Treviglio, Barney James, Neil Kennedy, Richard Warwick, Ken Hicks, Gerald Incandela, Christopher Hobbs.

Houston, Texas. 10 mins. (Shot in Super 8 mm.).

1977 *Jordan's Dance.* 1 min. (Shot in Super 8 mm. later used in *Jubilee*).

Every Woman for Herself and All for Art. 1 min. (Shot in Super 8 mm.).

1978 *Jubilee.* 103 mins. (Shot in 16 mm.). Whaley–Malin Production for Megalovision. SCRIPT: Derek Jarman. PRODUCERS: Howard Malin and James Whaley. ASSISTANT DIRECTOR: Guy Ford. PHOTOGRAPHY: Peter Middleton. PRODUCTION DESIGNER: Kenny Morris (from Siouxsie and the Banshees), and John Maybury. COSTUMES: Christopher Hobbs. EDITORS: Nick Barnard and Tom Priestley. MUSIC: Brian Eno. CAST: Jenny Runacre, Little Nell, Jordan, Toyah Willcox, Wayne County, Adam Ant, Ian Charleson, Karl Johnson, Neil Kennedy, Richard O'Brien, Jack Birkett.

1979 *Broken English.* 12 mins. (Shot in Super 8 mm. and 16 mm.). PRODUCER: Guy Ford, with Marianne Faithfull. MUSIC: Marianne Faithfull (three songs: 'Broken English', 'Witches' Song', 'The Ballad of Lucy Jordan').

The Tempest. 96 mins. (Shot in 16 mm.). Kendon Films for Boyd's company. SCRIPT: Derek Jarman (adapted from the play by Shakespeare). PRODUCERS: Sarah Radclyffe, Guy Ford, and Mordecai Schreiber. EXECUTIVE PRODUCER: Don Boyd. PHOTOGRAPHY: Peter Middleton. PRODUCTION DESIGNER: Yolanda Sonnabend. EDITOR: Lesley Walker. MUSIC: Wavemaker (Brian Hodgson, John Lewis). CAST: Heathcote Williams, Toyah Willcox, Jack Birkett, Karl Johnson, Claire Davenport, Elizabeth Welch.

1980 *In the Shadow of the Sun.* 51 mins. (Shot in Super 8 mm.). (Using material from *A Journey to Avebury, Tarot, Fire Island*, and footage from *The Devils*. Film edited in 1973/4. Sound added in 1980.) Dark Pictures with assistance from Freunde der deutschen Kinemathek (Berlin). PRODUCER: James Mackay. MUSIC: Throbbing Gristle. CAST: Gerald Incandela, Christopher Hobbs, Kevin Whitney, Luciano Martinez, Andrew Logan.

1981 *T.G. Psychic Rally in Heaven.* 8 mins. (Shot in Super 8 mm.). Dark Pictures with financial assistance from the Arts Council of Great Britain. MUSIC: Throbbing Gristle.

1982 *Pirate Tape.* 12 mins. (Shot in Super 8 mm.). Film of W. S. Burroughs's visit to England.

Pontormo and Punks at Santa Croce. 10 mins. (Shot in Super 8 mm.).

1983 *Waiting for Waiting for Godot.* 10 mins. (Shot in Super 8 mm.). From a RADA production of Beckett's play designed by John Maybury.

B 2 Film. 35 mins. (Shot in Super 8 mm. and video.). MUSIC: various, from 1920s and 1930s. CAST: Dave Baby, John Scarlett-Davies, Volker Stokes, Judy Blame, Hussain McGaw, James Mackay, Padeluun, Jordan.

1984   *The Dream Machine.* 35 mins. (Shot in Super 8 mm.). Arts Council. Made with John Maybury, Cerith Wyn-Evans, Michael Kostiff.

Catalan. 7 mins. (Shot in 16 mm.). TVE, Spanish Television. PRODUCTION DESIGNER: Christopher Hobbs. MUSIC: Psychic TV.

Imagining October. 27 mins. (Shot in Super 8 mm.). Picture by John Watkiss. PRODUCER: James Mackay. PHOTOGRAPHY: Derek Jarman, Richard Heslop, Cerith Wyn-Evans, Sally Potter, Carl Johnson. EDITORS: Derek Jarman, Cerith Wyn-Evans, Richard Heslop, Peter Cartwright. MUSIC: David Ball and Genesis P. Orridge. CAST: John Watkiss, Peter Doig, Keir Wahid, Toby Mott, Steve Thrower, Angus Cook.

1985   *The Angelic Conversation.* 78 mins. (Shot in Super 8 mm.). BFI/Channel 4. PRODUCER: James Mackay. SCRIPT: Shakespeare's sonnets numbers 27, 29, 30, 43, 53, 55, 56, 57, 61, 90, 94, 104, 126, 148. Read by Judi Dench. PHOTOGRAPHY: Derek Jarman, James Mackay. EDITORS: Cerith Wyn-Evans, Peter Cartwright. MUSIC: Coil (John Balance, Peter Christopherson). CAST: Paul Reynolds, Philip Williamson.

1986   *Caravaggio.* 93 mins. (Shot in 35 mm.). BFI/Channel 4. SCRIPT: Derek Jarman. PRODUCER: Sarah Radclyffe. EXECUTIVE PRODUCER: Colin McCabe. PHOTOGRAPHY: Gabriel Beristain. PRODUCTION DESIGNER: Christopher Hobbs. EDITOR: George Akers. MUSIC: Simon Fisher Turner. CAST: Nigel Terry, Sean Bean, Gary Cooper, Spencer Leigh, Tilda Swinton, Nigel Davenport, Robbie Coltrane.

The Queen is Dead. 13 mins. (Shot in Super 8 mm.). Made with Richard Heslop, John Maybury, Christopher Hughes. PRODUCER: James Mackay. MUSIC: The Smiths (three songs: 'The Queen is Dead', 'There Is a Light that Never Goes Out', 'Panic').

1987   *Aria.* 98 mins. (One of the 10 segments at 4½ mins.). (Shot in Super 8 mm. and 35 mm.). RVP Productions. PRODUCER: Don Boyd. ASSOCIATE PRODUCER: James Mackay. ASSISTANT DIRECTOR: Cerith Wyn-Evans. PHOTOGRAPHY: Mike Southon and Christopher Hughes. PRODUCTION DESIGNER: Christopher Hobbs. EDITORS: Peter Cartwright, Angus Cook. MUSIC: Charpentier's 'Depuis le jour' from the opera *Louise.* CAST: Amy Johnson, Tilda Swinton, Spencer Leigh.

The Last of England. 87 mins. (Shot on Super 8 mm.). British Screen/Channel 4/ZDF. SCRIPT: Derek Jarman. PRODUCERS: James Mackay and Don Boyd. PHOTOGRAPHY: Derek Jarman, Christopher

Hughes, Cerith Wyn-Evans, Richard Heslop. PRODUCTION DESIGNER: Christopher Hobbs. EDITORS: Peter Cartwright, Angus Cook, Sally Yeadon, John Maybury. MUSIC: Simon Turner, Andy Gill, Mayo Thompson, Albert Dehlen, Barry Adamson, El Tito. CAST: Tilda Swinton, Spencer Leigh (with the voice of Nigel Terry).

1988   *L'Ispirazione*. 2½ mins. (Shot on Super 8 mm.). As an introduction to the opera of the same name by Silvano Bussotti. PRODUCER: James Mackay. EDITORS: John Maybury and Peter Cartwright. MUSIC: Silvano Bussotti. CAST: Tilda Swinton, Spencer Leigh.

1989   *War Requiem*. 86 mins. (Shot on 35 mm. and Super 8 mm.). Anglo International Films for the BBC in association with Liberty Films. PRODUCER: Don Boyd. POEMS: Wilfred Owen. PHOTO-GRAPHY: Richard Greatrex. PRODUCTION DESIGNER: Lucy Morahan. EDITOR: Rick Elgood. MUSIC: Benjamin Britten's War Requiem: London Symphony Orchestra conducted by Benjamin Britten. CAST: Laurence Olivier, Nathaniel Parker, Tilda Swinton, Owen Teale, Patricia Hayes, Claire Davenport.

SET DESIGN

1970   Set designer for Ken Russell's *The Devils*.

1972   Set designer for Ken Russell's *Savage Messiah*.

1974   Set designer for Ken Russell's abandoned film *Gargantua*.

1976   Set designer for a Ken Russell Nescafé advertisement.

MUSIC VIDEOS

1982   *Diese Machine ist Meine Antihumanistiches Kunstwerk* for Psychic TV. 6 mins. (Shot on 16 mm.).

1983   *Dance With Me* for Lords of the New Church. Aldabra.

        Willow Weep for Me for Carmel. Aldabra.

        Dance Hall Days for Wang Cung. Aldabra.

        Stop the Radio for Steve Hale. Why/B2. PHOTOGRAPHY: Peter Middleton.

1984   *What Presence* for Orange Juice. (Shot on 16 mm.). Adalbra.

        Tenderness is a Weakness for Marc Almond. 6 mins. (Shot on 16 mm. and video.). Adalbra.

1985   *Windswept* for Bryan Ferry. Aldabra. CO-DIRECTOR: Marc Almond. PHOTOGRAPHY: Gabriel Beristain.

1986   *Ask* for The Smiths. 3 mins. (Shot in Super 8 mm.). PRODUCER: James Mackay.

Whistling in the Dark for Easterhouse. 4 mins. (Shot in Super 8 mm.). PRODUCER: James Mackay.

1969 for Easterhouse. 4 mins. (Shot in Super 8 mm.). PRODUCER: James Mackay.

I Cry Too for Bob Geldof. 4½ mins. (Shot in Super 8mm.). PRODUCER: James Mackay.

Pouring Rain for Bob Geldof. 4½ mins. (Shot in Super 8 mm.). PRODUCER: James Mackay.

1987   *It's A Sin* for The Pet Shop Boys. 5 mins. (Shot on 35 mm.). PRODUCER: James Mackay.

Rent for The Pet Shop Boys. 4½ mins. (Shot on 35 mm.). PRODUCER: James Mackay.

THEATRICAL DESIGNS

1967   Showed theatrical designs for Prokofiev's *Prodigal Son* at Biennale des Jeunes, Paris.

1968   Sir Frederick Ashton's production of *Jazz Calendar* for the Royal Ballet.

Don Giovanni for the Royal Opera House.

Thruway for the Ballet Rambert. Choreography by Stere Popescu.

1973   *Silver Apples of the Moon*. Choreography by Tim Spain. Banned from Coliseum in London. Showed once at Oxford.

1982   Ken Russell's production of Stravinsky's *Rake's Progress* at the Pergola Theatre, Florence.

1984   Ballet by Micha Berghese.

PAINTING EXHIBITIONS

1967   'Young Contemporaries' at the Tate Gallery.

Lisson Gallery (with Keith Milow).

1968   Lisson Gallery.

1976   'Six British Painters' in Houston, Texas.

1978   Sarah Bradley's Gallery.

1982   Edward Totah Gallery.

1984   Institute of Contemporary Art.

1987   Richard Salmon Gallery.

1988   Richard Salmon Gallery.

OPERA

1988   Directed Silvano Bussotti's *L'Ispirazione*, Florence.

ACTING

1986   *Ostia*. Directed by Julien Cole (Jarman plays Pasolini).

             Prick Up Your Ears. Directed by Stephen Frears (Jarman plays Patrick Procktor).

ABANDONED FEATURE SCRIPTS

1976   *Bible* for Ken Russell.

1979   *Neutron*, to have been produced by Don Boyd and to have starred David Bowie; *Bob Up A Down*.

1981   Plan for film of Steve Birkoff's stage play *Decadence*.

1983   *B Movie*.

1984   *Nijinsky's Last Dance*.

             Pier Paolo Pasolini.

1985   *Lossie Mouth*.

BIBLIOGRAPHY

Guardian, 12 Mar. 1969. Norbert Lynton reviews an exhibition of Jarman's paintings, drawings, and stage designs, at the Lisson Gallery.

Derek Jarman, *A Finger in the Fish's Mouth* (Bettiscombe Press, 1972). Described by Jarman as a 'book of adolescent poems'.

Daily Express, 3 Feb. 1976. William Hickey describes how the Marquis of Dufferin, Lord Kenilworth, and David Hockney put up the money for *Sebastiane*.

Time Out, 5–11 Nov. 1976. Interview about the making of *Sebastiane*.

Guardian, 21 Dec. 1976. Derek Malcolm on *Sebastiane*, future projects and Jarman's work for Ken Russell.

Sunday Times, 19 Feb. 1978. 'History is Punk'. Jarman interviewed by Philip Oakes on *Jubilee*.

Gay News, 23 Feb. 1979. Keith Howes talks to Jarman about *Jubilee*.

Screen International, 24 Mar. 1979. Comments from Jarman on the making of *The Tempest*, working on a low budget, his previous films, and his particular style.

Film Directions, No. 8, 1979. Interview on how he came to make *The Tempest* and his approach to Shakespeare's play, his earlier film *Jubilee*, and current project on the painter Caravaggio.

Screen International, 24 Nov. 1979. Note on *The Tempest*, and some of Jarman's comments on the film's showing in the 1979 London Film Festival.

Time Out, 2–8 May 1980. Interview with Jarman about *The Tempest*.

London Evening Standard, 9 May 1980. Charles Spencer on the favourable critical reception of *The Tempest* and Jarman's background.

Time Out, 9–15 May 1980. Brief listed preview of the forthcoming screenings of his work at the London Film-Makers Co-op Cinema.

London Magazine, 20 Oct. 1980. 'Talking to Derek Jarman', by Timothy Hyman.

The Times, 16 Apr. 1981. Short article on Jarman's Super 8 work shown at the ICA.

Continental Film and Video Review, May 1981. Note on his films and new projects.

London Evening Standard, 29 Oct. 1982.

Derek Jarman, *Dancing Ledge* (Quartet, 1984). First part of Jarman's autobiography.

Time Out, 2–8 Feb. 1984. Summary of his output and career on the occasion of the screening of his work at ICA.

London Evening Standard, 8 Feb. 1984. On the publication of Jarman's autobiography *Dancing Ledge*; Jarman briefly comments on US influence on British cinema.

City Limits, 17–23 Feb. 1984. Interview on his paintings, the first part of his autobiography *Dancing Ledge*, his projected film on the painter Caravaggio, and his views on world politics.

Monthly Film Bulletin (51), June 1984. Interview by Michael O'Pray on Jarman's Super 8 mm. film work, his experience directing pop promotional videos, and his proposed film on the artist Caravaggio. Also a review of three of Jarman's shorts: *Gerald's Film*, *Pirate Tape*, and *Waiting for Waiting for Godot*.

Movie Maker, July 1984. Jarman talks about his films and his work on Super 8 mm.

ICA Seminar, Dec. 1984.

BFI: Derek Jarman Special Collection, 1984. Scripts, notes, sketches, interviews.

Time Out, 3–9 Jan. 1985. Jarman's comments on British Film Year, voiced at a seminar at the ICA.

Time Out, No. 754, 1985. Jarman talks to dancer Lindsay Kemp about his career, influences on his work, and living in Spain.

Three Sixty, May 1985. Article about Jarman's career and the financial problems which he has had to overcome.

New Musical Express, 25 May 1985. Detailed discussion and interview on Jarman's career and his views on British Film Year.

City Limits, 18–24 Oct. 1985. Jarman comments on his career and his film, *The Angelic Conversation*.

Screen International, 19–26 Oct. 1985. On *Caravaggio*'s production problems, with Jarman's comments on his plans for future projects. Also details of a national tour of Jarman's films, organized by Films and Video Umbrella, with BFI and Arts Council backing.

Afterimage, No. 12, Autumn 1985. Special issue on Jarman. Articles on the place of 'Eros' and 'Thanatos' in Jarman's work; sequences from two of his projects: *Pier Paolo Pasolini in the Garden of Earthly Delights* and *Nijinsky's Last Dance*, both later abandoned; the characteristics of his films; his painting; and an interview with Jarman.

Derek Jarman, *Caravaggio* (Thames and Hudson, 1986). A beautifully produced script-book on the film.

Films and Filming, Jan. 1986. Article on the making of *Caravaggio*, with comments from Jarman and the cast.

Stills, Apr. 1986. Interview with Jarman on *Caravaggio* and his views on British films.

Monthly Film Bulletin, (53), Apr. 1986. Interview with Jarman about responses to his films.

NFT Booklet, Apr. 1986. Note of the forthcoming Guardian Lecture by Jarman at the National Film Theatre. Also article on Jarman's career and his film *Caravaggio*.

BFI Special Collection, 1986, on *Caravaggio*.

Guardian, 17 Apr. 1986. 'A Natural Obsession'. Nicholas de Jongh talks to Jarman about *Caravaggio*.

What's On, 24 Apr. 1986. Philip Bergson interviews Jarman on *Caravaggio*.

New Musical Express, 24 Apr. 1986. Interview with Jarman by Peter Culshaw on *Caravaggio* and Jarman's view of sex, politics, and censorship.

Scotsman, 8 May 1986. Interview by Ian Bell on *Caravaggio*.

New Musical Express, 9 Aug. 1986. About Jarman directing a music video for The Smiths.

American Film, Sept. 1986. Article on *Caravaggio* and Jarman's career.

Observer Magazine, 22 Feb. 1987. 'Painting the Big Screen'. Maureen Cleave talks to Jarman about his career and *The Last of England*.

Screen International, 6–13 June 1987. Jarman comments on the reception of his film *Caravaggio* by Italian art historians on its release in Rome, and his next project.

Cinema Papers, Sept. 1987. Interview with Jarman about British Cinema, the themes of his films, and *Caravaggio*.

City Limits, 15–22 Oct. 1987. Extract from Jarman's book chronicling his feelings about the making of *The Last of England*.

London Evening Standard, 29 Oct. 1987. Jarman talks to Neil Norman about his views on British society and about *The Last of England*.

Guardian, 29 Oct. 1987. Derek Malcolm reviews *Aria*.

Marxism Today, Oct. 1987. Review of *The Last of England*.

Derek Jarman, *The Last of England* (Constable, 1988). The second part of Jarman's autobiography.

Sunday Times, 10 Jan. 1988. 'Through a Lens Darkly', by Norman Stone.

Sunday Times, 17 Jan. 1988. 'Freedom Fighters for a Vision of the Truth', by Derek Jarman. (Reply to criticisms of British films by Norman Stone.)

Art Monthly, Mar. 1988. Article by Jarman on the importance of toleration and equality for homosexuality, especially in the arts.

Art Monthly, June 1988. 'Fierce Visions: Derek Jarman'. Thoughtful article on Jarman's work, by Michael O'Pray.

Sunday Times, 23 Oct. 1988. Anthony Holden on Sir Laurence Olivier's involvement in *War Requiem*.

Independent, 3 Nov. 1988. Location report by Christopher Cook on *War Requiem*.

Screen International, 10–17 Dec. 1988. Production report on *War Requiem*.

Observer Magazine, 1 Jan. 1989. Article and interview by Janet Watts, on *War Requiem* and the effect being HIV positive has on Jarman's work.

What's On, 4 Jan. 1989. Philip Bergson interviews Jarman on *War Requiem*.

Radio Times, 18–24 Mar. 1989. Extract from *War Requiem*.

7 KENNETH LOACH

Kenneth Loach on location of *Kes*, 1969. © MGM/UA. Photograph: BFI.

KENNETH LOACH

KENNETH LOACH is one of Britain's world class directors; someone who has significantly contributed to the development of the art forms of film and television. He has a reputation for being one of the most affable directors in the business; he is certainly one of the most unassuming. He might be described as retiring, shy, and introverted, but then he is also a crusader who vehemently proclaims his political views. The writer Trevor Griffiths[1] remarked, 'Ken is a very secretive, intense man, who demands total commitment from everyone around him.'[2] Loach's cameraman for most of his films since *Kes*, Chris Menges,[3] has an immense respect for him: 'Ken is the man I admire most.'[4] Other British directors have voiced similar sentiments. To Roland Joffé, 'The first time that I saw *Cathy Come Home*[5] I was ravished because there were "people". This is a man who allows people to be. I wanted to embrace everything that Ken stood for.'[6] For Alan Parker, the film represented 'The single most important reason why I wanted to become a film director.'[7] Attenborough and Frears regard him with similar esteem. Yet for all this, he has been, from the mid-1970s, something of a giant in the wilderness. He is one of the few film-makers in this country who has faced direct political censorship, and even when he is free to work he has found it increasingly difficult to raise finance for his films. He quotes from his film *Fatherland*: 'Political folk are never safe this side of the grave,' and points out, 'I've spent as much time defending my films as I have making them.'[8]

The reason for this is that under the guiding influence of his Marxist friend, and for many years his producer, Tony Garnett, he has made

---

[1] British socialist writer for stage, film, and television. He has scripted *Reds* (1981, dir. Warren Beatty) as well as Loach's *Fatherland* (1986).

[2] *Time Out*, 25 Mar.–1 Apr. 1987.

[3] Now a director in his own right, who made *A World Apart* (1987).

[4] David Chell, *Movie Makers at Work* (Microsoft Press, 1987).

[5] Directed by Loach in 1966. It is a documentary style fiction examining the plight of homelessness in London.

[6] Sir Richard Attenborough's contribution to *British Cinema, a Personal View* (Thames Television, broadcast on 19 Mar. 1986).

[7] Parker's contribution to *British Cinema, a Personal View: A Turnip-Head's Guide to the British Cinema* (Thames Television, broadcast on 12 Mar. 1986).

[8] *Observer*, 22 Mar. 1987.

himself through his work a champion of the Left and the British working class. His views of class itself are orthodox Marxist (he supports an exclusively economic interpretation), dismissing wider sociological explanations. 'The working class is organised labour. Whether organised labour wears a white collar or not is really irrelevant. Class isn't defined by people's habits, and what they want to acquire. It's actually defined by their relationship to the productive forces.'[9] As such it is, for him, 'the only class worth connecting with'.[10] His broad aims seem to be threefold.

First, he wants to give a voice to the working class, one which he feels is never heard. But he does not want to be patronizing. He wants to show how the working class really lives, along with the sorts of everyday problems that they face. As Garnett points out, 'The Left often forgets to attend to the living details of working class life. In our films, however, we see a fidelity to the texture of the everyday as an act of political respect and solidarity . . . we try to give a sense of the complexity of the working class, their social conditions and context.'[11] Loach feels that, particularly under the influence of Hollywood, the actual range of topics tackled in film has become extremely narrow—'The whole spectrum of human experience has been reduced to Westerns and Romances.'[12] By addressing social issues such as homelessness, the family, or mental hospitals, he hopes to counteract this trend. These themes are very much rooted in the tradition which traces its way back to the post-war Italian 'Neo-Realist' movement. But although Loach plays down any direct influences, the similarities are striking. Pasolini,[13] commenting on *Poor Cow*,[14] said, 'Even a child could see that it is a product of Italian neo-realism which has moved into a different context.'[15]

Secondly, Loach wants to change society with his films. He wants to proselytize politically in the interests of the working class. This radical-ism distances him from the relative passivity of the Free Cinema direc-tors of the fifties, with their own, more lyrical belief in the 'significance of the everyday'. 'We (in Britain) have an appearance of democracy but we don't have the substance. We can't democratically decide many of the things which govern our lives.'[16] Loach wants to articulate the pre-dicament of the working class in order to help them break out of their

---

[9] *MFB*, Jan. 1983.
[10] *The Times*, 15 Jan. 1972.
[11] *Cineaste*, Autumn 1980.
[12] *The Times*, 21 Feb. 1980.
[13] Iconoclastic Italian director. His films reflect his uneasy allegiances to Marx, Freud, and Catholicism, and include: *Accattone!* (1961), *The Gospel According to St Matthew* (1964), and *Salo* (1975).
[14] Directed by Loach in 1967.
[15] Oswald Stack, *Pasolini* (Thames and Hudson, 1969).
[16] *Cinema Papers*, Apr. 1977.

traditional restraints; the first step towards the ultimate goal of a more equitable distribution of power. But he mourns the way that the working class blindly turn their frustrations in on themselves, rather than directing it at their political adversaries, through a lack of any true political consciousness. Some of his films are certainly more openly political than others, but he claims not to have a rigid political position. The political meaning especially in the fiction, rather than the documentaries, tends to be implicit, and not directly presented. The viewer, he hopes, will extract the meaning. He pessimistically admits, however, that the most that can be hoped for is the opening of a discussion, and that one is lucky if one does more than just rattle the status quo.

Thirdly, in the process of doing this he also hopes to sustain the Left 'by letting them see their own ideas reflected back at them, which is something that very rarely happens in the media'.[17]

Although some of Loach's political views might be seen as oversimplified, he would never claim to be a theorist, or intellectual. Anderson, one of the leaders of the 'Free Cinema' movement, and Britain's most caustic social satirist, sees Loach as having 'allowed himself to become obsessed with other people's political ideas. He's a sort of sentimental leftist who's got mixed up with a group of Trotskyists.'[18] The problem is that Loach's political ideas appear too polemic when presented in the form of documentaries. By contrast, his films, through their emphasis on characterization, represent a personalization of his political views. Powerfully humanist, they reflect a social morality which has a strong effect on the audience. He sees the difference between his documentaries and his films as 'the difference between a pamphlet and a novel. Sometimes it's quite useful to write a pamphlet, but it can never dig as deeply as a novel whose aim is to reconstruct something which has the stamp of authenticity.'[19] Nevertheless he remains, as Parker points out, 'interested in politics first, and film-making second'.[20] When asked what he felt about people discussing the merits of his cinematic techniques, he replied, 'I'm very happy that people should be discussing these issues, but I'd rather they were discussing the role of trade union leaders. These are issues which it seems to me are very much more important than the role of naturalism in film—that's quite interesting but marginal.'[21] Yet it is, ironically, in the actual development of ways of showing everyday life with veracity and sensitivity that his greatest achievements lie.

[17] *Time Out*, 17–30 Dec. 1982.
[18] Eva Orbanz, *Journey to a Legend and Back: The British Realistic Film* (Edition Volker Spiess, 1977). Interview with Anderson.
[19] *MFB*, Jan. 1983.
[20] *British Cinema: A Personal View* (Thames Television, broadcast on 19 Mar. 1986).
[21] *Framework*, No. 18, 1982.

Loach explains why he places such stress on this approach:

If you're trying to communicate the experiences of the people in the film to the audience then it has to really look as if they're actually having those experiences, and so it has to be authentic—that's a less loaded word than 'naturalistic'. If it doesn't, then people who are not particularly interested in film won't believe, so you don't even begin to get through.[22]

The ways in which he has tried to do this have seen some development over his career. Initially, the aim was to be more persuasive by stylistically imitating documentaries or the television news alongside which many of his dramas were scheduled. He hoped to achieve a new synthesis of documentary and fiction, and succeeded in films such as *Cathy Come Home* and *In Two Minds* (1967). He was prepared to let the camera follow improvised action and dialogue, and even let actors talk directly to the camera. But his style from *Kes* onwards, until recently, left much of this behind because 'in the end it seemed to inhibit the development of characters and their relationships'.[23] He was no longer interested in these devices, which he felt too often hid the absence of style. He disliked unusual camera angles or compositions, found zooms to be too assertive, and, instead, looked for 'quiet' shots. Chris Menges describes this from the point of view of a cameraman: 'The thing about working with Ken is that you learn very, very quickly that he wants a very sensitive quiet camera that isn't going to impose a style on the actors or the script. It should quietly observe.'[24] At the moment, Loach feels that this approach has become somewhat lethargic, and in his most recent works he has returned to a more assertive and energetic style.

It is however, the position and importance of the actors which has always been his first priority. Trevor Griffiths believes that, 'If Loach could make a film without a camera he would. He wants the actors to just be themselves so that everything looks as though it has just happened.'[25] Loach tries to make the actors forget the camera and make the whole filming process seem as unassuming and ordinary as possible. He uses a small, dedicated crew so that the actors will feel less conscious of the whole technical process. Garnett recognizes Loach's strength: 'Ken is very good with actors. They respond to him as a human being, not as a film director. Gradually, if that trust is there, they come up with things that they didn't know they could come up with.' Loach feels that it is

[22] *Framework*, No. 18, 1982.
[23] *MFB*, Jan. 1983.
[24] David Chell, *Movie Makers at Work*.
[25] *Observer*, 22 Mar. 1987.

Loach discussing a scene with Carol White in *Poor Cow*, 1967. © Weintraub Screen Entertainment. Photograph: BFI.

crucial to allow and encourage the actors to react spontaneously to both their words and their movement. Sometimes, he does not ask them to learn their lines, but to improvise based on the script. The inspiration for the acting of a scene must come from within them. To foster this, he likes to shoot in continuity (that is, in the order which the events happen, which, for logistical reasons is rarely the case in most films) and on the real locations so that the actors are taken as far as possible into the situations of their characters.

The writer puts down dialogue which is accurate up to a point, and often ends up being said. But words are the tip of the iceberg—it's everything else that you've got to get right. When you've got actors, their personalities are obviously involved. You've got to build it all up so that the exact words don't matter. You've got to have three dimensional characters, which don't exist in the script as paper.[26]

The consequence of this approach is that compared to the Hollywood style of acting, or that of the English theatrical tradition, and the resulting expectations these have created in the audience, Loach's characters seem very subdued. But then, that *is* real life.

This encouragement of a non-self-consciousness in the acting is supplemented by his preference for casting non-professional actors even for lead roles. He feels that with suitable casting he is able to get people to draw on their lives and experiences, so that what is seen on the screen is as much a reflection of themselves, as their characters. To Chris Menges, 'Ken's ability to choose actors from among real people is phenomenal.'[27] But this use of non-professional actors has caused Loach considerable problems with Equity, the actors' union. He has occasionally had to use performers, entertainers, or semi-professional actors because they have union cards without being the sort of 'full blooded' actors he wishes to avoid. For *Fatherland* Loach was fortunate enough to be able to cast the part of the East German singer with a real East German who had actually been forced to leave for the West, and who was himself a singer/writer of protest songs.

In pursuing these aims, Loach has been creatively and politically out on a limb.

I am alienated from most of the films that are being made now. I find American films exploitative—they both use and take from people, and the English film industry is dominated and colonised by American film companies. Of course, the American audience is not interested in unsensational, quiet films that ask

---

[26] Interview with the authors.
[27] David Chell, *Movie Makers at Work*.

them to draw general conclusions from the nuances of English working class life.[28]

He believes that, 'Traditionally British Cinema hasn't developed the expectation that there will be ideas in films. If you're interested in ideas, you don't tend to go to the cinema here—British Cinema is modelled on American movies and West End Theatre.'[29] He has, especially in recent years, found it very difficult to raise money for his films,[30] and believes that today he would not even have been able to get films like *Kes*, or *Family Life* (1971) off the ground. 'Perhaps I'm just not the man to arouse interest, unless any story I dealt with was very contemporary and pretty sensational—but the system seems pretty inadequate too.'[31] One must agree with him. Of the films which he has raised money for, many have in turn faced stormy receptions, and some even actual political censorship. He is, however, proud of the fact that his BBC television dramas *Up the Junction* (1965) and *Cathy Come Home*, were amongst the first things that Mary Whitehouse[32] ever complained about. Early on in his career, the pressure on him was rather half-hearted and never particularly direct; the criticisms from BBC management about his plays tended to be directed at their aesthetic quality rather than explicitly at their politics. In particular, they felt disturbed at the realist/documentary approach to fictional work. Garnett commented that the management once told him, 'The trouble with your films, and it's very worrying, is that people might believe them. Look, we want actors to look like actors, we want it to be clear.' In *Rank and File* (1971) the BBC persuaded Loach to change the ending, but half an hour before transmission, he replaced it with the original, and comments, 'no one complained, no one probably even noticed.'[33] Garnett got *The Big Flame* (1969) shown on the BBC by making an announcement to the *Radio Times* before anyone had seen it, so militating against any attempts to stop it because of the attention focused on it. However, with *Questions of Leadership* (1983),[34] *Which Side Are You On?* (1984),[35] and the play *Perdition* (1987)[36] at the

[28] *Cineaste*, Autumn 1980.

[29] *Stills*, May–June 1986.

[30] For many years he tried to make a film scripted by Jim Allen set in the Ireland of 1918–21, and also had plans to make a film about the English Civil War to 'expose the myth that Cromwell was on the side of the people'.

[31] *Guardian*, 16 Feb. 1980.

[32] Since the early 1960s she has been a self-appointed voice of the public conscience.

[33] Alexander Walker, *Hollywood, England* (Harrap, 1974).

[34] A four-part examination of Trade Union leadership.

[35] An anthology of striking miners' songs and poems.

[36] A highly controversial play written by Jim Allen. It is an examination of the collaboration of a group of Hungarian Zionists with Eichmann in 1944, in an attempt to win moral justification for the foundation of the state of Israel. It claimed to address the question of 'How were over half a million Jews sent to the Death Camps by a few hundred German SS, when

Royal Court, his work was actually banned for openly political reasons, and few people were prepared to stand up for him.

Ken Loach started work at the BBC in 1962 as a trainee director, although he only got six weeks of actual training. He really began to learn two years later, when he directed several episodes of *Diary of a Young Man*. He was working with the writers, Troy Kennedy Martin and John McGrath, and became deeply involved in the whole movement of taking the television play out of the studio, and onto location. But in some senses it was a very experimental, not very naturalistic, movement. Loach's television plays of this period use voice-overs, songs, location filming, monologues to the camera, factual information on titles, disjunctive episodes, and even some live studio sequences. Loach was open to a lot of influences, but most of all he was rebelling against the traditions of filmed theatre, studio drama, and overacting. He tried to make his work cinematic and visually exciting. While this was happening he became increasingly interested in politics. He was influenced by Garnett—a passionate Marxist, who had started as an actor for theatre and television and became a producer for the BBC's Wednesday Play series which began in 1965. The gritty plays that they were to make together, each within a mere seven-week period, are remarkable—some of the best dramas of the sixties. *Up the Junction* (1965), is based on a book by Nell Dunn which Loach admired for its 'accurate, sympathetic portrait of a group of people. It had a great sense of life, and a terrific energy.'[37] It is an anecdotal account of the lives of three working class girls in Clapham (south London), and tackled many issues including abortion, commercialism, love, sex, and youth. *Cathy Come Home* (1966) focused its sights more concisely and more profoundly on the problems of homelessness and overcrowding. It is the desperate story of a young but very close family destroyed by the poverty trap. They are faced by rent that they have to pay, which seems always to be higher than their wages, and by the callousness of a petty bureaucracy unsympathetic to their plight. It was a highly emotional film which sent reverberations across the country. Everybody discussed it. It even inspired the creation of the charity 'Shelter'. But Loach and Garnett both felt the film had failed: 'If the inference was that the difficulties of the homeless could be solved by forming a charity, it was a backward step.'[38] He felt that energy had been stimulated, but that he had failed to channel it, or use it politically, by offering a specific solution to the

Germany was facing defeat?' It was criticized for factual inaccuracies, implying generalizations from one incident, and blurring the distinction between Zionist and Jew. The director of the Royal Court, Max Stafford Clark, lacking a deep commitment to the play, cancelled the production.

[37] Interview with the authors.
[38] *Observer*, 22 Mar. 1987.

problems. The final titles stated, 'All events in this film took place in Britain within the last eighteen months; four thousand children are separated from their parents, and taken into care each year because parents are homeless; West Germany has built twice as many houses as Britain since the War.' In retrospect, Loach wishes the film had managed to indicate the causes of homelessness, and to show that it was an example of how the free market does not work.

In Two Minds (1967) is another brilliant and astonishing documentary fiction. Scripted by David Mercer, it is about a girl suffering from the diagnosis of schizophrenia, whose identity has been repressed by her parents, and later by the mental institution in which she is placed. Her predicament is revealed through the questions of an unseen therapist, and progress seems to be made until the girl decides to admit herself to hospital. The power of the film lies in its subtlety, the slow disclosure of the real sources of the girl's troubles: her apparently well-meaning, but in fact, totally uncomprehending parents; her genuine sense of bewilderment, so dependent on them that she cannot leave them; and her conviction that because everyone, even the staff of the mental institution, talks about 'her problem', she must in fact be mad. The camera style is suitably claustrophobic, concentrating on facial close-ups, which, because of the intensity and naturalness of the acting, is particularly effective. The film is also a plea on behalf of family therapy as opposed to physical treatment as the remedy for such problems, suggesting that here a cure for the girl's schizophrenia stems from a sociological understanding and does not come through the use of drugs and shock treatment.

However, during this time Loach was eager to move into feature films. He was given his opportunity by Joseph Janni,[39] who agreed to put up the £210,000 for *Poor Cow* (1967). Like *Up the Junction*, and *Cathy Come Home*, it stars Carol White. She plays a down-trodden mother whose petty-criminal husband is sent to prison, and who sets up temporary home with another lover, who also ends up in gaol. Although it was a commercial success, Loach now regards the film with misgivings:

It was a good book. The problem was that I just made a bad film. I framed the shots badly, timed them badly, cut them badly. It's the equivalent of bad prose. I just wasn't good enough, and I suppose that it was partly because it was my first film, but also because the story hadn't the same drive as *Cathy*, and *Junction*. I didn't think through the content sufficiently. It was a mess, and it was my fault.[40]

[39] Janni, an independent producer, working with Anglo-Amalgamated, had previously also given Schlesinger his first chance to direct.
[40] Interview with the authors.

Loach is rather unfair on himself. Even if it is a rather self-conscious film, it did successfully catch the raw tone of the heroine's situation.

Loach's next film was *Kes* (1969), a project that he had been trying to get off the ground for two years. It was eventually made for £157,000 for United Artists largely due to the positive intervention of Tony Richardson and Woodfall Films. But the film was very nearly never distributed; it was felt to be too depressing, to be neither an adult nor a children's film, and to have northern accents which were too strong for people to understand. As a concession, Loach redubbed a small amount of the film, but it was largely through the pressure exerted by certain British critics on the distributors that it was given a release, although one which was limited in scope and rather unimaginative. Despite this, it was ultimately a great success in Britain, though not in America.

Loach believes that *Kes* had more emotional commitment than anything else that he has ever done.

Tony Garnett, who had a great creative input as producer, Barry Hines as writer, and I were very much of one accord. It wasn't a big money film involving too great a consideration of production issues. We were doing everything ourselves on the real locations, tackling problems which if you were sitting in London, thinking about how to make a film about a kestrel, would have seemed too difficult. Instead it was cast in Barnsley, shot there at the school which Barry used to teach in, and his brother, Richard, even trained the kestrel. It was a very straightforward film to make, and a much happier experience than *Poor Cow*.[41]

The story, scripted by Barry Hines from his book, is set in a Yorkshire mining town. It is about Billy (brilliantly played by David Bradley), a 15-year-old school boy who lives with a worn-down mother and tough half-brother who works down the pits. This is an uncaring world, and the educational system offers no escape, merely holding the children down until they are old enough to start work in the mines. Billy is teased by the other children because he is bad at sport and too dreamy. But he keeps a beautiful untamed kestrel, which becomes an important symbol in the film, highlighting Billy's own plight (the kestrel was the only hunting bird in medieval feudal ranking that the lowest social orders could possess). In the end it is killed by Billy's half-brother in a moment of fury. The film poignantly shows how the repressed working class turn their frustrations in on themselves, and how the vitality and energy of youth is so often snuffed out.

The next film which Loach and Garnett made was *Family Life* (1971). Scripted by David Mercer, it was an adaptation of their previous television play, *In Two Minds*, and in many ways is as successful a film as *Kes*. Garnett, who had studied psychology and who had very personal

---

[41] Interview with the authors.

David Bradley with his pet kestrel in *Kes*, 1969. © MGM/UA. Photograph: BFI.

reasons for wanting to make the film, contributed a great deal to both pieces. The adaptation into a feature film naturalized the play's tight, fragmented narrative form. Mercer and Loach 'wanted to let the characters breathe a little more, give them more space to reveal themselves, and I also think perhaps that our own feelings about pace had changed'.[42] Loach argues that the only reason that he was able to get the £190,000 budget from MGM-EMI was that it starred a 19-year-old girl. As with the play, the film looks at the plight of a suicidal schizophrenic girl whose own family shows itself to be the real cause of the disturbance in her life. Based on the theories of the psychiatrist R. D. Laing,[43] mental illness is seen as largely family induced. In this case her schizophrenia is partly triggered by her uncomprehending parents' destruction of her individuality, and their demands that she have an abortion. 'It is an attack on a social system which produces families like the one we show, families who are led to believe that life consists of working, bringing up your children, and not complaining. This is the great working class cliché.'[44] Not only is it an indictment of the family unit itself, but also of the practices of psychiatry, especially of the inhuman use of drugs and shock treatment, which have little concern for care or understanding of the patients themselves. While all the hospital's doctors are portrayed as callous and insensitive, all the Laingian doctors are seen as humanitarian. Although this is rather distorted, Loach and Garnett wanted to make sure that the audience would draw only one conclusion.

Loach's next major project, *Days of Hope* (1975), initially started out as a feature film. Lack of money, however, resulted in it being made as a four-part drama for BBC television. Along with *The Big Flame* (1969), *Rank and File* (1971) (both scripted by Jim Allen), and to some extent *After a Lifetime* (1970) (scripted by the actor Neville Smith), it showed Loach's new obsession with the ability of the working class to free itself from the constraints of capitalism by withdrawing labour. *The Big Flame* is about the occupation of the docks in Liverpool; *Rank and File* is about the Pilkington glass workers' strike (although it is here called Wilkinson's); and *After a Lifetime* is a sensitive description of the days leading up to the funeral of an old militant of the General Strike.[45] *Days of Hope* directly tackles the years of social conflict between the First World War and the collapse of the General Strike in 1926. Although it was Loach's first film not set in contemporary Britain, it did have specific contemporary significance relating to the labour situations under the

[42] *Framework*, No. 18, 1982.
[43] R. D. Laing wrote books such as *The Divided Self* and *Self and Others*.
[44] *The Times*, 15 Jan. 1972.
[45] For the writer, Neville Smith, it was, like his script for Loach's *The Golden Vision* (1968), a very autobiographical film.

Loach's four-part television series, *Days of Hope* (1975) about labour relations from the end of WW1 to the General Strike. Copyright © BBC.

Heath and Wilson governments. 'The big issue which we tried to make plain to ordinary folk was that the Labour leadership had betrayed them fifty years ago, and were about to do so again.'[46] The four parts follow the life of a working class family during the period, but individual characters are generally secondary to the portrayal of political meetings and arguments, and Loach clearly felt problems in balancing the fictional and factual. The last of the films was an attempt to be more measured and thoughtful in tone so that the audience would be able to view at a greater distance issues like the rights and wrongs of gradual constitutional reform versus 'more major upheaval' to achieve socialism, or the historical positions the TUC or Labour Party have taken in this context. Historically it did over-simplify events, and its accuracy has been questioned, but it still remains, in many ways, his most ambitious project.

Apart from one play for the BBC, *The Price of Coal* (1977), Loach made nothing for around four years. He returned to feature films with *Black Jack* (1979) and *The Gamekeeper* (1980), both highly lyrical, rural pieces, marking a significant move away from the polemic of his more recent works. *Black Jack* was financed by the NFFC with help from France and Germany, after it had been turned down by every major British and American company because Loach and Garnett[47] refused to use stars. The story was based on a children's historical novel by Leon Garfield. 'I found *Black Jack* on one of my children's shelves, and I thought it was the perfect answer to all the sex and violence on TV today. It is set in 1750 and the background intrigued me.'[48] In the sense that it explored issues of childhood and madness, Loach was returning to familiar themes and beliefs. He did the adaptation himself: 'It is something I don't like doing, but fortunately *Black Jack* is a very articulate book.'[49] Tolly, a draper's apprentice and mild young hero, saves the thief and cut-throat, Black Jack (the Frenchman, Jean Franval, was cast as a condition for the French money), only to be forced to accompany him on various adventures through Yorkshire. On the way Tolly falls for Belle, a young girl who has escaped from a madhouse where she was secretly placed by her family to maintain their good name. Like *Days of Hope*, the film was under-budgeted. Loach made it for £400,000, largely on Super 16 mm., on a short six-week shoot. He commented that in

---

[46] *MFB*, Jan. 1983.
[47] Except for *Black Jack*, the Loach–Garnett collaboration ended after *Days of Hope*. Garnett subsequently produced the controversial *Law and Order* (1979), about the London police force and court system; directed *Prostitute* (1980), about a provincial girl hoping for better things by becoming a high class prostitute; and directed *Handgun* (1983) for Thorn-EMI. He worked in America for several years, but in 1990 returned to the BBC.
[48] *Films Illustrated*, Dec. 1978.
[49] Ibid.

some places, he did not even have enough shots to cut together. But he does capture the pace of life at that time, and the quality of the pre-industrial landscape. He has been criticized by some people for not being able to handle historical subjects, but this is not true. The real problem with this film is its pacing. Moments of tension and excitement are positioned with little thought beside those of quiet beauty, leaving the audience a little unsteady on its feet.

The Gamekeeper, made for ATV, is more successful in maintaining the lyricism, although unfortunately it lacks some of the charm given to *Black Jack* by the children, and the historical setting. Nevertheless, it is one of the films Loach has most enjoyed making. On one level it is about class in the British countryside, about a steelworker who leaves his work to become a gamekeeper. His job largely involves keeping old friends, and even children, off the duke's estate, and he is unable to perceive the underlying class reality of the situation. On a non-political level it is a simple account of the day-to-day running of an estate throughout the four seasons of the year.

The following year Loach made the disappointing *Looks and Smiles* (1981). He recognized that one of its problems was that 'we simply didn't have such a strong image as the bird in *Kes*, or the countryside in *Gamekeeper*.'[50] The story was originally about a teenager getting his first job — 'until we went to Sheffield, and then it was obvious that there weren't any jobs'.[51] It charts the impact of the recession on two school leavers. One decides to go into the army, tempted by the financial security and the sport he would be able to play. The other grows desper-ate as his job applications are turned down. Things look more optimistic for a while, when he becomes involved in a relationship. But ultimately that does not work either. It was conceived by Barry Hines and Loach as a sort of sequel to *Kes*, using older characters. 'We were anxious to make a film which said that most kids are not involved in riots in city centres. They lead quiet lives — quiet despair.'[52] In retrospect he sees the film as being too quiet: 'I'm not sure if the anguish the kids undergo is adequately represented, or if it kicks you hard enough in the stomach.'[53] As with *Cathy Come Home*, fifteen years earlier, he felt that the film had not sufficiently created the desperately needed outrage in the audience.

It was five years before Loach was to make his next feature film, *Fatherland*. Unable to raise the finance for his projects, he made several forays into documentary work. Stirred up by the strikes of the early

[50] *Stills*, May–June 1986.
[51] *Time Out*, 17–30 Dec. 1982.
[52] *The Times*, 13 Nov. 1981.
[53] *Time Out*, 17–30 Dec. 1982.

1980s, he was eager to tackle trade union issues head on. *A Question of Leadership* (1980) was about the 1980 steel strike; the similarly titled *Questions of Leadership* (1983) looked in four parts at trade union leadership as a whole; and *Which Side Are You On?* (1984) was an anthology of striking miners' songs and poems. They all faced politically motivated opposition. The former was cut, and a studio discussion tacked onto the end for 'balance' by Central Television; *Questions of Leadership* was banned by the board of Central Television; and *Which Side Are You On?* was not shown by the South Bank Show, allegedly because it did not have enough artistic content (it later won several awards at Film Festivals, and was eventually shown on Channel 4).

Fatherland (1986), like *Black Jack*, received much of its finance from France and Germany. Although originally budgeted at £1,500,000, it had to be cut back to £800,000—a very small sum, considering that it was shot in Berlin and England. The film marked Loach's departure from the straightforward camera style he had used since *Kes*, and which he felt had, to a degree, let him down in *Looks and Smiles*. Once more we see the titles on the screen like, 'Stalinism is not socialism. Socialism is not capitalism.'

The film is about Klaus Drittemann, an East German political singer and song-writer. Having displeased the authorities he is forced to leave his country. He has few illusions about the West and he finds new and different repressions; he sees money-obsessed, exploitative music company executives who want to make him into a marketable commodity; and he is followed and questioned by the CIA rather than the KGB. He is, however, far more interested in finding his father; a former communist and freedom fighter, who has been missing since the war. This search, in which he is joined by a French journalist, Emma, gives the film certain aspects of a thriller, although the slow pace and general absence of melodrama distance it from the genre. When he ultimately finds his father, he discovers the reality of his past: a Stalinist in the Spanish Civil War who had been ordered to wipe out 'the enemy within'; then blackmailed into working for the Nazis; and after the war, an American agent.

It is amusing to see how the financial pressures which Loach had to contend with affected the film. Emma, the journalist, was originally Dutch, but on the insistence of the French financiers became French (as happened with *Black Jack*). However, Loach refused to alter—and thereby further compromise—crucial parts of the storyline to meet this new situation, even though it makes the film more difficult to understand. Likewise the German backers, threatening to withdraw their money, tried to stop a line in the script which mentioned the continuing presence of Nazis in Germany. In making the film, Loach also found

Gerulf Pannach in Loach's *Fatherland*, 1986 © Channel 4.

some problems in working with Griffith's script because it was so specific that it worked against his own methods for encouraging spontaneity. He jokes, 'Trevor ought to direct. His scripts do.'[54]

But the film remains a very personal work for Loach. 'It's about someone coming from East to West, but in a biographical sense, it's about where I am now, and where others like me are now.'[55] In particular, the press conference in the film vocalizes strong complaints about television censorship, the financial exploitation of art, and the problems of not being able to work in one's own country. This scene was, however, substantially cut by the German financiers for German television, and they refused to show it in German cinemas.

It is an odd fate that a film-maker so acclaimed in the late sixties and early seventies, and still with so much enthusiasm, should be ignored by both critics and financiers for so long. His new film, *Hidden Agenda* (due to be released at the end of 1990), has received considerable attention since it won the Special Jury Prize at the 1990 Cannes Film Festival.[56] Hopefully the film is a sign that he has found a new sense of direction and a topic through which he can express his compassion. However, the film has not yet been released and we must wait to see its impact on the audience and how it will affect his subsequent career.

Under the pressure of opposition, Loach has considerably intensified his emphasis on the higher political issues which concern him. Unfortunately, at the same time the tastes of the public for controversial, political drama in film or television are changing. There has been a profound shift in Loach's work. As Michael Winner wrote,

I'm a great friend of Ken's, but when I think of the man who made *Kes*, which tells us more movingly about the disinherited than any other film I've seen, I wonder what has happened. *Poor Cow*, *Up the Junction*, and *Cathy Come Home* were all films of great humanity, and were probably political films in their way, but the compassion conquered all. He seems to be moving away from that, and becoming more politically motivated, and less interesting. It's a pity.[57]

There has been a shift of emphasis in the relationship between compassion and the political undertones in Loach's work. The years of struggling have sapped some of his vitality, and one senses that in his frustration, the direct political content, latent in all his work, has risen to the surface. Subsequently, his films have become more polemic, and in

[54] *Observer*, 22 Mar. 1987.
[55] *Time Out*, 25 Mar.–1 Apr. 1987.
[56] Scripted by Jim Allen, and based on the Stalker inquiry, it is a thriller tackling the controversial issues of Northern Ireland head on: alleged malpractices of British Intelligence and the Ulster Security Services, and establishment cover-ups.
[57] *Observer*, 22 Mar. 1987.

the process, more detached from the very people they were claiming to represent—authentic working class characters. What is quite remarkable is that, despite the tremendous opposition he has faced, neither Loach nor his films have been corrupted by political or economic forces.

DISCUSSION WITH

KENNETH LOACH

As a child, were you conscious of having a working class background?

Well, my childhood was very ordinary. My father was an electrician at a machine tools factory. I went to a grammar school; we lived in a semi-detached suburban house—as I guess most people do. So I wouldn't say that I was conscious of having anything other than a suburban and pleasant upbringing in a small town in the Midlands.

Were your parents political?

No, not at all.

So when and how did your interest in politics arise?

Well it was after I left Oxford University. I had worked in the theatre for a bit, having a very pleasant time, and then joined the BBC. It was through the influence there of friends, producers, and writers with whom I worked—people like Tony Garnett and Roger Smith[1]—that I started reading politics. The novels of the late fifties particularly influenced me; writers like Alan Sillitoe,[2] John Braine,[3] Stan Barstow,[4] and David Storey.[5] They had provincial settings in common with my background, and I became interested in reflecting that non-metropolitan life. The actual subjects we were all trying to tackle in films made us political . . . But it was quite a politicizing decade wasn't it?—all those public events of the sixties.

Were you influenced by Free Cinema or the Social-Realist movement of that time?

---

[1] Roger Smith is a British playwright whose work includes *Duet for One* and *Steaming*. Loach began to work with Garnett and Smith on the Wednesday Play series which began in 1965.
[2] British north-country novelist. Scripted the celebrated social-realist *Saturday Night and Sunday Morning* (1960, dir. Karel Reisz), and *The Loneliness of the Long Distance Runner* (1962, dir. Tony Richardson).
[3] His novels include *Room at the Top*.
[4] British playwright and novelist, whose work includes *A Kind of Loving*. He was one of the founders of a school of working class northern writers and unlike many of this group has remained firmly rooted in that society.
[5] British novelist and playwright. Author of *This Sporting Life* and *In Celebration* both of which were adapted into feature films, and directed by Lindsay Anderson. During the 1960s and 1970s he worked closely with Anderson at the Royal Court Theatre.

No, I don't think so. I didn't see any films which were part of Free Cinema until a few years ago. I don't wish to denigrate them because they were very interesting, and, I am sure, very important. But personally they didn't affect me at all.

You were committed to a career in the theatre before you decided to switch to television.

Yes, I had terribly unrealistic ideas about setting up a provincial theatre. My move into television was a desperate bid to earn money and pay the rent! I felt then, that theatre was where art was, and television was where the money was. I managed to get the job at the BBC because they were employing more people for the start-up of BBC 2.

Did you feel that you were abandoning your ideals or hopes for a lifetime in the theatre by changing tack in this way?

Well, I don't think I would be that pretentious, even at that age. It was just a question of where you could find work.

Your film and television work has maintained a characteristic interest in, and even championship of, the working class. Why do you want to limit yourself to these topics?

It's not only because I believe that the working class are very interesting. I also believe that, however much thirty years in the film industry detaches yourself from your roots, it's where my loyalty is. I also believe that one of the things which is good to do is to give a voice to those who don't have one, and who in film and television generally become stereotyped—that is the working class. The prevailing ideology sees the working class as victims, or as the deserving or undeserving poor, who must adapt to market forces. In fact, I see the working class as the vehicle for change. It is the class from which you have to hope for the future. I don't expect change to be imposed from the top. The analysis that change will come from the working class means that it is there that you have to aim your work, whether you are working class or not.

Could you describe your political inclinations more precisely?

I really prefer not to be labelled—the more that happens the more likely your work is to be dismissed.

You have been accused of being close to a position of anarcho-syndicalism in so far as you have stated that you affirm 'the power and will of the people to control their own lives'.[6]

I wouldn't accept that at all. I would stress the need for leadership.

[6] *Cineaste*, Autumn 1980.

Questions of Leadership is about this problem which we have had for generations. It was an analysis and strategy for change instead of accommodating to the way things are.

Of course, not all your films are so overtly political. What factors do you bear in mind when deciding what sort of film you will make?

Human beings are infinitely varied and infinitely absorbing, and that's what makes you want to make films. That's what really draws you to the medium. But you can't separate it from the way you view the world, and that is what colours the sorts of films you want to make. Yet you can't go on making the same film again and again. So just because I believe that there is a crisis in the leadership of the labour movement, doesn't mean that every film I do will be about that. It would be very boring if it were. You look broadly. You say, 'That's an interesting story and I think I can raise money for it.' So the films all start out in different ways, and you must proceed in a fairly pragmatic way.

Have your political opinions changed much since the 1960s—any more or less extreme as you have grown older?

No, I've never been an extremist [laugh]. Extreme depends on where you stand. Your opinions are always pretty central from your own point of view. Nobody is an extremist to themselves . . . I was once close to having a Trotskyist stand, and I am still anti-Stalinist, but I find the label revolutionary to be rather embarrassing.

Have any of your views on topics like the family and mental illness altered?

Yes, developed, I hope. I hope they have deepened. You often start out with rather two-dimensional views which need qualifying. But, for example, I still think that families can be very destructive as well as supportive. So I think that the family parts of *Family Life* are still valid.

Would you say that you should be angry or have strong political views about a topic, if you are to make a film on it?

If you ever lost your sense of anger that would really be quite disabling. You need a fund of anger from which to work—it cannot be artificially created. It has to be an anger at the way things are. My prime anger now is really against those who have disabled the cause of labour—from Wilson through to Kinnock.

So in a sense rather than criticize the government, you are criticizing the opposition for failing to oppose?

Yes, one always knows what the right wing will do—they are ideologically committed to sustaining capital, and creating an economy fit for capital to thrive in. The trouble with the Labour movement is that it has a leadership which neither understands nor represents its interests. The result is that all the strength, and all the power, is bled away by these careerists whose analysis really leads them to prop up the same system which exploits those people whom they are supposed to be representing . . . Increasingly you realize that the system cannot keep on going for ever, that it must collapse at some point.

Collapse?

Political and social collapse. More and more people are redundant to the process of production. Fewer and fewer people produce everything we need. Yet there is no way of harnessing those resources to satisfy people's *real* needs. The 'market' system does not work. Production centres on whichever working class can be most easily exploited; so if Taiwan manages to get organized and put its rates up, the exploitation will switch somewhere else. In the end, however, it just won't work. Production itself becomes concentrated in fewer and fewer hands. Whatever happens in the short term, the army of the unemployed will just increase because the production process is geared that way. Even in terms of something like transport or the infrastructure, the market cannot deal with the problems that arise. It's all cracking up.

Do you see significant changes in the nature of class divisions in 1980s Britain, as old working class institutions like the council estate or trade unions are being destroyed? Do you think that Britain is becoming more socially homogeneous?

Well, old working class patterns change and new ones arise. But I think the fact that workers might be being given a small share of the money will not alter their position as workers, because a company will always have to respond to needs of capital rather than the needs of the workers. So if market forces say 'lay people off', the company will have to do it. The fact that the workers might have nominal shares doesn't greatly affect them. In the long run this cannot work.

What about the tradition of working class conservatism in Britain?

Yes, it is a problem. But the issue comes back to questions of leadership. What is conservative one day may easily be forced into collision with the forces of capital, just through the particular circumstances involved. But the system will crumble . . . It has to crumble, doesn't it?

Do you see any danger of succumbing to a form of emotional blackmail in portraying political arguments through fiction?

Well, it's never a danger with yourself because you hope you've taken the right decisions about the characters. But yes, I know what you mean.

It is a danger but if they are your own views then you must be satisfied that you use emotion in a justifiable way. You mustn't get the audience to sympathize with things that are untenable. The danger is inherent in fiction.

Do you find it hard to tackle some of these broad and complex political and social issues in the two hours of a movie?

Yes, film is obviously a different medium from a book, and you can't try and do those same things as you would in a book. But there are other compensations. You can, for instance, show nuances of behaviour which can be very indicative, and can take the place of pages of writing, or which, in fact, can never be shown in a book.

Do you find that you have to simplify things for the audience?

Yes, certainly. You've got to start by judging what the audience will be. Sometimes I think that we've misjudged it. Certainly, in the last part of *Days of Hope* we tried to do too much, far too much. Or in *Fatherland*, we assumed that the audience know more about the Spanish Civil War than they plainly do.

Do you ever have to make compromises for the financiers to make the film more commercial?

Yes, I think you probably have to, although I don't make commercial films. The trick is to make those compromises without spoiling the film, so that the film can, at least, be made. The biggest problem is the American connection because they subvert films by constantly demanding Americans in the cast. They feel that if the good guy is American then there can't be too much wrong with it. It's a form of cultural imperialism isn't it? Look at Denzil Washington playing Steve Biko in *Cry Freedom*. He's a good actor but Africa is full of black actors who could have played him.

How do you try and balance these different responsibilities which you feel to the audience, the political ideas, and the financiers?

Well, primarily you feel a responsibility to the ideas in the film, and secondly to the crew and people involved. Then to the audience. But that does not mean that I would ever try to pull a stroke on the financiers because it would, of course, inhibit what you could do the next year. When a film is successful then all the problems which you've had with the financiers—there is always that bit of tension involved—is soon forgotten.

The real problem is that the commercial decisions of the financiers are based on preconceptions about the audience, which may not be right. The films they want to distribute and show properly are those which made money last year. Generally, they don't perceive that they have, in fact, created the tastes of the audience by what they have financed and shown in the past. It's a self-perpetuating thing and hard to break out of. Moreover, I'm not optimistic about the future for film, because the majority of finance will come from TV, or will be linked to television finance, and market pressures are going to drive financiers into making things which are, in their eyes, even more commercial.

Have you found the whole production process harder since you stopped working with Tony Garnett?

Yes, it is obviously much more difficult. I've worked with good friends since then. But I think that Tony had a very particular contribution. He was creatively involved, and also created a very good atmosphere to work in.

How important to you, is the whole collaborative/co-operative side of film-making?

You always have to get the best out of the people you are working with. But how exactly you do that, I don't know, because every job is different.

Is it hard to balance this satisfactorily with the desire to maintain a personal style?

Yes, that's the trickiest bit. The contributions of the people you work with have to fit into a central template. It has to knit together to become a personal work, while still not less than the sum of many diverse talents. In particular, the cameraman, designer, and editor are really crucial in combining to make what appears to be a very personal, separate, individual statement and approach. If it does not, it's a sort of mass production; everybody putting their contribution according to their own judgement and taste without it relating. By and large, that's what television is—an industrial process. To avoid that happening, you have to make certain that the people that you work with share the same vision and taste and are excited by the same things. Sharing taste is a most important thing. You therefore try and work with the same people—to build up a backlog of shared experience. It then takes less long to get inside each other's minds, and decide simple things like what camera movements you want. That's why I've used the same designer for twenty years—Martin Johnson. I've used the same cameraman—Chris Menges. I've worked with him since 1968—no before that.[7] I've worked with three or four editors over a long period. I have worked with my current editor, Jonathan Morris, for around eight or nine years. It is, in large part, a matter of very practical, technical questions.

What about working with the script-writer? You haven't mentioned how he fits in?

We would work through it all together, but I can't write. I'm not a writer. When I work with Jim Allen[8] and Barry Hines,[9] I work through with them, closely, and right from the beginning. Trevor Griffiths[10] is a different sort of writer, so I didn't do that with him. As a whole we would always talk about the ideas, discuss how the characters should develop, and what the resolution of the film should be. But they write it; although it's very hard to say who does what in a collaboration, I would never claim the writer's credit.

Your style has undergone some quite profound changes since you started making films at the BBC. But you have always maintained a search for 'naturalness', haven't you?

Yes, initially it was a question of making the fiction have the same feeling as documentaries; it meant taking the camera off the tripod, and taking it into places where you don't normally get a heavy 35 mm. camera, and so generating the excitement that you only get with a hand-held running camera. Then from *Kes*, Chris [Menges] and I developed a style where you sort of sat back, and showed what happens, so giving greater freedom for the things that happen in front of the camera.

Why change style bearing in mind the success and critical acclaim that you were getting with films like Cathy Come Home?

It was a conscious decision. Everybody started copying that early documentary style, and I felt that this chasing after events would very quickly become a mannerism. It became plain that really, if what was in front of the camera was important, you needed to photograph it as

---

[7] Chris Menges started working with Loach on *Poor Cow* (1967), as the camera operator.

[8] Script-writer of *The Big Flame* (1969), *Rank and File* (1971), *Days of Hope* (1975), and the play *Perdition* (1987).

[9] Script-writer of *Kes* (1969), *The Price of Coal* (1977), *The Gamekeeper* (1980), and *Looks and Smiles* (1981).

[10] Script-writer of *Fatherland* (1986).

simply and economically as you could. I wanted to put forward more sympathetic arguments, more observant; more thoughtful. When we started this with *Kes*, it was all quite viable. It was OK. The problem now is that I've spent too long doing that. After a decade it has become lethargic, which is a pity. A film like *Looks and Smiles*, I would describe as lethargic, and to some extent *Fatherland* as well. I now feel that I need to jack up the pace again; take the camera off the tripod again; put a bit more energy into it. It's a throw-back to the sort of things that I was doing with *Cathy*. I want to do this with my new film, but I don't know quite how I'm going to do it yet. *Hidden Agenda* is scripted by Jim Allen; style is also dictated by the writer you are working with, and the argument of his scripts comes through the confrontation and battle between people. In those circumstances you don't run around with the camera.

How do you react to the comments of praise and esteem from fellow British directors like Frears, Joffé, or Parker?

[Laugh] . . . I don't know about that . . . I don't know . . . they're just being kind, being generous. In some respects everybody has to work on their own . . . in others you have to work . . . together. But, by and large, I think that we all have to plough our own furrows. You have to work out your own artistic strategy as directly from life as you can, and say, 'This is what I observe. How can I transform this into a film?', or 'How can I bring the various elements of reality which I see in the streets onto film?', or 'How can I make a film which breathes of the outside world, rather than of other films, theatricality, or old techniques or methods?'

In trying to work out this artistic strategy for a film, do you storyboard your work?

No, I find that very sterile. In a storyboard, you are stuck with what you draw. I would never do that. I think it's very sterile and it works against the actors and against improvisation.

You seem to spend a great deal of energy and thought in working with the actors.

Yes, I shoot in sequence so that the actors, and, in fact, everybody else as well, experiences the story as it unfolds. The rehearsal for one scene will be the scene that you did yesterday. That way the actor will come into the scene in the right emotional condition. In fact, as far as possible, you should even give the actors the script a little bit at a time, so that they won't know what will happen. How the actors come into the scene is very important.

How much improvisation do you allow the actors?

Well, it depends entirely on the sort of script and type of film. In a film like *Family Life*, we did improvise quite a lot. So a large percentage of the scenes with the analyst were improvised, based on the family knowledge which everybody had. There is a lot of variation in the amount of improvisation even within parts of one film. If the scene is carefully plotted—say in terms of the dialogue of an argument between two people—then there won't be much. But if it is an open press conference like the one in *Fatherland*, you can improvise quite a lot. It's important not to make a fetish out of it. It's only one tool out of many.

But naturalistic performances are always very important to you. How else do you try and obtain them?

Yes, it's very different to a theatrical technique. I never work closely on the words, because then the words would become stale. You need to see the thought come into the brain of the actor. You see it behind his eyes; the real uncertainty. The camera can pick out the lie very quickly. In most films you can see that the thought hasn't just come into his mind— that he is just standing on a mark saying his lines on cue. But for the kind of things I do, you've got to try and prevent that premeditated quality so that you feel that it really has just happened.

You use a lot of non-professional actors in your films. Do you find it hard to find really good ones?

No, not really. In fact, it's much easier. But you do have to see lots and lots of people. Just recently I have started working with a casting director. Before that I did it on my own, and just saw lots of people.

Does the use of non-professional actors imply making more than average number of takes?

It does mean using a lot of stock, but not a lot of takes. After a few takes you use up the store of emotion anyone has got at any one time. It gets worn out. You really can't go beyond six or eight takes, unless there is some technical thing that you just have to get right. If you do, the words become mechanical and lose their meaning.

When you are casting actors for these parts, what do you look for?

If a part has to carry a lot of information, then you need someone who can learn their lines well, yet deliver them with spontaneity. If it is a part which involves a primarily emotional exchange with somebody, then you want someone whose emotions are very available, and who can

From left to right, Loach discussing a scene from *Family Life* (1971) with Sandy Ratcliffe and Grace Cave. Actors' improvisation is one of the hallmarks of Loach's films. © Weintraub Screen Entertainment. Photograph: BFI.

respond in that way. You also cast for authenticity of age, of class, of region.

Of looks?

Not too much. What is crucial is the experiences people have had in common with the part they must play in the film, or else the ability to project themselves into it.

Two rather different ways of acting?

Yes, yes, but there's always got to be something; some of your own experience that you can refer to; something, somewhere.

As a director do you have to draw on your own experiences in the same way?

Well, yes, but no more than just your own experience of living in the world. I want to make films which are real, which correspond faithfully to experience; to describe what is going on between people; to be authentic about the world.

The IBA wouldn't broadcast Questions of Leadership *because they said that it did not satisfy the criteria of objectivity. Do you take on board any of their criticisms?*

That notion of objectivity is just unsound because every question you ask, every camera position you adopt, depends on what you think is important. There is a decision behind it and consequently it isn't objective. Everyone has their own conceptions. In the case of these documentaries our understanding of what had happened is different, and therefore seemed prejudiced. They react by saying, 'This isn't objective by our standards, and therefore we have to discriminate.' If they said, 'Yes, what we have is a point of view, as well,' then that would be fine. They hypocritically defend this notion of objectivity. There are certain issues which Channel 4 are more tolerant of than others—such as feminism, or apartheid. In the broadcasting organization the senior appointments are political. You can't get a film made about Frank Chapple,[11] because Frank Chapple is big mates with the head of the IBA, and the Chairman of Channel 4. The same happened with *Perdition* because it was openly critical of Israel. There's a power network of contacts to stop these sorts of things. You can trace them.

Is there nothing equivalent in left wing circles, particularly in theatre and film?

[11] General Secretary of the Electrical, Electronic, Telecommunication, and Plumbing Union, 1964–84.

No, I don't think so. Certainly, it has never appeared when my work has been at issue. Yes, there are left wing people, but they are not in positions of power. They have no access to newspapers, to broadcasting, or to money. I don't think that there was ever a golden age of tolerance. I was once even instructed in the 1960s to take out a rather bland quotation from Trotsky, even though it was just saying that the world should be a cleaner and better place. But today the manipulation is more apparent, because it is also more dangerous. We are increasingly manipulated by the media, day-time and night-time television, more and more channels. There has also been an ideological shift as the media have become more and more commercially orientated, so narrowing the variety of programming. It's an invasion of the mind which just disables people's ability to draw independent conclusions about their lives. They become like chickens in a battery-hen system.

You sound very pessimistic about the possible benefit which a film can give to the public?

Yes, I don't think that there is much that you can do. I think that my broad aims are not achievable for someone who is a freelance director. The possibilities for a film-maker are very limited, and you shouldn't exaggerate what you can do, because you really cannot do much at all. Film is plainly so very transient. I am not optimistic at all in the short or medium term. . . . But people's capacity to fight back is inexhaustible. The tide will turn sooner or later . . . I'm optimistic about that.

FILMOGRAPHY

BRIEF BIOGRAPHICAL DETAILS

Born in 1936 in Nuneaton, Warwickshire. Attended King Edward VI school, Nuneaton, and then St Peter's Hall, Oxford, where he read Law, and became president of OUDS (Oxford University Dramatic Society) and secretary of the Experimental Theatre Club. He served his Military Service as a typist with the Royal Air Force. He then worked as an actor at various repertory companies. In 1961 he received a year's sponsorship from ABC TV as a director at Northampton Repertory Theatre. He was accepted as a trainee television director at the BBC in 1963.

FILM AND TELEVISION WORK DIRECTED BY KEN LOACH

1964   *Teletale* (Episode: *Catherine*). BBC TV.

Z Cars (various episodes) 60 mins. each. BBC TV. SERIES FORMAT: Troy Kennedy Martin.

Profit by their Example. BBC TV.

The Whole Truth. BBC TV.

Diary of a Young Man. 45 mins. each episode. (Episodes one, three, and five out of six.) BBC TV. SCRIPT: John McGrath and Troy Kennedy Martin.

1965   *Tap on the Shoulder*. 70 mins. BBC TV (Wednesday Play). SCRIPT: James O'Connor. PRODUCER: James MacTaggart. STORY EDITOR: Roger Smith. CAST: Lee Montague.

Wear a Very Big Hat. BBC TV (Wednesday Play). SCRIPT: Eric Coltart. CAST: Neville Smith.

Three Clear Sundays. BBC TV (Wednesday Play). SCRIPT: James O'Connor. CAST: Tony Selby, Rita Webb, Finnala O'Shannon.

Up the Junction. 72 mins. BBC TV (Wednesday Play). (Precursor of the theatrical film, dir. Peter Collinson). Based on the book by Nell Dunn. PRODUCER: James MacTaggart. PHOTOGRAPHY: Tony Imi. PRODUCTION DESIGNER: Eileen Diss. EDITOR: Roy Watts. STORY EDITOR: Tony Garnett. CAST: Carol White, Geraldine Sherman, Vickery Turner.

The End of Arthur's Marriage. BBC TV (Wednesday Play). SCRIPT: Christopher Logne. PRODUCER: James MacTaggart.

The Coming Out Party. BBC TV (Wednesday Play). SCRIPT: James O'Connor. PRODUCER: Tony Garnett. CAST: Dennis Golding.

1966 *Cathy Come Home*. 80 mins. BBC TV (Wednesday Play). SCRIPT: Jeremy Sandford. PRODUCER: Tony Garnett. PHOTOGRAPHY: Tony Imi. PRODUCTION DESIGNER: Sally Hulke. EDITOR: Roy Watts. CAST: Carol White, Ray Brooks, Winifred Dennis, Wally Patch.

1967 *In Two Minds*. 85 mins. BBC TV (Wednesday Play). SCRIPT: David Mercer. PRODUCER: Tony Garnett. PHOTOGRAPHY: Tony Imi. PRODUCTION DESIGNER: John Hurst. EDITOR: Roy Watts. CAST: Anna Cropper, George Cooper, Helen Booth.

Poor Cow. 101 mins. VIC Films. SCRIPT: Ken Loach and Nell Dunn (from the novel by Nell Dunn). PRODUCER: Joseph Janni. PHOTOGRAPHY: Brian Probyn. PRODUCTION DESIGNER: Bernard Sarron. EDITOR: Roy Watts. MUSIC: Donnovan. CAST: Carol White, Terence Stamp, John Bindon, Queenie Watts, Kate Williams.

1968 *The Golden Vision*. 75 mins. BBC TV (Wednesday Play). SCRIPT: Neville Smith and Gordon Honeycombe. PRODUCER: Tony Garnett. CAST: Alex Young, Neville Smith, Billy Dean.

1969 *Kes*. 113 mins. Woodfall Films/Kestrel Films. SCRIPT: Ken Loach, Tony Garnett, Barry Hines (from the novel *A Kestrel for a Knave* by Barry Hines). PRODUCER: Tony Garnett. PHOTOGRAPHY: Chris Menges. PRODUCTION DESIGNER: William McCrow. EDITOR: Roy Watts. MUSIC: John Cameron. CAST: David Bradley, Colin Welland, Lynne Perrie, Freddie Fletcher, Brian Glover.

The Big Flame. BBC TV (Wednesday Play). SCRIPT: Jim Allen. PRODUCER: Tony Garnett.

In Black and White. London Weekend Television (not transmitted).

1970 *After a Lifetime*. 80 mins. Kestrel Films for London Weekend Television. SCRIPT: Neville Smith. PRODUCER: Tony Garnett. PHOTOGRAPHY: Chris Menges. PRODUCTION DESIGNER: Andrew Drummond. EDITOR: Ray Helm. MUSIC: John Cameron. CAST: Edie Brooks, Neville Smith, Jimmy Coleman, Billy Dean, Peter Kerrigan.

1971 *Family Life*. 108 mins. Kestrel Films for Anglo-EMI distributors (feature film version of his TV film *In Two Minds*). (In the US a.k.a. *Wednesday's Child*.) SCRIPT: David Mercer. PRODUCER: Tony Garnett. PHOTOGRAPHY: Charles Stewart. PRODUCTION DESIGNER: William McCrow. EDITOR: Roy Watts. MUSIC: Mark Williamson. CAST: Sandy Ratcliffe, Bill Dean, Grace Cave, Malcolm Tierney.

Talk About Work. Central Office of Information.

Rank and File. 80 mins. BBC TV. SCRIPT: Jim Allen. PRODUCER: Graeme McDonald. PHOTOGRAPHY: Charles Stewart.

PRODUCTION DESIGNER: Roger Andrews. EDITOR: Roy Watts. CAST: Peter Kerrigan, Billy Dean, Tommy Summers.

1973   *A Misfortune.* BBC TV (Second House) (based on a short story by Anton Chekhov). CAST: Lucy Fleming, Ben Kingsley, Peter Eyre.

1975   *Days of Hope.* 95 mins. each episode. BBC TV. SCRIPT: Jim Allen. PRODUCER: Tony Garnett. PHOTOGRAPHY: Tony Pierce-Roberts and John Else. PRODUCTION DESIGNER: Martin Johnson. EDITOR: Roger Waugh. CAST: Paul Copley, Pamela Brighton, Nicholas Simmonds, Helen Beck. Four episodes: *Days of Hope 1916: Joining Up*; *Days of Hope 1921*; *Days of Hope 1924*; *Days of Hope 1926: General Strike.*

1977   *The Price of Coal.* 75 mins. BBC TV (Play for Today). SCRIPT: Barry Hines. PRODUCER: Tony Garnett. CAST: Bobby Knutt, Rita May, Paul Chappell, Jayne Waddington, Jackie Shinn. Two episodes: *Price of Coal: Meet the People*; *Price of Coal: Back to Reality.*

1979   *Black Jack.* 109 mins. Kestrel Films in association with NFFC. SCRIPT: Ken Loach (from the novel by Leon Garfield). PRODUCER: Tony Garnett. PHOTOGRAPHY: Chris Menges. PRODUCTION DESIGNER: Martin Johnson. EDITOR: Bill Shapter. MUSIC: Bob Pegg. CAST: Jean Franval, Stephen Hurst, Louise Cooper, Andrew Bennett.

   Auditions. 60 mins. documentary. ATV. PRODUCER: Ken Loach. PHOTOGRAPHY: Chris Menges. EDITOR: Jonathan Morris.

1980   *The Gamekeeper.* ATV. SCRIPT: Barry Hines (based on his book *The Gamekeeper*). PRODUCTION TEAM: Ashley Bruce, June Breakell, Julie Stoner. PHOTOGRAPHY: Chris Menges and Charles Stewart. PRODUCTION DESIGNER: Martin Johnson. EDITOR: Roger James. CAST: Phil Askham, Rita May, Andrew Grub, Peter Steele.

   A Question of Leadership. 60 mins. documentary. ATV. PHOTOGRAPHY: Chris Menges and John Davey. EDITOR: Roger James.

1981   *Looks and Smiles.* 104 mins. Black Lion Films (an ITC subsidiary) in association with Kestrel Films for Central TV. SCRIPT: Barry Hines. PRODUCER: Irving Teitelbaum. PHOTOGRAPHY: Chris Menges. PRODUCTION DESIGNER: Martin Johnson. EDITOR: Steve Singleton. MUSIC: Mark Wilkinson, Richard and the Taxmen. CAST: Graham Green, Carolyn Nicholson, Tony Pitts, Roy Hardwood, Phil Askham.

1983   *The Red and The Blue.* 90 mins. (a.k.a. *Impressions of Two Political Conferences*). Central TV. PRODUCER: Roger James. PHOTOGRAPHY: Chris Menges.

   Questions of Leadership. Central TV (not transmitted). Four Episodes.

1984   *Which Side Are You On?* 50 mins. London Weekend Television (for the South Bank Show). (Due to the considerable controversy that it aroused its transmission was postponed from November 1984 to January 1985.) PRODUCER: Ken Loach. EXECUTIVE PRODUCER: Melvyn Bragg. PHOTOGRAPHY: Chris Menges. EDITOR: Jonathan Morris.

The Coal Dispute (for Diverse Reports for Channel 4).

1986   *Fatherland*. 111 mins. Film Four International (London)/Clasart Film (Munich)/MK2 (Paris)/A Kestrel 2 Production with the participation of the French Ministry of Culture. SCRIPT: Trevor Griffiths. PRODUCER: Raymond Day. PHOTOGRAPHY: Chris Menges. PRODUCTION DESIGNER: Martin Johnson. EDITOR: Jonathan Morris. MUSIC: Christian Kunert, Gerulf Pannach. CAST: Gerulf Pannach, Fabienne Babe, Christine Rose, Sigfrit Steiner.

1990   *Hidden Agenda*. Initial Film and Television. CAST: Mai Zetterling, Brian Cox, Brad Dourif, Francis McDormand.

He has directed two commercials: one for potatoes in the mid-sixties, and one for the *Glasgow Herald* in 1986.

OTHER WORK

1969   Founded the production company Kestrel Films.

1985   Produced *End of the Battle . . . Not the End of the War* (for Diverse Reports for Channel 4, credited as 'Edited by K.L.')

1987   Was to direct the stage play *Perdition* by Jim Allen for the Royal Court, but it was withdrawn a few days before opening.

BIBLIOGRAPHY

Financial Times, 10 Nov. 1965. 'Experimental Slot', by T. C. Worsley. Essay on the Wednesday Plays, with particular reference to *Up the Junction*.

Sight and Sound, Winter 1968–9. 'Case Histories of the Next Renaissance'. Article and biofilmography on Loach by David Robinson.

Guardian, 30 June 1969. Article by Stacy Waddy.

Sight and Sound, Summer 1970. 'The Kes Dossier', by John Taylor. An analysis of the production of *Kes*.

Today's Cinema, 11 Sept. 1970. Notes on a forum at the NFT at which Loach talked about distribution problems.

G. Roy Levin, *Documentary Explorations* (New York, 1971). Interview with Loach.

Sunday Times, 9 Jan. 1972. Article by Philip Oakes.

The Times, 15 Jan. 1972. Article by Barry Norman.

Films and Filming, Mar. 1972. 'Spreading Wings at Kestrel'. Article by Paul Bream in which he interviews Ken Loach and Tony Garnett.

Observer, 8 Feb. 1976. Article by Ian Mather.

Jump Cut, June 1976. A comparison of the films of Ken Loach and Tony Garnett, including an interview with the two film-makers.

Cinema Papers, Apr. 1977. Interview about the series *Days of Hope*.

Films Illustrated, Dec. 1978. Interview in which Loach discusses his return to feature film-making with *Black Jack*; and his involvement in film and television.

Guardian, 16 Feb. 1980. Article by Derek Malcolm.

The Times, 21 Feb. 1980. Article by Glenys Roberts.

Cineaste, Autumn 1980. A look at the work of Garnett and Loach, followed by an interview with them, in which they discuss their relationship with the BBC, the terms in which they see their work and their audience, and their thoughts for the future.

Time Out, 12–18 Dec. 1980. Discussion of Loach's recent work, particularly *Auditions*, and *The Gamekeeper*, and also *A Question of Leadership*, which he made for ATV, but which has been shelved because of its political bias.

Tony Bennett, Susan Boyd-Bowman, Colin Mercer, and Janet Woollacott, *Popular Television and Film* (Open University, 1981), 297–318, debate on *Days of Hope*.

Screen International, 30 May–6 June 1981. Profile of Loach, and his comments on the making of *Looks and Smiles*.

Broadcast, 17 Aug. 1981. Loach criticizes ATV and the IBA for their treatment of his documentary *A Question of Leadership*, which was substantially altered and transmitted only locally.

Guardian, 21 Aug. 1981. Article by Stephen Cook.

The Times, 13 Nov. 1981. Article by Nicholas Wapshott.

Framework, No. 18, 1982. Interview with Loach on working at the BBC during the 1960s, his reasons for moving to ATV, his experiences there, and how he got into feature film-making.

Time Out, 17–30 Dec. 1982. Loach talks about *Looks and Smiles* and reflects on his career.

Monthly Film Bulletin (50), Jan. 1983. Interview on *Looks and Smiles* and his documentary work for Central TV.

Tribune, 25 Nov. 1983. Article by Loach.

Television Weekly, 3 Aug. 1984. Brief article on the political censorship of his series *Questions of Leadership*.

Broadcast, 14 Mar. 1986. A note that Loach is to direct commercials for Scope Picture Productions, advertising the *Glasgow Herald*.

Stills, May–June 1986. Article tracing Loach's career, with comments from the director on his work.

Stills, Nov. 1986. Article on *Questions of Leadership*.

Film Dope, Feb. 1987. Biofilmography.

Guardian, 18 Feb. 1987. Article by Loach.

Monthly Film Bulletin (54), Mar. 1987. Biofilmography.

City Limits, 19–26 Mar. 1987. Notes on the Ken Loach season at the ICA.

Observer, 22 Mar. 1987. Article by Christine Aziz.

Time Out, 25 Mar.–1 Apr. 1987. 'No Place like Home', by Steve Grant. Interview with Trevor Griffiths on *Fatherland*.

What's On, 26 Mar. 1987. Article by Phillip Bergsun.

City Limits, 26 Mar.–2 Apr. 1987. Interview with Loach and Trevor Griffiths about *Fatherland*.

Video Business, 18 May 1987. Loach discusses the making of *Fatherland* and the reason for his interest in the subjects in the film.

Screen International, 27 Feb.–5 Mar. 1988. Letter from Tony Garnett, Ken Loach, and David Puttnam, paying tribute to Nat Cohen following the news of his death.

Film Comment, Mar.–Apr.1988. Career interview.

8 ALAN PARKER

ALAN PARKER

ALAN PARKER was the first of the group of British directors who had learned their craft in the commercials industry to go and live and work in America,[1] a country and an ethos with which he strongly identifies. His great talent has been his ability to make films which appeal to a mass audience, are emotionally compelling, and employ evocative visual imagery. His trademark, which has hit an extremely responsive chord with the audience, is power rather than subtlety. The commercial success of his films has bought him considerable creative freedom, as well as a steady stream of potential backers. But despite this success, Parker remains defensive and truculent about his work and beliefs. He sees himself as a maverick, at odds with both the reviewers, who often find the emotional anger and intensity of his films distasteful, and the British film establishment for what he sees as their dogged and misguided attachment to the ideal of an indigenous cinema. Trusting in the principle that attack is the best form of defence, he arms himself with a cutting sense of humour and a cartoonist's pen.[2] This combination of diligence and belligerence stems from Parker's early years in Islington. It was in this working class area of north London that the more powerful aspects of his personality were moulded; particularly his frank-talking, fighter's instincts and a loathing of all forms of intellectualizing. He is keen to convey an image of himself as someone who escaped from his environment by hard work and determination, but also with a fair bit of good fortune—'I'm a yobo who got lucky,' he likes to joke.

Parker has always been a man of ambition and of action. He is apt to take the initiative, because he is rarely prepared to take 'no' for an answer. Leaving school with 'A' Levels in Pure and Applied Mathematics and Physics, he was ill prepared for his future career. He got his first job through the influence of his father. 'My dad worked at the *Sunday Times*, in the transport department. He used to accost people in the lift, anyone who looked important, and ask them to give me an interview.'[3] Initially obtaining a job as a dogsbody at one of their

---

[1] Others include Ridley Scott, Tony Scott, and Adrian Lyne.
[2] His scathing sketches, until recently, appeared regularly in industry magazines and journals.
[3] Interview with the authors.

Alan Parker relaxing at Cannes Festival.

obscure publications, the *Hospital Equipment News*, he soon got work elsewhere, at a small advertising agency in Holborn. Desperate to write the commercials himself, but confined to the mailroom, he started working at it in his spare time, eventually becoming a full time copy-writer[4] for Collett, Dickenson, Pearce.[5]

Parker thrived in a business where merit was the deciding factor, and relished the competitive atmosphere. By 22 he was one of the highest paid writers in the industry, and two years later was directing commercials. It was at Collett, Dickenson, Pearce that he met many of the people with whom he would work closely during his film career: David Puttnam,[6] Alan Marshall,[7] Ridley Scott, Adrian Lyne,[8] and Charles Saatchi.[9] Television commercials were just beginning to appear in Britain and Parker got a budget to experiment on 16 mm. film in the agency basement. Alan Marshall produced and edited them, and 'at a loss for a role, having written the commercials, I got to say "action" and "cut". It was at this point that the megalomania set in.'[10] They began to transform television commercials by using evocative atmosphere and story-telling techniques. This period, at the end of the sixties and beginning of the seventies, was a prolific time for Parker. In 1970, together with Marshall, he formed the *Alan Parker Film Company*, a commercials production house partially funded by Collett, Dickenson, Pearce, who became their largest client. Making about six hundred advertisements in six years,[11] Parker became, along with Ridley Scott, one of Britain's top commercials directors.

But Parker had his eyes firmly fixed on feature films. 'True, we learned our craft in thirty second bursts, but secretly we only dreamed of movies. Well, I did anyway.'[12] Puttnam feels that 'What hasn't been

[4] A writer of advertisements.

[5] One of the most progressive agencies in the 1960s.

[6] Independent British film producer who produced *Midnight Express* and Parker's script *Melody*, and was executive producer on *Bugsy Malone*.

[7] Producer of all Parker's feature films until *Mississippi Burning* (1989). Marshall also produced *Another Country* (1984) and in 1987 was executive producer for another British commercials director—Paul Weiland—on *Leonard: Part IV*.

[8] Scott and Lyne are both commercials directors who went into feature films and now work in Hollywood. Ridley Scott directed *Alien* (1979), *Blade Runner* (1981), *Someone to Watch over Me* (1988), and *Black Rain* (1989). Adrian Lyne directed one of the most successful of all American films, *Fatal Attraction* (1987).

[9] He now owns one of the largest advertising companies in the world, Saatchi & Saatchi, which includes the Conservative Party among its clients. Parker believes that of all the people he was working with at this time, 'Charles was always the most gregarious and astute. A lot of his wisdom rubbed off on all of us' (interview with the authors).

[10] Alan Parker, interview with the authors.

[11] Including the famous series for Cinzano with Leonard Rossiter and Joan Collins, the Silk Cut advert set during the Zulu War, and the series for Slumberland beds where Fred and Mabel Pottle have romantic dreams.

[12] *A Turnip Head's Guide to the British Cinema*, 60 mins. For Thames Television series *British Cinema: A Personal View*.

realized is that there was a period in the early seventies when commercials were the only place an aspirant film director could become trained in his craft.'[13] Shooting around two a week for years on end, Parker was able to explore the camera's possibilities; to experiment dramatically, technically, and of course, visually.

It is in visual terms that Parker speaks most eloquently and personally, and his training in advertising equipped him with an ability to create a visual shorthand in order to achieve an immediate sense of atmosphere or emotion. 'It's very unfashionable to talk about visual style, but I really do care very much how my films look. Technically, I don't think I could do better. The surface would be hard to improve, but I must go deeper.'[14] This ability is exemplified by the early montage scenes in *Birdy* (1984) which capture childhood friendship so well and so concisely. The two boys, Birdy and Al, are seen catching pigeons from the metal girders of a railway bridge, riding off on their bicycles with the birds they have caught, and then happily building a bird-house in Birdy's backyard. Another particularly concise and simple example is at the beginning of *Mississippi Burning* (1989). Two drinking fountains stand next to each other in a seedy old room, a gleaming silver one under a sign saying 'Whites' alongside a broken down porcelain wash basin beneath a placard saying 'Coloreds'. An ugly white boy comes and drinks from one, and then a handsome black boy drinks from the other. By means of this motif Parker swiftly and unmistakably establishes the tone of the film. Certainly this ability is one of Parker's great strengths, but at times it also proves to be a weakness when his scenes border on, or actually become, clichés and stereotypes. In *Midnight Express* (1978), the gaoler's fat and sweaty appearance predictably forebodes his viciousness. To Parker, cinema is predominantly an emotional art form, for which the creation of such forceful images is crucial. Puttnam sees this as the strength of all directors with a background in commercials, 'They have been trained to deal in images and this [cinema] is an imagist business.'

Parker places a strong emphasis on lighting to help create the desired mood. His gothic imagination often manifests itself in his use of strong beams of light which fall through windows, often at acute angles, diffused by bars and grills. This is used in *Birdy*, together with distorting perspective through wide angled shots, in order to capture the atmosphere in the cell of the mentally disturbed central character. Similar images can be seen in the archaic prison in *Midnight Express*, or the seedy, colour-drained environment in *Angel Heart* (1987). Supporting this is the choice of locations, for Parker prides himself on finding the

[13] Ibid.
[14] *City Limits*, 1–8 Oct. 1987.

unusual even in familiar settings, such as the eerie back streets of New Orleans in *Angel Heart*. Indeed, he will often work the script around stimulating locations. For one scene in *Mississippi Burning* he found an authentic shoe repair shop. 'No one could ever dress a set like that. In the ground there were embedded nails from many, many years. I wrote a whole scene there because of finding the place.' Consequently, as Marshall claims, 'Whether our films are disliked in terms of their storylines, nobody could ever criticise them for what they look like . . . I'm not saying that we're the greatest film-makers, but when we make a film, we make it with a quality that we believe is as good as can be made for the amount of money we have to spend.'[15] This degree of sophistication and the polish of Parker's films owe much to the fact that since he started to make commercials he has worked frequently, and sometimes uninterruptedly, with the same crew; particularly his cameraman, Mike Seresin; production designer, Geoffrey Kirkland; editor, Gerry Hambling; and producer, Alan Marshall. So important is it for him to work with these people that on several occasions locations have been chosen, partly at least, to allow access to his British crew. Such means as these have equipped Parker to make films which attract a world audience through their visceral appeal.

However, the subject and style of the films he makes now are very different from the first scripts he wrote in the early 1970s and the films he admired at that time. Parker was greatly influenced by the works of Loach and Garnett.[16] Commenting on his reaction to Loach's *Cathy Come Home* (1966), a documentary-style fiction examining the plight of the homeless in London, Parker calls it 'the single most important reason I think why I wanted to become a film director.'[17] His first scripts, *Melody* (1971), *Our Cissy* (1973), *Footsteps* (1973), and *No Hard Feelings* (1974), the last three directed by him, were written while he was making commercials for his own company. They were all set in London, and reflected personal British subjects. In their approach, most were close to the working-class realist traditions which had so dominated British cinema in the 1960s. But it proved quite difficult to get such work made, and they were not popular with the mass audience. *Melody*[18] was Parker's first script and was produced by Puttnam (also his first film). Set in a south London comprehensive school, it is the story of a love affair between two children and captures the feeling of childhood autobiography. 'We had also acquired the rights to seven Bee Gees songs, and I had

---

[15] *Stills*, June–July 1985.
[16] Pioneering and politically motivated director and producer of many of the 1960s' and 1970s' most important British television films concentrating on social issues.
[17] *A Turnip Head's Guide to the British Cinema*.
[18] a.k.a. *S.W.A.L.K.*

Mickey Rourke in Parker's colour-drained 'film noir', *Angel Heart*, 1987. © Courtesy of Columbia Pictures Industries Inc. Photograph: BFI.

to incorporate them into the story.'[19] Although it proved to be a sensational hit in Japan, it was a disappointment in the UK and the US. *Our Cissy* was a much more sophisticated, gritty, 'realist' film, about an elderly worker from Preston who comes to London to find out the sordid truth about his daughter's suicide. Similarly his sinister short story, *Footsteps*. Part social realism and part mystery adventure, it focuses on the plight of a lonely old woman, frustrated by the unsympathetic world around her and paranoid about her own safety. *No Hard Feelings*, funded by Parker's own company, was, like *Melody*, a love story with a similar sense of autobiography, recounting the pain of adolescence set against the working-class, East End of London during the blitz. The £30,000 it cost him was money well spent because it led to the BBC producer Mark Shivas's entrusting Parker to direct Jack Rosenthal's script *The Evacuees* (1975).[20] This humorous tale, told with a superb sense of atmosphere, received tremendous critical reviews, and won an international Emmy for Best Drama and a British Academy Award. Yet, after this, nearly all Parker's efforts to get his scripts made came to nothing. 'The industry was run by Wardour Street spivs; cigar chomping Lew Grade types producing mid-Atlantic pap. The scripts I wrote at the time were based on my background in Islington, real "angry young man" stuff, and got the rubber rejection seal—"too parochial"—stamped on them by the potential backers.'[21]

His first feature film *Bugsy Malone* (1976), a musical spoof of American gangster films performed entirely by children, represented a calculated volte-face.[22] 'It was a totally pragmatic exercise to break into film because the TV work I had been doing was in a Ken Loach vein.'[23] In the film, the child actors imitate their adult models in dress, manner, and voices. To an extent the novelty of this whole idea wears a little thin towards the end, but the film has charm, visual grace, and great humour, particularly in the songs[24] mouthed by children and sung by deep voiced adults. The film achieved international acclaim. But its success changed Parker. 'With everyone giving us a standing ovation at Cannes, and every single American studio after us to do stuff, I looked over those early screen-plays again but I didn't want to do that kind of film any

[19] Parker in Andrew Yule's, *Puttnam: A Personal Biography* (Mainstream, 1988).

[20] A warm story, also set during World War II, about two Jewish boys who are evacuated to Manchester and have to stay in a non-Jewish home.

[21] *Face*, July 1985.

[22] It began as a story Parker told his children. 'I had four young children, and we had a house in the North of England which we used to drive to each weekend. That drive was a nightmare. I couldn't bear driving three and a half hours with four kids every week, so I invented this story to keep them quiet. All of the characters in the film were in these stories.' (*Cineaste*, No. 2, 1986).

[23] *Cineaste*, No. 2, 1986.

[24] All composed for the film by Paul Williams.

Parker directing Scott Biao and Florrie Dugger in *Bugsy Malone*, 1976. Like Frears', Parker's first film was a parody of the American gangster genre, but unlike Frears he chose to work with an exclusively child cast. Courtesy of the Rank Organisation plc. Photograph: BFI.

more. Something had happened—the seduction of world cinema.'[25] The experience also proved to him the need to make films with international appeal. Despite its modest budget of £575,000,[26] and although it ultimately netted some £2 million, it did not break even in Britain alone. 'You cannot make a personal British picture anymore. *Grease* is the only word.'[27] Since then, Parker has been an ardent champion of the idea of world cinema, as only a convert can. Never again would he return to Britain for the subject of his films.[28] As a result, he is intolerant of those critics who, as he sees it, narrow-mindedly champion parochial British themes, while denigrating the importance of films which are able to break into the international arena. Not only, he says, are such critics intellectually isolated from the international mass audience, but they are allowed to condemn films, particularly his, with impunity.

The problem with the Channel 4 idea of British film, which the critics regard as the real British film culture whilst relegating the work of Puttnam and myself to the status of Hollywood pap, is that it has more to do with anthropology than entertainment. It's too self-consciously British. The culture, the milieu, the society, should filter through the films not, as is so often the case, be its *raison d'être*.'[29]

Although Parker, in his feature films, chooses predominantly American subjects with mass appeal, he does retain many of his previous interests and attitudes. He remains concerned about the backgrounds, motivations, and problems of working class people. Few of the central characters in his films are middle or upper class; and *Birdy* and *Fame* (1980) have strong working class themes. *Birdy* is set in the 'blue-collar' district of Philadelphia, which reminded Parker of 'the working-class terraced streets and backyards of Islington in London, where I grew up.'[30] *Fame*, set in the tough world of New York City, is about predominantly working class youths with burning aspirations, tremendous energy, and varying degrees of confidence; just like Parker himself. Similarly, while filming *The Wall* (1982), he got on well, up to a point, with the skin-head extras.

I really loved them because they were *my people*. In between shots they used to make me laugh. Until they started to laugh about how they'd seen this Pakistani kid and had tried to throw him off a moving train. I thought, 'How can people

[25] *Cinema Papers*, July 1985.
[26] Roughly one-third financed by the National Film Finance Corporation, which proved to be one of the best investments they made.
[27] *What's On in London*, 13 Oct. 1978.
[28] Except *The Wall*, which he did not intend to direct when he originally got involved with the project.
[29] *Face*, July 1985.
[30] *Literature/Film Quarterly*, No. 3, 1987.

whose sense of humour is exactly the same as mine on every level, suddenly hit something which isn't?' I found that very disturbing.[31]

Something else which has not changed is the intensity and aggression of Parker's work. 'I think that anger never leaves you, and is in all my films. Sometimes I look at them and think they're too angry.'[32] The morality reflected in his work is that of 'rough justice'. Middle class liberalism is replaced by retribution as the guiding tenet. *Midnight Express* and *Mississippi Burning* are both deeply pessimistic in that the tension in both is released through the revenge the central characters exact on their adversaries. Commenting on Anderson's (Gene Hackman) torturing of a Klan member to get information in *Mississippi Burning*, Parker admits that 'it is old testament justice', but sees it as an emotional lever which is universal in its impact. Violence, or the constant threat of it, is a feature of many of his films, particularly *Midnight Express*, *The Wall*, and *Angel Heart*. 'To me, arguing passionately, sometimes violently, is not such a terrible thing. That's probably why I have become so argumentative in my work. I enjoy seeing what kind of a reaction I can get from someone.'[33]

Parker strives for a seriousness in his work, and sees his films as political in that they express 'some facet of the human condition'.[34] Certainly there is 'injustice' in *Midnight Express*, the hardship of the struggle for success in *Fame*, the necessity and warmth of companionship in *Birdy*, marital problems in *Shoot the Moon* (1982), the sense of individual isolation and attendant frustrations in *The Wall*, as well as racism in *Mississippi Burning*. Yet this seriousness is subordinated to 'entertainment' largely because of his desire to reach the international audience; which implies the predominance of the less sophisticated elements to achieve popularity. But his approach and skills are well suited to this desire to reach a mass market: his visual style, his energy and his clearly defined — if simplistic — sense of morality. Parker reacts against the intellectual and literary traditions of film-making, and is without question a master at arousing an emotional intensity in the audience. Yet it is an ability which must be treated with the utmost care. In *Midnight Express* and *Angel Heart*, and to some degree in *Mississippi Burning*, he shows that it is easier to arouse a superficial gut feeling about a subject than anything more profound; the reactions he calls for have an intensity and anger which can leave little room for rational judgement. Indeed, it is Parker's ability to arouse emotions which has been at the

[31] *Time Out*, 29 Mar.–5 Apr. 1989.
[32] Interview with the authors.
[33] *Cineaste*, No. 2, 1986.
[34] Interview with the authors.

root of those of his films which have achieved critical praise, as well as those singled out by some for condemnation.

Midnight Express (1978) thrust Parker into the mainstream of world cinema. It was a far cry from *Bugsy Malone*: a harrowing account of the actual experiences of Billy Hayes (Brad Davis), a young American imprisoned for life in Turkey for drug smuggling.[35] Subjected to appalling treatment and on the verge of insanity, he is determined to escape—in prison jargon, to take the 'Midnight Express'—which he ultimately does.

It is a harsh, violent, and compelling film, with Hayes starkly portrayed throughout as the victim, subjected to the sadism of his Turkish guards. It lacks both profundity and characterization, but not power. While the opening credits run on a black screen, we hear the rattle of a machine gun and then the laments of a woman in a foreign language—emotive triggers which have nothing to do with the story. No one can deny the brilliance of the work as a piece of film-making. The sets and locations, the old buildings of the prison with its decaying masonry and dark, squalid catacombs, combined with Parker's dramatic lighting techniques, conjure up an atmosphere so far removed from our Western world that the audience instantly empathize with Billy's sense of isolation. Yet, at the same time, viewers could not escape from the fact that they had been emotionally manipulated, that animal feelings had been aroused in them. To take one example, Rifki, a Turkish helper in the prison, lies in order to secure the brutal punishment of Billy's friend, Max (John Hurt). Such is Billy's eruption of blind rage that he bites the Turk's tongue out. Puttnam, the co-producer, was shocked at the audience's reaction to this:

When we made the film it seemed as powerful a way as any of illustrating just how far the human spirit can descend under those conditions. But in the cinema it had (at least to me) a quite unforeseen effect. Some people were so infected by the cliché of brutal retribution for brutal treatment that they leapt to their feet, cheered and applauded.[36]

The film also provoked a public outcry against the apparent blanket condemnation of the Turkish race for its treatment of Hayes. The film's polarity was partly due to exaggerations made by Hayes and partly the result of the script-writer Oliver Stone's concern for a tighter story. The savaging which the film received from many critics, despite popular enthusiasm, surprised Parker. He felt he had made a film about injustice and prison conditions. He said that his ignorance of the racist implications of his work may have reflected his political naïvety, but were in no

---

[35] Based on the book of the same name by Hayes, ghost written by William Hoffer.
[36] *Sight and Sound*, Spring 1989.

Parker working with Billy Hayes on location of *Midnight Express* (1978), the powerful and controversial film which established his international reputation. © Courtesy of Columbia Pictures Industries Inc. Photograph: BFI.

way intentional. 'I'm immensely proud of the film. It's one of the most important films in many people's lives, and I stand by every frame of it.'[37] In contrast, for Puttnam the film represented an important turning point. Through it he saw the impact that films could have and the damage that negative messages could bring, and recognized the need for responsibility on the part of the film-maker. Subsequently, he has always addressed the wider moral issues raised by the films he has chosen to make. Parker acknowledges this change in Puttnam, and sees it as marking a major difference between them. 'David's tended to become more pompous and self righteous because he's gone evangelical. He's an optimist and I'm a pessimist. He wants to make films which will show how human beings can aspire to being better, while I try to find the things that are wrong in the human psyche and explore what's rotten with the world in order to make it better.'[38]

Nevertheless, *Midnight Express* was an immense success. Costing $2.8 million it has grossed about $40 million, winning Academy Awards for the Best Screenplay (Oliver Stone) and Best Original Score (Giorgio Moroder), and Parker himself was nominated for Best Director, although he did not win.[39] The film also won six Golden Globe awards.

Fame (1980) represented another unpredictable change of subject matter for Parker, but an equally successful one,[40] spawning the long-running television series. It is a highly entertaining combination of music and straight drama, telling the story of a group of eager young hopefuls during four years of tough training at Manhattan's High School for the Performing Arts. It shows their enthusiasm and vitality, as well as the personal problems they have to face in their struggle for success—cloying parents, pregnancy, illiteracy, and homosexuality. Parker noted that 'as usual in our work, we have given everything a theatrical edge. We have pushed reality as far as we can push it. On one occasion we have what I call a sort of "choreographed bedlam" out in the street. It's not really real but it's done in a way so you think it could happen.'[41] The contrast between the largely ethnic composition of the school and the classical arts they are studying is shown by means of an effective sequence in which a black student bursts out of a classroom and unleashes his anger on the glass bookcases in the corridor. Suddenly choral music is heard. 'I wanted to counterpoint the raw black anger of the kid. So against that scene we had a piece of choral music by Rossini, "Stabat Mater"—very white classical music which is absolutely the

[37] Andrew Yule, *Puttnam: A Personal Biography.*
[38] Ibid.
[39] This was the first time a British director had been nominated since John Schlesinger won the award for *Midnight Cowboy* in 1969.
[40] Winning Academy Awards for Best Original Score and Best Original Song.
[41] *Screen International*, 16–23 Feb. 1980.

opposite of what you would expect to hear in that situation. But immediately after we finish the scene, we cut to the voices, who are singing it, and they are not white, but black faces.'[42]

The film is less successful when it tries to explore the psychological problems of the students in anything beyond a superficial way. This is done in a theatrical manner reminiscent of the Broadway musical, A Chorus Line—all the main characters are given one scene in which to reveal themselves to the others, telling of their hopes and fears. The problem is that these characters do not evolve, they suddenly burst forth. It is a little too crude. The film also says something about the nature of success, of single-mindedness which all too often becomes selfishness, and of the hardships and humiliations which must be endured to achieve it. One student, Leroy, tries to persuade his English teacher not to fail him, while she is visiting her dying husband in hospital. Eventually she explodes, deploring his lack of consideration. Another student, Coco, is approached by a bogus Frenchman who tries to flatter her into starring in a pornographic film. She apparently accepts this humiliating proposition as a necessary step on the way to success.

Shoot the Moon (1982) marked another change of subject and, more importantly, of pace. It is a marital drama focusing on the acrimonious, and at times violent, break-up of a fifteen-year-old marriage and the subsequent tentative affairs of both partners, complicated by their four young daughters. Parker liked the idea of such a change from his previous work. 'With Midnight Express and Fame we sort of grabbed people by the lapels and dragged them through the film at 100 mph without their feet touching the ground. I wanted to slow down, do something delicate, and show some kind of maturity.'[43] He worked on the script with the writer, Bo Goldman, and the two of them became amateur psychoanalysts, examining each other's feelings in similar emotional situations. 'It was a very personal film for me, about my views of marriage. It was the first time I ever did a film that was close to my own life. There is a lot of me in that story, and making the film served to exorcise some of the things I wanted to say about my life . . . although afterwards I was very angry for revealing so much of myself.'[44]

Starring Albert Finney and Diane Keaton it marked the first of Parker's films in which the main characters were truly believable. The audience could not only identify with an emotion or statement, but with a whole situation. Here the characters determined the development of the plot, and not the plot the development of the characters. It is a stark,

[42] American Film, Apr. 1980. A similar use of counterpoint was used to equal effect in The Mission (1981), directed by Roland Joffé and produced by David Puttnam.
[43] Photoplay (UK), Aug. 1982.
[44] Cineaste, No. 2, 1986.

striking film—a sincere attempt to find deeper, more authentic feelings. 'I believe the film's strength is that it does hit certain truths about all our relationships.'[45]

The film deserved the praise it received from Paulene Kael, until then a vehement critic of Parker's work, who believed that Finney and Keaton gave 'the kind of performances that in the theatre become legendary'. This film would have been his most complete but for its unsatisfactory ending. Finney is beaten up by Keaton's lover, after destroying her new tennis court. The final shot is a freeze-frame of Keaton standing over Finney, whose hand is outstretched. Will they get back together? Who knows? Parker said that his original intention was for Keaton to walk away, to leave him, but decided that this was too cold and callous. Yet she would not really take him back. So he polled the crew. All the men said she should take him back, all the women said she should not. So he hedged his bets. Such indecision seems surprising from a man who usually expresses his opinions with such confidence.

Parker's next film, *The Wall* (made in the same year, 1982) was based on the best-selling album of the same name by Pink Floyd. It was his last film to be made in Britain.[46] It follows the psychedelic story of a rock star's mental breakdown, played by the lead singer of the pop group The Boomtown Rats, Bob Geldof. Over the years he has isolated himself from his wife and fans, and, sitting in a Los Angeles hotel room, he begins to dream of the past. His mind wanders to various incidents in his life, including the tragic death of his father during the war, all of which, in nightmare fashion, build up on top of each other like bricks, which form an enclosing wall from which he must escape. 'It is my angriest film to date. In fact it's almost too angry. When I look at it now, I'm almost embarrassed. It's like when you're in an argument with somebody and you raise your voice and later you wish you hadn't. With *The Wall* I feel I was raising my voice for an hour and a half.'[47] It was not the easiest film to make, as the collaboration between Parker, Roger Waters (the main force behind Pink Floyd), and Gerald Scarfe (responsible for the film's superb animation sequences) seems to have been quite an ordeal. As Parker remarked, 'Can you imagine, three megalomaniacs all used to getting their own way.' After this, Parker took a well-earned sabbatical, returning to feature films the following year with *Birdy*.

Birdy (1984) is undoubtedly the most accomplished film of Parker's career. He turned the book down when it was first offered to him in

[45] *Cineaste*, No. 2, 1986.
[46] Parker and Marshall kept on their offices at Pinewood Studio, where they had worked since *Bugsy Malone*, until 1985 when they left the UK, transferring their base to the US. Marshall returned to the UK in 1987, after the break-up of their long-term partnership.
[47] *Cineaste*, No. 2, 1986.

1978, but after it had been rewritten,[48] took it up in 1983. It is a sensitive and moving account of how two teenage boys, Birdy (Matthew Modine) who is obsessed with birds, and his closest friend Al (Nicholas Cage), grow up in a working-class district of Philadelphia and are changed by their involvement in the Vietnam war. Both are traumatized in different ways. Al is badly burned, while Birdy, suffering from shock, is confined to a mental institution, where he retreats into a private world in which he becomes one of the birds which had so fascinated him as a teenager.

The film has a warmth and sense of humanity lacking in most of Parker's previous work. The characters are well developed and believable. Parker is able to show that his great ability is to construct a visual atmosphere attuned to the emotions he is trying to evoke—in this case a feeling of compassion. He recognizes this, believing that 'Birdy is my most complete film.'[49]

After the problems caused by the ambivalent ending of *Shoot the Moon*, Parker vacillated over *Birdy*'s. 'Everyone from my children to my agent had begged me not to follow my normal macabre route and kill them. But the dramatic line that we were following seemed, partly at any rate, to lead to this conclusion.'[50] As with *Shoot the Moon* the ending remained undecided until about two weeks before it was due to be shot. 'Films have a way of taking on a life of their own and consequently endings tend to present themselves.'[51] The final sequence shows the two boys escaping from the mental institution. They reach the flat roof, and as Al blocks the door against the attendants who are chasing them, Birdy is seen on the outer wall of the building with his arms in the air. From Al's point of view it looks as if he is about to leap to a certain death. His shouts of restraint go unheeded and Birdy jumps. Al runs up and peers over the wall to see that Birdy has, in fact, only jumped to a lower level. The film ends with the bemused look on Al's face. This final moment of warmth, given the charm of their friendship at the beginning of the film, is particularly appropriate. *Birdy* received the best reviews of any of Parker's films and won the Grand Prix from the Cannes Special Jury, to a standing ovation.

Angel Heart (1987) is as violent as *Birdy* was gentle. It is a synthesis of the genres of the classic *film-noir* detective story and the Faustian tale. A private detective, Harry Angel (Mickey Rourke) is hired by Louis Cyphre (Robert de Niro) to find someone who disappeared without paying a debt. As Harry delves into the case, encountering devil worship

---

[48] By Sandy Kroopf and Jack Behr.
[49] *Cineaste*, No. 2, 1986.
[50] *Screen International*, 4–18 May 1985.
[51] Ibid.

and a series of brutal murders, he is ultimately struck by the revelation that both the missing person and the elusive murderer are, in fact, a subconscious side of himself. He owes the devil his soul, and Parker uses the image of a huge, descending, metal-grilled elevator to conjure up the final price he must pay.

'I was interested in doing a film-noire, because it was what people like me were brought up on. It gives you the chance to play around with atmospherics . . . I wanted to make a black and white movie, but had to shoot it in colour, so what we did was shoot with regular stock and drain the colour out of what we were filming.'[52] The visual virtuosity, particularly in terms of camera movement, is astonishing. The menacing atmosphere is created by moody images using subdued lighting. Some of these are the most striking of Parker's career. All provide a visual and emotional intensity which temporarily blinds the audience and delays any qualms about whether the story actually fits together. The tortuous plot is difficult to take seriously, and Harry Angel is essentially a two-dimensional vehicle for the action scenes, stunning as they are.

In *Mississippi Burning* (1989), Parker has combined his talents for creating atmosphere and emotion with a serious political subject—a fictional account of the FBI investigations into the murder of three civil rights activists by the Ku Klux Klan in Mississippi in 1964. The result is a work of power and substance. As usual there is a strong visual style, full of evocative images. However, they are sometimes a little too dogmatic, too consciously 'set up', detracting from the impact of the film (for example, the sequence when a black father is hanged by the Ku Klux Klan outside his burning house).

For added appeal, there is the classic struggle between the two principal detectives. Ward (Willem Dafoe) is a fervent young officer determined to play by the book, while Anderson (Gene Hackman), with the experience of middle age, realizes that the rules must occasionally be bent. Although Parker handled these topics well, the film aroused controversy on the grounds of historical distortion. Anderson terrorizes a Klan member to extract information, which in reality was gained by a bribe. The work of black activists in tackling racism is not shown, but the role of the FBI is emphasized as is the passivity of the Mississippi blacks. But to Parker, 'It is a polemical film; which is one to create debate, and it has done that'[53]—not least by stressing economically motivated racial hatred. In one scene, Anderson explains to Ward his own redneck background. His father had lived by the dictum, 'If you ain't better than a nigger, who are you better than?' 'Where does that leave you?' Ward asks. 'With an old man so full of hate he didn't know

[52] *City Limits*, 1–8 Oct. 1987.
[53] *Film*, Apr. 1989.

Parker rehearsing with Matthew Modine for the final scene from *Birdy*, 1984. © Courtesy of Columbia Pictures Industries Inc. Photograph: BFI.

Gene Hackman and Willem Dafoe in *Mississippi Burning*, 1988, Orion Pictures Corporation © 1988. Photograph: BFI.

that being poor was what was killing him', Anderson replies. Parker actually rewrote much of Chris Gerolmo's script and one senses that there is much of himself in Anderson. As he says of his own working class background, 'class bigotry and economic inequalities were very much part of my life, and something about which I felt strongly.'[54]

With *Mississippi Burning*, Parker has developed a potent style. He has synthesized greater sophistication—in terms of weightier subject matter handled with maturity—with his visceral approach, already tremendously popular with the audience. This seems to be the direction Parker will take in the future. He has just completed *Come See the Paradise*, about the Japanese interned in America during World War II. He is now (July 1990) working on *The Commitment* (due to start filming in Dublin in the summer of 1990) about a group of unemployed young people who decide to form a rock band. After that he plans to make a film version of *Les Misérables*.

Mississippi Burning marked the end of Parker's long-term partnership with his British producer, Alan Marshall.[55] Early in 1989, Parker also signed a deal with an American studio, Tri-Star Pictures, agreeing to give them a first option on projects developed and directed by him over the next three years.[56] Basing himself permanently in Los Angeles, Parker has effectively broken his remaining ties with Britain. As with so many British directors, the appeal of America has proved irresistible.

[54] Parker's production diary.
[55] The film was produced by Robert Colesberry, who had incidentally been assistant director on *Fame*.
[56] These films are to be produced by Colesberry.

ALAN PARKER

Would you say that your early childhood was a particularly formative period?

I find it so boring talking about my childhood because it always comes out like that Monty Python sketch, the one where they say, 'We were so poor we used to have to lick the gravel off the road for breakfast.' You see I came from a working class background; my father worked for the *Sunday Times*, first as a store-man and then in the transport department, and my mother was a dress-maker. Our politics were strongly working class socialist. I grew up in post-war London—in a council flat in Islington. I got a scholarship to the local grammar school, Owen's, which was fortunately the best school in the area. With both parents working and as an only child I probably spent a lot of time on my own and in the streets, although in no way would I describe myself as a tearaway—my parents were too strict for that, and the work ethic in our family too strong.

How did this upbringing affect your personality?

I think it was the basis of almost all of my subsequent attitudes. On the Islington street corners I was always the one with the big mouth which invariably got me into, and by necessity out of, trouble. I'm a pretty 'up front' character; outspoken, vulnerable, volatile. Sometimes I wish I could be more guarded, but it's not my nature.

I feel that at this time I was scarred by the odious British class system, and the scars have taken a long time to heal. But I was well prepared for the future. I grew up among a very articulate and angry bunch of people who were a lot smarter than anyone I've encountered since. Sadly, they couldn't escape.

Was cinema important to you at this time?

Well, I was always obsessed with cinema. My influences were anything and everything I could devour at the local Odeon and the many fleapits in north London, and my greatest love and fondest memory was Saturday morning pictures. On television my influences were Garnett and Loach.

Were you influenced by 'The Angry Young Men' of the early sixties—Lindsay Anderson, Tony Richardson, and Karel Reisz?

No. People think that I must have been, but it's not true. My work is much more influenced by the generation *before* them, which includes Carol Reed[1] and David Lean.[2] I know every shot of *The Third Man*[3] and *Great Expectations*[4] by heart.

Is your anger still there, and has it affected your films?

Well, I think the original rage has calmed down to a more controllable anger these days. But it does tend to break out at the slightest scratch. That kind of anger never leaves you, and I think that it's in all my films. Sometimes I look at them and think they're too angry.

Do you see yourself as a rebel?

Not consciously. I do feel very strongly about things that I believe in, and it's true that I can be difficult. But a rebel is how other people see me, not really as I see myself.

Would you say that there are any continuous themes running through your films?

I've never understood the philosophy that a director should make twenty versions of the same film throughout his career. But, every country I go to, and every journalist or critic I talk to, seems to have a different theory on the thematic constancy of my films. One of them is that people are often trapped in situations from which they are trying to escape, whether trapped by marriage in *Shoot the Moon* or by an environment in *Birdy* or by real prison walls in *Midnight Express*. But of course it's not in all my films, so I don't know how important it is to me. I never look at different stories and consciously think of thematic continuity. In the end it can only be subconscious.

What then has to appeal to you in a potential project?

Good question, but I don't know the answer. All I know is that when an idea or a book or a script comes along with the magic ingredient, you know it immediately. But it's an elusive quality which I have great difficulty describing.

[1] Renowned British director. He reached his peak in the late 1940s with films such as *Odd Man Out* (1948), *The Fallen Idol* (1948), and *The Third Man* (1949). His later work, generally large projects undertaken in Hollywood, were disappointing.
[2] Extraordinary British narrative director. For most of the 1940s and 1950s he directed intimate dramas; *Brief Encounters* (1945), *Great Expectations* (1946). With *Bridge on the River Kwai* (1957) he abruptly moved onto more epic productions with which he has since been identified.
[3] Made by Carol Reed in 1949.
[4] Made by David Lean in 1946.

Why is it so important to you that all your films should reach a mass audience?

There is undoubtedly a thrill about seeing a whole bunch of people lining up for your films round the block in Rio or Tokyo or Paris. For me a film doesn't exist in a can or as a column in a newspaper. It lives only when an audience reacts to it. And to have as wide an audience as possible, from many different backgrounds and sensibilities, go through this experience is the very reason that I make films.

I feel that the film-maker has a responsibility to find a large audience, and certainly this is true in America. Maybe this means you have to change the way in which you make the film, but personally I don't think there's anything wrong with this, because I think that what you do is so hard and you spend so long on it that if you don't find an audience it's kind of irrelevant. Besides, there are many directors like Woody Allen[5] and Martin Scorsese[6] who work under that commercial umbrella and who do very individual and very good work.

How do you aim to do this? There does seem to be a marked development through many of your films, from the aggression of Midnight Express *to a more subtle and effective mix in* Birdy.

Well, I think my films often operate on an emotional or visceral level, which obviously opens up the audience possibilities. In *Midnight Express* I tended to grab the audience by the lapels, drag them through the movie, and chuck them out of the cinema at the end, drained and exhausted. But I'm very proud of that film and knowing what I know now, which I might not have known then, I don't think that I would have done it very differently. The film has certainly had its effect.

With *Birdy* maybe I didn't shout so much, but I suppose that as you get better at what you do, you learn when to shout and when not to. There's nothing wrong with raising your voice now and again, it depends very much on whether it matters to you if people are listening or not.

One also senses that the need for recognition is very important to you?

Well, I'm pretty proud of what I do. The emotional and physical commitment is so great for all film-makers that it's nice to have a pat on the back every now and again. To make a film you have to operate on an overdrive of self-confidence, which I've always had. But it probably masks all of your insecurities, of course—and recognition helps this. I've

---

[5] While many of Woody Allen's works reflect his particular brand of Jewish humour, he has developed into a sophisticated psychological film-maker with films such as *Hannah and Her Sisters* (1986), and *Crimes and Misdemeanors* (1989).

[6] Scorsese's films include *Mean Streets* (1973), *Taxi Driver* (1976), *Raging Bull* (1980), *After Hours* (1985), and *The Last Temptation of Christ* (1988).

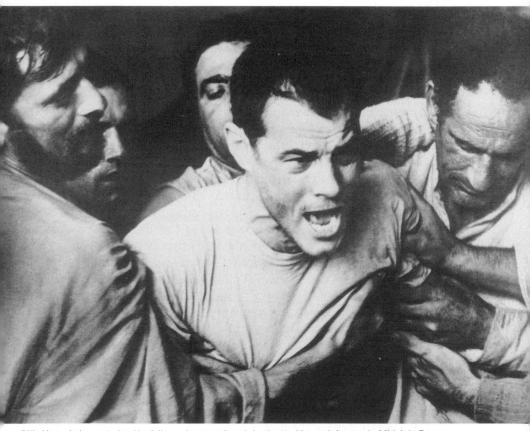

Billy Hayes being restrained by fellow prisoners after violently attacking an informer, in *Midnight Express*, 1978. The preview audience's unexpected applause for Hayes's action deeply affected the film's producer, David Puttnam. © Courtesy of Columbia Pictures Industries Inc. Photograph: BFI.

also been pretty fortunate in that recognition has allowed me to continue making films. A lot of fine directors haven't been so lucky.

Yet you give critics quite a rough ride. Why despite your remarkably successful career are you so concerned with what they say?

I think that it matters less now than it did when I started. My films are seen in forty or fifty countries, and there's a hundred critics in every one. Some like what you do, some don't. So I've learned to be philosophical and pragmatic about it now. As I said in *Fame*: 'No true artist should be concerned about what other people think of him or her, a custard pie comes with the job.' But I wrote that long before I was able to trust in it.

You see, with the early films we had no support from anyone in England. I wanted the creative integrity of our work to be recognized in England as it had been elsewhere, and as a hooligan from Islington I was programmed to throw a brick at anyone who threw one at me. Also I think that time has proven that critics weren't really ready for what was going to happen to cinema. Over the last fifteen years a revolution has occurred. My generation of film-makers were suddenly able to make films which broke the chauvinistic, nationalistic, British mould. We made films to be seen all over the world, and we suffered for it, because it was somehow vulgar to do that. But film criticism, like the films, has of course changed drastically since.

You've used your sharply humoured cartoons to attack your pet hates, including the critics. How much are they a means of getting something out of your system, and how much are they quite simply a tease?

I think the cartoons are partly a wind-up. I tend to get a perverse pleasure out of putting the boot into certain established film preconceptions. But they do also allow me to articulate certain opinions I have without resorting to a long polemical essay, that I may have trouble reasoning out anyway.—One-liners are much easier, and they cut quicker although not deeper.

But some of them, while cutting and humorous, are quite cruel.

They're only spindly little scribbles on a page with a speech bubble, they're not meant to be cruel—just a gentle nudge at the critics, wafflers, spivs, and academic clerks. I just used humour to make a point; to prick the balloons of pomposity and dishonesty that were floating all over the place. They were my mouthpiece for a while. Really, during the days when the opinions and views of non-film-makers were all you could read or hear, I thought it's about time that the people who actually made the films spoke up, and I had this facility, which I used. It was just throwing back a few bricks, which isn't so necessary now, not for me anyway. I haven't done any for a year or two, except in personal letters to people, which is how it started originally.

The British film industry has come in for numerous attacks by you over the years. What problems do you see plaguing it?

Well, I got fed up going to seminars on 'The future of the British film industry', and 'How to improve the British film industry', it's gone on for as long as I've been directing. It seemed everyone talked too much and did very little in practical terms. There's still a very narrow, parochial attitude left over from the years of academic posturing and general myopia.

Our problem is one of leadership, because God knows there's enough talent in England. This continual conflict between Wardour Street bookies and the BFI clerks has confused any government help that might have occurred, and stifled any real progress. The truth is that the talent has gravitated to television, and the TV 'drama on film' that Loach and Garnett pioneered now gets a limited theatrical life in a few cinemas around the world. Anyway, with all those satellites buzzing around in space, soon Britain might not even need a film industry, so hopefully then they'll stop wondering or caring where it went.

Which new British directors' work impresses you?

I'm greatly impressed by the work of David Leland[7] and Stephen Frears, but of course Stephen started before I did so he's hardly a new director, although it seems that the world has just discovered him. But the trouble with many British film-makers is that while they may have great intellectual depth they have no cinematic breadth. They have little understanding of what cinema can do because they are hidebound by an intellectual and literary tradition. I think that Peter Greenaway's films are the biggest con in a long time.

How detailed a vision do you have of a film? Some directors see their role not really as creators, but facilitators, providing a framework flexible enough for others to do the actual creating. Stephen Frears would be a British example.

I think a director should have a video tape of the finished film in the back of his head before ever beginning. Stephen is a magnificent director, and a refreshingly modest one. His contribution is total, but he recognizes honestly that no film in history was made by one person, and that the work of cinematographers, editors, production designers, has been given

[7] Writer/director/actor. He wrote *Personal Services* (1987, dir. Terry Jones), co-wrote *Mona Lisa* (1987, dir. and co-written by Neil Jordan), wrote and directed *Wish You Were Here* (1987), directed *Checking Out* (1989, his first film made in America), and has made *The Big Man* (1990) for Palace Pictures.

a raw deal by film writers who are forever on the periphery of the creative process. But I would say that the major difference between directors is in their involvement with the writing of a film.

You write a lot of your own material. When did you start and why is it so appealing?

I love writing more than anything because it's just you and the page. I tend to get more satisfaction out of writing a good scene than from shooting it.

   Actually, this all started because David Puttnam and Charles Saatchi took me into the 'Kebaba-Mousse' in Charlotte Street[8] in about '68, and asked me to write a film script.[9] I'd written a few things before but never had the courage or self-confidence to do this. So I had a go. There's a great strength in naïvety.

You've also worked with other script-writers: Oliver Stone on Midnight Express *and Bo Goldman on* Shoot the Moon. *Has this kind of collaboration been effective?*

Working with Oliver wasn't a bundle of laughs. I wasn't sad to see the back of him and I'm sure that the feeling was mutual, but he did write a great script. Bo Goldman is also a wonderful writer, and my collaboration with him was I think, the closest and among the most enjoyable I've ever had. I learned a lot from him about my craft and myself.

You seem to remain remarkably flexible while shooting your films; the endings of both Shoot the Moon *and* Birdy *remained fluid for a long time. How important is such flexibility to you?*

Once I've written 'Final Shooting Script' on the first page, I stick pretty close to it. But a film is an organic process, and often things tend to be revealed to you as the story evolves and the actors seek out their own truths. I think you must be able to think on your feet, and to have an open mind to changes right up until you send the negative off to be cut; although there is of course the danger of improvising to death.

Do you use a storyboard?

No I don't. I think that's the way you'd make movies in Dagenham. I very rarely miss a shot. I can shoot a very complicated sequence and hold every shot in my head, because a lot of the film is at the back of my eyeballs before I start. You must have a clear view of your intentions, but a lot of magic does occur at the time, which mustn't be dismissed just because it's different to how you originally conceived it.

[8] A restaurant in Soho, London's film district.
[9] *Melody* (a.k.a. *S.W.A.L.K.*). Made in 1971, directed by Warris Hussein.

Does this apply to the acting as well? How do you get the best out of the actors?

Well, all actors are different, so it's misleading to generalize too much. I think that the important thing is to create an atmosphere of trust, where they can feel free to give of their best, to explore different avenues without fear, and to reveal themselves and their characters without me being too judgemental. Their eyes always flicker towards you after a take and you have to let them know you're on their side. It's too bloody hard for them otherwise. I rehearse wherever possible, but in an exploratory way, discussing every scene and problem which might ensue many weeks later. Although if you rehearse too much it becomes a theatrical repetition and the very spontaneity of the film process is lost.

Do you prefer working with American or British actors?

Well again, it's dangerous to generalize, but I enjoy working with American actors because I feel their approach is more suited to film, they tend to bring the part they're playing much closer to themselves and their own personalities. British actors of course hate this because its not really acting, they think that they can play anything as long as they stick on a false nose or have a club foot. The theatrical cleverness of British actors tends not to suit the camera which is able to reveal the difference between truth and artifice. I think with many British actors you do tend to admire the performance rather than believe it, which is often not the case with American acting.

Why on most of your films have you consistently used your British crew?

I've grown comfortable working with certain people because I happen to believe that they're the very best at what they do. It's so hard making a film that it's nice to be among friends. It also means that the creative process is a little more fluid because of the rapport—we all speak the same language.

But you retain firm control?

I'm very strong when it comes to making a movie, and *nobody* tells me what to do. It's *my* film and no one else's. I'm very egocentric and megalomaniac about that. I believe that the director has the most important part to play, although of course, only one part. Michael Winner[10] once said that democracy on a film set is sixty people doing what he tells them to do. Michael may be wrong about virtually every-thing else, but he might be right about this.

[10] British director who made TV films for the BBC in the early 1960s. Since the early 1970s he has worked frequently in Hollywood. His films include *The Big Sleep* (1978), and the first three *Death Wish* films (1974, 1981, 1985).

Your partnership with Alan Marshall, your producer, lasted a long time.

He didn't produce *Mississippi Burning*, but our partnership did last eighteen years, which is extraordinary by film industry standards where people's relationships tend to self-destruct after one film. We stayed together so long because we were very compatible. We grew up making films together, there was no conflict of interest or of egos. His great common sense and basic wisdom were very important to me, and his abilities freed me to concentrate on the creative side of things, secure in the knowledge that he was always there to back me up and make things happen.

Do you still find it hard to raise the money for your films?

It's not a big problem, but asking anyone to part with the many millions of dollars that a film takes to make is never easy, especially large budgets. Big budget films become double-edged swords. They do allow you more to spend on the production, and most importantly, time to plan and shoot correctly. If you look at the work of Bertolucci, Storaro, and Scarfiotti[11] on *The Last Emperor* (1987), which is in my opinion the very finest film-making, it just cannot be achieved inexpensively. Of course if you spend more, there's a certain pressure on your films finding an audience if the backers are going to get a return. But just because you spend more money on a film doesn't mean that you automatically compromise its creative integrity.

Why in recent years have you worked almost solely in the States, and now seem to be set to stay there for the foreseeable future?

I know it's fashionable to dislike America, but I like working there, it's more in tune with my personality. I find it stimulating and have always had encouragement, which I've rarely had in England. I do miss my friends in England, and the irreverent sense of humour that kept us all sane. But, ultimately we're gypsies. It's a big world and I'll go wherever I think I can do my best work.

Do you have any plans to branch out into any other media apart from film?

Well, Placido Domingo asked me this week to do an opera at the Los Angeles Opera and I'm very tempted, although terrified. I don't think I've got enough 'O' Levels. I'd like to do more documentaries[12] and perhaps eventually write another novel. I get asked to do music videos every week, but *The Wall* was enough for me. So, although I do think

---

[11] Director, cameraman, and designer, respectively.

[12] Parker's first was *A Turnip Head's Guide to the British Cinema*, 12 Mar. 1986 (Thames Television).

it's important to try as many different things as possible, it's hard when a film takes so much of your time.

Would you say that you were happy with your work to date?

It's foolish to say that you couldn't go back and do something better the second time around. But overall, when I look back on the last twenty years I'm pretty pleased with what I've done and *how* I've done it. That's not to say that I don't pinch myself every day at how lucky I've been. But I've got rid of the fear of someone tapping me on the shoulder and whispering in my ear 'the game's up son, get back to Islington.'

Any regrets?

Having thrown back so many bricks.

FILMOGRAPHY

BRIEF BIOGRAPHICAL DETAILS

Born 14 February 1944. He was brought up in the working-class area of Islington, north London. Educated at Owen's school, Islington. He was an advertising copy-writer from 1965 to 1967, and a commercials director before he made his first feature film in 1976. He has published a novel of his film, *Bugsy Malone* (1976), and a book of cartoons, *Un Cartoon de Alan Parker* (1982).

FILMS DIRECTED BY ALAN PARKER

1973   *Our Cissy.* 30 mins. EMI/Alan Parker Productions. SCRIPT: Alan Parker. CAST: John Barrett, Ian East, Ellis Dale, Michael O'Hagen, Graham Ashley.

   Footsteps. 33 mins. EMI/Alan Parker Productions. SCRIPT: Alan Parker. CAST: Gemma Jones, Rose Hill, Robert Bridges.

1974   *No Hard Feelings.* 55 mins. Alan Parker Productions. (Shown on BBC TV in 1976.) SCRIPT: Alan Parker. CAST: Anthony Allen, Mary Larkin, Joe Gladwin, Kate Williams.

1975   *The Evacuees.* 75 mins. BBC TV. SCRIPT: Jack Rosenthal. PRODUCER: Mark Shivas. PHOTOGRAPHY: Brian Tufano. PRODUCTION DESIGNER: Evan Hercules. EDITOR: David Martic. CAST: Maureen Lipman, Ray Mort, Gary Carp, Margery Mason, Steven Serember.

1976   *Bugsy Malone.* 99 mins. Bugsy Malone Productions in association with the National Film Finance Corporation and Rank. SCRIPT: Alan Parker. PRODUCER: Alan Marshall. EXECUTIVE PRODUCER: David Puttnam. PHOTOGRAPHY: Michael Seresin and Peter Biziou. PRODUCTION DESIGNER: Geoffrey Kirkland. ART DIRECTOR: Malcolm Middleton. EDITOR: Gerry Hambling. MUSIC: Paul Williams. CAST: Scott Baio, Florence Dugger, Jodie Foster, John Cassisi.

1978   *Midnight Express.* 120 mins. Casablanca Record and Filmworks Production for Columbia Pictures. SCRIPT: Oliver Stone (based on the book of the same name by Billy Hayes and William Hoffer). PRODUCERS: David Puttnam, Alan Marshall. EXECUTIVE PRODUCER: Peter Guber. PHOTOGRAPHY: Michael Seresin.

PRODUCTION DESIGNER: Geoffrey Kirkland. EDITOR: Gerry Hambling. MUSIC: Giorgio Moroder. CAST: Brad Davis, Randy Quaid, John Hurt, Bo Hopkins.

1980   *Fame.* 134 mins. MGM. SCRIPT: Christopher Gore. PRODUCERS: David de Silva, Alan Marshall. PHOTOGRAPHY: Michael Seresin. PRODUCTION DESIGNER: Geoffrey Kirkland. EDITOR: Gerry Hambling. MUSIC: Michael Gore. ASSISTANT DIRECTOR: Robert Colesberry. CAST: Lee Curreri, Paul McCrane, Irene Cara, Antonia Franceschi, Gene Anthony Ray, Laura Dean, Barry Miller, Maureen Teefy.

1982   *Shoot the Moon.* 124 mins. MGM. SCRIPT: Bo Goldman. PRODUCER: Alan Marshall. PHOTOGRAPHY: Michael Seresin. EXECUTIVE PRODUCERS: Edgar J. Scherick, Stuart Miller. PRODUCTION DESIGNER: Geoffrey Kirkland. ART DIRECTOR: Stu Campbell. EDITOR: Gerry Hambling. CAST: Albert Finney, Diane Keaton, Peter Weller, Diana Hill.

      The Wall. 95 mins. Tin Blue Ltd./Goldcrest Films. SCRIPT: Roger Waters. PRODUCER: Alan Marshall. EXECUTIVE PRODUCER: Stephen O'Rourke. DIRECTOR OF ANIMATION: Gerald Scarfe. PHOTOGRAPHY: Peter Biziou. PRODUCTION DESIGNER: Brian Morris. EDITOR: Gerry Hambling. MUSIC: Pink Floyd. CAST: Bob Geldof, Kevin McKeon, Bob Hoskins, James Laurenson, Eleanor David.

1984   *Birdy.* 120 mins. A and M/Tri-Star-Delphi 3/Columbia. SCRIPT: Sandy Kroopf and Jack Behr (based on the novel of the same name by William Wharton). PRODUCER: Alan Marshall. EXECUTIVE PRODUCER: David Manson. PHOTOGRAPHY: Michael Seresin. PRODUCTION DESIGNER: Geoffrey Kirkland. ART DIRECTORS: Armin Granz, Stu Campbell. EDITOR: Gerry Hambling. MUSIC: Peter Gabriel. CAST: Matthew Modine, Nicholas Cage.

1986   *A Turnip Head's Guide to the British Cinema.* 60 mins. Thames TV. SCRIPT: Alan Parker. PRODUCERS: David Gill and Kevin Brownlow. EXECUTIVE PRODUCER: Catherine Freeman. PHOTOGRAPHY: Ted Adcock, Chris Ward.

1987   *Angel Heart.* 113 mins. Winkast—Union/Carolco International. SCRIPT: Alan Parker (based on the novel *Fallen Angel* by William Hjortsberg). PRODUCER: Alan Marshall, Elliot Kastner. EXECUTIVE PRODUCERS: Marrio Kassai, Andrew Vajna. PHOTOGRAPHY: Michael Seresin. PRODUCTION DESIGNER: Brian Morris. EDITOR: Gerry Hambling. MUSIC: Trevor Jones. CAST: Mickey Rourke, Robert de Niro, Lisa Bonet, Charlotte Rampling.

1989   *Mississippi Burning.* 121 mins. Orion Pictures. SCRIPT: Chris Gerolmo. PRODUCERS: Robert Colesberry and Fred Zollo. PHOTOGRAPHY: Peter Biziou. PRODUCTION DESIGNER: Phillip Harrison, Geoffrey Kirkland. EDITOR: Gerald Hambling. MUSIC: Trevor Jones. CAST: Gene Hackman, Willem Dafoe.

FILMS SCRIPTED BUT NOT DIRECTED BY PARKER

1971   *Melody* (a.k.a. *S.W.A.L.K.*). 107 mins. Sagittarius/Hemdale/Good-time Enterprises. DIRECTOR: Warris Hussein. PRODUCER: David Puttnam. EXECUTIVE PRODUCER: Ron Kass. PHOTOGRAPHY: Peter Suschitsky. SECOND UNIT DIRECTOR: Alan Parker. CAST: Anthony Allen, Mary Larkin, Joe Gladwin.

THEATRE

1983   *Alfie*. Liverpool. SCRIPT: Bill Naughton. CAST: Adam Faith.

BIBLIOGRAPHY

Cinema TV Today, 8 Mar. 1975. Article on Parker's work for television.

Screen International, 11 Oct. 1975. Article on Parker's film background and his first feature film *Bugsy Malone*.

Daily Express, 12 June 1976. On *No Hard Feelings*.

Sunday Telegraph, 13 June 1976. On *No Hard Feelings*.

Time Out, 23–9 July 1976. Article/interview about his start in advertising, and about *Bugsy Malone*.

Guardian, 27 July 1976. Career article by Derek Malcolm.

Film Review, Sept. 1976. Interview about the problems encountered during the making of *Bugsy Malone*.

Alan Parker, *Bugsy Malone* (Armada, 1976). A novel of the film.

Screen International, 18 May 1978. Parker talks about *Midnight Express*.

BFI: Special Collection on Midnight Express, 1978.

Time Out, 11–17 Aug. 1978. Interview in which Parker discusses the reasons why his film *Midnight Express* has been misunderstood by the critics.

Daily Express, 16 Feb. 1979. On the making of a TV commercial.

Film and Television Technician, Apr. 1979. Interview with Parker on the move from making TV commercials to feature films, the adjustments which have to be made in that transition, the racist charges against *Midnight Express*, and his next project, *Hot Lunch* (*Fame*), to be made in America.

Daily Mail, 5 Apr. 1979. 'Why is Britain's Oscar hope getting the cold shoulder?', by Margaret Hinxman.

Screen International, 16–23 Feb. 1980. Parker talks about *Fame*.

American Film, Apr. 1980.

Focus on Film, Apr. 1980. Interview on *Fame* and *Midnight Express*.

The Times, 23 July 1980. Nicolas Wapshott on *Fame*, and Parker's relationship with the US.

Films Illustrated, Aug. 1980. Parker talks about the making of *Fame*.

BFI: Fame special collection, 1980. US reviews and interviews.

New York Times, 22 Jan. 1982. Chris Chase. 'Director with a whole world to explore' on *Shoot the Moon*.

Screen International, 13–20 Feb. 1982. Parker talks about *Shoot the Moon* and *The Wall*.

Design and Art, 21 May 1982. On the current state of advertising directing.

Observer Magazine, 30 June 1982. Ray Connolly on the autobiographical nature of *Shoot the Moon*.

NFT Booklet, June 1982. Note on Guardian lecture and a retrospective on his films.

Daily Mail, 12 July 1982. Margaret Hinxman, 'Another Brick in a Wall of Anger', on *The Wall*.

New Musical Express, 24 July 1982. Richard Cook, on his career, film critics, and *The Wall*.

Screen International, 17–24 July 1982. Parker talks about making *The Wall*.

Cinema (UK), Aug. 1982. Parker talks about the influences on his career, how he started as a director, and the films he has made.

Photoplay (UK), Aug. 1982. Parker talks about the autobiographical nature of *Shoot the Moon*.

American Cinematographer, Oct. 1982. Parker gives his personal view of the Pink Floyd LP *The Wall*, and an account of the making of *The Wall*.

Un Cartoon de Alan Parker (Alan Parker Film Company, 1982). A collection of Parker's cartoons.

AIP & Co., June 1983. Parker's view of the 1983 Cannes film festival.

Sight and Sound, Summer 1983. Selection of cartoons by Parker on aspects of the film business.

Guardian, 6 Oct. 1983. Robin Thornbes, 'What's it all about Alan?'—about Parker's stage debut with *Alfie*.

Screen International, 18–25 Aug. 1984. Parker talks about *Birdy*.

Alan Parker, *A Film Diary* (Rank Film and Television Services, 1984). A diary featuring Parker's cartoons.

Screen International, 4–18 May 1985. Parker gives a step-by-step account of the making of *Birdy*.

Screen International, 1–8 June 1985. Parker's anger at leading UK's film critics' inaccurate reporting of the prize *Birdy* won at Cannes.

Face, July 1985. Parker on British film and his career.

Cinema Papers, July 1985. Parker talks about his career.

Encore, 1–13 Aug. 1985. Parker talks about the problems encountered in using a Skycam during the filming of *Birdy*.

Sunday Times, 9 Mar. 1986. 'Parker's Poison Pen'. Parker on his first documentary, *Turnip Head's Guide to the British Cinema*.

Televisual, Mar. 1986. Comment on Parker's contribution to the Thames TV series *British Cinema: Personal View*.

Screen International, 3–7 May 1986. Parker talks about the production of *Angel Heart*.

Stills, Nov. 1986. Location report on *Angel Heart*, with comments from Parker.

Film Directions, No. 31, 1986. Michael Open defends Parker's work in the face of general hostility from many critics.

Film Directions, No. 32, 1986. Sinead Jones praises Parker's work and discusses *Angel Heart*.

Cineaste, No. 2, 1986. Excellent interview on Parker's films, background, and anger.

Screen International, 20 Dec. 1986–3 Jan. 1987. Parker and Marshall decide to leave Pinewood, where they have worked since 1975.

Films and Filming, Sept. 1987. Part 1 of a production diary by Parker on the filming of *Angel Heart*.

Films and Filming, Oct. 1987. Part 2 of Parker's diary on the filming of *Angel Heart*.

City Limits, 1–8 Oct. 1987. Interview with Parker about the making of *Angel Heart*, and his reaction to the cutting of one scene in America to get it past the censors.

Literature/Film Quarterly, No. 3, 1987. *Birdy: The Making of the Film—Egg by Egg*. Production notes by Parker.

American Film, Jan.–Feb. 1988. Interview with Parker about his career.

Screen International, 28 Jan.–4 Feb. 1989. Parker signs his first contract with an American studio, a three-year 'first option' deal with Tri-Star. Films are to be produced by Robert Colesberry.

Time Out, 29 Mar.–5 Apr. 1989. Interview.

Film, Apr. 1989.

Sight and Sound, Spring 1989. An interesting review of *Mississippi Burning* by Sean French.

9 NICOLAS ROEG

Nicolas Roeg at the time of editing *The Witches*. © H.P. Productions.

ESSAY ON
NICOLAS ROEG

NICOLAS ROEG is a sad-eyed man, who chain smokes his Gitanes cigarettes and talks about himself in a soft-spoken, polite, English tone. He expresses himself cautiously and without great self-confidence, tending to say 'one does . . .' when he means 'I do . . .', and not finding it easy to discuss his movies—'It's hurtful to expose your genuine thoughts.'[1] He is rather a loner: 'I don't have a very big social life, I don't like meeting a lot of people and sometimes I reveal myself through my work more than I mean to.'[2] His films have never made much money and some have in fact lost a great deal. He has never won a single award (although as he once pointed out, 'I won a gym prize at school—I don't know how or why because there was no gym in the school'). Yet in the sixties he was one of Britain's major cinematographers and over the last twenty years he has made a full length feature film on average every two years. Although his first film, *Performance* (1970), had some good reviews, it was greeted by *Variety* with the damning condemnation that it was a substandard British thriller with sadism and sex hypes, which should have been shot on 16 mm.[3] Yet today many critics would vote it one of the best British films ever made.[4] All his films are very personal in nature and raise more than their fair share of controversy—people eager to take almost a political stand over what they think 'film' should be about are quick to label him as 'pretentious', or alternatively as 'a genius'. Meanwhile Roeg continues to make his films.

His interest in film began at an early age, although it was a long time before he would direct. At school he had attempted to run film societies, in the army he was the unit projectionist, and when he left in 1947 he got a job, using a contact of his father's, dubbing French films into English in the cutting room at Marylebone Studios. In 1950 he moved to working on the studio floor as a clapper boy, and started the long apprenticeship up to lighting cameraman. He knew that he wanted to make films and admired the gumption of his colleague

---

[1] *Screen International*, 27 Mar. 1976.
[2] *The Times*, 15 Feb. 1987.
[3] *Variety*, 31 Aug. 1970.
[4] See the survey in Gilbert Adair and Nick Roddick, *Night at the Pictures* (Columbus Books, 1985).

Walter Lassally[5] who worked his way up to cameraman much more quickly and more forcefully. But he was quite content with the slow and steady progress which he was making, interestingly having no great creative force within him driving him on. 'I never felt a burning ambition and just enjoyed being part of the film which was being made.'[6] By the 1960s he had become a lighting cameraman and came to work closely with several distinguished directors, and gradually showed greater discrimination over which films he wanted to be involved in. He was cinematographer on Truffaut's *Fahrenheit 451* (1966), which had an important influence on him and which he greatly enjoyed, partly 'because it was a film very much to be "read" in terms of images'[7] . . . 'I thought that Truffaut was extraordinary and for the first time in all those years, I felt that I didn't mind making movies for someone else.'[8] The following year he worked for John Schlesinger on *Far from the Madding Crowd* (1967)—'I think that that film was underestimated. John tried to capture the feeling of the seasons through a rather leisurely pace, at a time when audiences were accustomed to another pace.'[9] The look of the film was however greatly praised. Other significant films which he shot included *Nothing But the Best* (1964) for Clive Donner, and *A Funny Thing Happened on the Way to the Forum* (1964) and *Petulia* (1968) for Richard Lester.

Since then Roeg has directed ten feature films, a television film (*Sweet Bird of Youth*, 1989), a concert movie (*Glastonbury Fayre*, 1971), a segment for the operatic collection *Aria* (1987), as well as numerous commercials. 'What I am trying to do, as is anyone who works in any form of art, is to express an emotion.'[10] But his underlying aim is a deeper one—that greater 'authenticity', or 'truth', or 'reality' can be achieved through emotional rather than, for example, chronological continuity. He hopes to penetrate the complexity of human beings. 'Usually producers read scripts and they want something rooted in the reality they know. I'm more anxious to look for what we don't know'[11] . . . 'My ambition is to go on looking at human relationships. We're all very complex, all of us, desperately complex. And I want to get closer in

---

[5] Walter Lassally started as a clapper boy at Riverside studios in 1946. He filmed his first feature *Another Sky* in 1954, and has since then photographed over fifty films including *Loneliness of the Long Distance Runner* (1962), *Zorba the Greek* (1965), and *Heat and Dust* (1982), as well as numerous documentaries (especially for the Free Cinema directors of the 1950s).

[6] Interview with the authors.

[7] *Sight and Sound*, Winter 1984/5.

[8] *Sight and Sound*, Winter 1973/4.

[9] *Films and Filming*, Jan. 1972.

[10] *The Times*, 2 May 1983.

[11] *Guardian*, 22 Mar. 1976.

a visual way, to how human beings actually behave in front of each other. I'll never get that one straight. I'd stop if I did—I'd have run out of breath.'[12]

His work, highly individual in both form and content, has made a significant artistic contribution to British film, yet as he said in 1985, 'There is a lot of nationalism about the place right now. It's just a cock crowing on its own dung heap . . . I'm just not interested'[13] . . . 'I suppose that really I am outside the mainstream of British cinema.'[14] This is largely because he has felt it to be too conservative and that until recently the British have not sufficiently accepted film as a fully independent cultural form.

His attitudes towards film form or film grammar have been motivated by a desire to break down barriers and push forward its possibilities — 'Film is one of the newest arts, it should also be one of the freest . . . I would like to have written on the front of all cinemas "Abandon all preconceptions, ye who enter here".'[15] The preconceptions and barriers which he objects to are those of the literary and theatrical tradition, which only the younger generation are now fully escaping. What he hopes to replace this with is derived from his desire to gain a 'reality' which must be reflected in the grammar of the film. He looks to the oral tradition of story-telling, where tales have no particular beginning or end. His attitude is a little like Goddard's answer when challenged that a film should have a beginning, a middle, and an end—'Yes, but not necessarily in that order.' He believes that, 'Life isn't like a Galsworthy story or a Priestly play, that flows, on and on, like this: a, b, c, and then she marries him, and so on. It happens in little jagged moments. Accidental encounters and random fortuitous events guide our lives.'[16] Moreover a finished story, he believes, leaves out so much. He wants the characters in his story to be real in that they are haunted by a past and the film is, in a sense, not their final resting place. The story should be unfinished and its development should not be linear.

This attitude is also reflected in the way that he frequently cuts back and forth, through time and place, within the story itself, in a style that was described by Peter Cowie as 'idiosyncratic'.[17] But Roeg has defended himself against criticism. He is trying to capture the random, jagged quality of life, and the way that people's thoughts and ideas can leap around—'I don't think my films are messy. To me they are quite

---

[12] *Independent*, 18 Feb. 1987.
[13] *Time Out*, 8–14 Aug. 1985.
[14] Interview with the authors.
[15] Interview with the authors.
[16] *Interview*, Mar. 1976.
[17] Peter Cowie, *International Film Guide* (Tantivy Press, 1981).

precise.'[18] In a justly famous scene in *Don't Look Now* (1973), Roeg cuts together shots of John and Laura Baxter making love with shots of them dressing for dinner afterwards, in order to convey their marital love. Another example is in the opening credits of *Performance* where the car taking the lawyer—whom the underground criminal, Chas, will later intimidate—to the courts, is intercut with scenes of Chas and his girlfriend making love and then leaving for work. The brilliance with which these scenes are cut together almost suggests a background in editing rather than camera. Curiously, the occurrence of this sort of intercutting seems to have decreased in his more recent films.

Film for Roeg is an independent art form, and he is acutely aware of its visual language—'I've always wanted to get my thoughts over in films visually.'[19] Certainly his films are well framed, as is to be expected of someone who spent twenty years behind the camera. But it goes deeper than that—and the camera work itself is generally not primary in his mind. Pauline Kael commented on him that he had more visual *strategies* than any director she could think of. So although he spends a great deal of time and effort in pre-production, deciding on the details of how the film will look, he doesn't storyboard much. He prefers to remain flexible, often re-shooting extensively from different angles 'so as to get as much truth out of the situation as possible'.[20] When editing, if he sees an expression on an actor's face which he likes, he may decide to re-shoot or try and adapt the character in the story to suit the new reality—changing the dialogue to suit the look.

Likewise, a great deal of attention is spent on props: 'Props are terribly important, they are part of us all. You are rarely surprised with people's houses although they are always fascinating to go into.'[21] So, for instance, in *The Man Who Fell to Earth* (1976), the professor is a simple but ambitious man, and into his study Roeg put a molecular model—'You knew he would have it out there to prove to himself that he was a professor and to let visitors think that he must be very clever to understand all that.'[22] Another example is the very specific demands he had with John and Laura's house in *Don't Look Now*: 'The house is half wood and half brick; it was right in line with the idea that this couple hadn't finally made up their minds about how they wanted to live . . . They were half way all the time.'[23]

Roeg considers it highly important that every line of dialogue, piece of music, and every object relate to the actual purpose of the film as well as,

[18] *The Times*, 10 Apr. 1980.
[19] *American Film*, Jan.–Feb. 1980.
[20] *Show Biz*, 21 Mar. 1980.
[21] *Time Out*, 12–18 Mar. 1976.
[22] Ibid.
[23] *Sight and Sound*, Winter 1973/4.

in the case of props generally, to the internal logic of the scene. In *The Man Who Fell to Earth* we see a picture of Breughel's *Fall of Icarus*—a metaphor for the lead character's own dilemma. In *Performance* the camera zooms through the gunshot wound in the pop star's head and we see a picture of the writer, Borges—a statement about the crisis of identity faced by the characters in the film. In *Track 29* (1988) the TV cartoons playing in the background of many of the scenes all seem to relate to themes in the film, such as motherhood. In *Eureka* (1982) when Gene Hackman discovers his gold in the snow wastelands, the music that we hear is, appropriately, Wagner's *Ring*.

In terms of content, Roeg starts, not with a plot but with a premiss— some aspect of the human condition. Again and again two fundamental themes crop up—the quest for self-identity, and the nature of love between man and woman. Most often he likes to place the characters involved against a foreign or abnormal background. 'Drop someone into an alien environment and their predicament stands out so much more clearly. When they have no one else to lean on, no friend to turn to, or equivalent thing of daily routine, it suddenly outlines how particularly strange most people's relationships are.'[24] So, for example, in *Performance* Chas, an East End criminal, finds himself trapped in the very different world of the house of the pop star Turner. In *The Man Who Fell to Earth* Newton is an alien unable to return home. *Castaway* (1987) looks at a man and a woman isolated on a desert island. This technique also gives Roeg an effective way of expressing his views on the way commercial society encroaches and imposes itself on human beings.

The theme of 'identity' and 'image' is an important one for Roeg. He described his own life as 'crawling from one shell to another, each time undergoing an emotional change with the new shell'.[25] *Track 29* (1988) explored the idea of the difference between how people should act and how they really do, with particular reference to our perceptions of what is proper adult behaviour, and why people deviate from it. Martin is a young man who behaves like a baby, Linda wears childish clothes, sucks her thumb, and cuddles her dolls, and Henry is a doctor obsessed by toy trains. As Roeg comments,

The film is about how we become the person others want us to be although we stay the child that we originally were. You can become a great brain surgeon or politician—which are intervening titles given to us by others—but immediately the door shuts, out come the model aeroplanes.[26]

Insignificance takes four people who happen to look like Einstein,

[24] *Time Out*, 8–14 Aug. 1985.
[25] *NME*, 14 May 1983.
[26] *Listener*, 26 May 1988.

Marilyn Monroe, McCarthy, and the baseball player DiMaggio, and explores their real identities.

The second major preoccupation is that of love, about which he has a very pessimistic attitude. He believes that 'Passionate love and sensuality are not easy to rationalise. Man and woman would have to be born under exactly the same circumstances—probably incestuously—to have that total understanding.'[27] He has even gone so far as to say, 'I don't think there is any relationship which is comfortable, do you? I mean what would that be? Dead. Or else so deeply secret that it would be unfathomable.'[28] The sex scenes in his films are intimate yet somehow cold. None of the relationships shown in his films ends with man and woman happily together. At best they are physically parted by circumstance or death. More often the ending involves rejection or shows the incompatibility of the partners. Even if the endings are on an upbeat of optimism, as in *Walkabout* (1971), *Don't Look Now*, or *Track 29*, the overall feeling in the films is of the failure of love. 'People are always accusing me of not making optimistic films. They say they never have a happy ending. I don't see that at all. I think all my endings are happy. No one ends up in a mad house. Death, of course, is not an unhappy ending. If we really thought that, then we would be in deep shit.'[29]

Roeg's first film was *Performance* (1970), which he co-directed with Donald Cammell (who had a background in painting and also in script-writing[30]), and which was to become something of a cult movie in later years. The nature of the partnership has been much talked about. But out of loyalty to each other neither of them has ever really disclosed exact details of their contributions: 'The main thing about any collaboration is secrecy. Happily the secrecy has been maintained against all attempts at division. What would it gain people to know who did what? What is on the screen is on the screen.'[31] It was a unique collaboration; after the film they discussed working again but decided against it.[32]

The film's production history is complex. Creative Management Associates, an agency for writers, producers, actors, and so on, had decided in 1967 to go into producing films. Sanford Lieberson was given the task of creating a film for their client, Mick Jagger, to star in. He

---

[27] *Time Out*, 4–10 Apr. 1980.
[28] *Time Out*, 8–14 Aug. 1985.
[29] Ibid.
[30] He scripted *Duffy* (1968, dir. Robert Parrish). In 1977 he directed *Demon Seed* (starring Julie Christie).
[31] Interview with the authors.
[32] The line in *Eureka*, 'There are no partners', was a personal reference by Roeg to this point.

Albert Einstein (Michael Emil) and Marilyn Monroe (Theresa Russell) reverse roles in *Insignificance*, 1985. © Zenith Productions Limited. Photograph: BFI.

invited Donald Cammell to write a script, and a story outline was drawn up. James Fox, who was to play the best role in his career, was given the other lead part; Roeg came in to co-direct, and Warner Brothers–Seven Arts to finance and distribute it (they were ultimately swayed by Mick Jagger's involvement, whom they saw as a guide to the youth market). It was a film specifically written with the social and psychological changes of mood of the 1960s in mind. Shooting began in Autumn 1968 with £400,000, and was done on location over eleven weeks. Problems arose with Warner's when they saw some of the more permissive rushes, and feared an 'X' certificate. They objected to its lack of story-line and the decadent behaviour of the characters, and the film itself was seized on completion of the shooting. But when Warner's itself was taken over by Kinney National Services, Roeg and Cammell were given the go-ahead for editing. Money dried up again as Warner's once more got cold feet, but fortunately not too many changes were demanded by them before it was released in 1971—nearly two years after it had been shot. Even then more minutes were cut by the censor. Roeg commented on these years, 'It was frightening—as if one had committed a terrible crime.'[33]

The film, whose original title was 'The Performers', deals with the interchangeability of image and identity whilst looking at drugs, sexuality, big business, and crime *en route*. It brings together two characters and two worlds—Chas (James Fox), a criminal from the underworld, and Turner (Mick Jagger), a bizarre, extravagant superstar. Chas is a mixture of strengths and weaknesses, terrorizing anyone who crosses his gangland boss. But they fall out and Chas flees to the basement of a house, where he hopes to wait, find a disguise, and escape. But Turner, a misfit and recluse from society, who lives there with two bisexual girl-friends, is also in search of a new identity. Bored with his lifestyle he begins to question its decadence and function. Turner becomes fascinated with Chas and his self-confident criminal image. Chas, initially disgusted by Turner's 'freak show', becomes more and more trapped as Turner tries to 'get inside his mind'. He is disorientated and drawn into their world, his old personality is bled from him, and his repressed self comes to the fore. Eventually the gang catches up with him. But before they take him away he shoots Turner dead. Yet, when we see the final image of the film it shows Turner, not Chas, being driven away; the change has been completed. Roeg had in this story the perfect vehicle to exhibit many of the attitudes and preoccupations visible in his later films.

The film's references and its influences, and the topics discussed, are wide ranging. When questioned why he thought it was so popular today,

---

[33] Interview with the authors.

he replied, 'Probably because of the issues which it had a blunt look at.'[34] There is calculated confusion and ambiguity, and at times it certainly does seem blunt and self-conscious. But this is perhaps to be expected from two directors making their first film. The first half hour, which shows Chas at work, gives a superb sense of the nature of organized crime in London. However, the constant cross-cutting between a court adjudicating on a commercial fraud case and Chas intimidating a bookmaker so that his boss can have a 'business merger' with him, in order to portray the parallels between the criminal world and legitimate society, is a little heavy handed. Likewise various scenes in Turner's house showing his decadence are more over the top than is necessary. However, Roeg's and Cammell's style was mature. Every scene is dense with meaning; there are numerous metaphors and symbols for all the changes of identity and the transformations that are occurring (hence, for example, the constant references to mirrors). Moreover, the symmetry of the elements, and the exchange of opposites on all their different levels, make it a film worthy of being watched and rewatched numerous times.

The collision and interchange of different worlds is also the dominant element in Roeg's next film *Walkabout* (1971);[35] as the copyline on the poster read—'The Aborigine and the Girl 30,000 years apart . . . together.' The story is a simple one. A girl (Jenny Agutter) and her young brother are left stranded in the outback after their father kills himself. They survive and get back to civilization because of the help of a young aborigine man with whom they are unable to talk, neither speaking the other's language. The girl, very much from the civilized and urban environment, a primeval man, and a child so young that he does not fully belong to either, act together and so, in a sense, become the most basic family structure of man, woman, and child. Roeg wanted the film to be a fable for life; to show the 'principles of life which somehow get obscured by civilisation under the guise of sophistication'—that man's sole duty to the world is survival and the continuation of mankind. The aborigine fails in his attempts at courtship, which are just met by the girl with total incomprehension, and he kills himself. The final shot shows the girl in her middle class house fantasizing about swimming naked back in a natural pool in the desert, while her husband recounts his minor business triumphs.

Roeg felt that there were more surfaces to *Walkabout* than to *Performance*, and it is clearly more concerned with love than the quest for identity. Its easily understood story made a good foundation for him to

[34] Interview with the authors.
[35] He had in fact tried and failed to get it off the ground before *Performance*.

build these other levels and contributed to making it his best film. It shows the indifference of one person to another, using the clash of cultures as an evocative setting. It is beautifully filmed; its beginning has the camera taking us from a busy city, tracking across a blank brick wall to the open desert that is all around. The end of the film has elements of civilization creeping gradually into the wasteland—discarded equipment, an old house, and finally the city itself is shown.

Don't Look Now (1973), based on the Daphne du Maurier short story, is Roeg's warmest view of the possibilities of love between two people. Shot over eight weeks, it tells of a devoted couple, John and Laura Baxter, played by Donald Sutherland and Julie Christie. They go to Venice, portrayed as a decaying city, to get over the death of their daughter, but meet someone who claims to be 'in touch' with the dead girl, and the couple try to bring her back to life through the paranormal. The film is full of enigmatic portents, premonitions, and flashbacks, which both John and Laura misinterpret. The final mistake is lethal; John becomes haunted by a figure in a red coat that he believes might be his daughter. When he finally confronts it, however, it turns out to be a manic dwarf, who like something from a nightmare, stabs him to death. The ambiguity—of the unexplained events themselves, and of the uncertain involvement of the supernatural—imbues the film with a sense of foreboding and menace. As one line in the film said, 'nothing is what it seems.' But despite the fact that it was a critical success, and was felt by many people to be one of the most original and exciting films of the last twenty years, it failed financially.

The Man Who Fell to Earth (1976) is about Newton (David Bowie), a stranger from another planet. Like *Performance*, it is concerned with a man outside society, who, as the film progresses, grows more like the people around him, more corrupted. He becomes trapped on earth, unable to return home, although never giving up hope. He has a futile relationship with an earth girl in which the barriers of understanding shown in *Walkabout* once more come to the fore. Ideas of time are also explored again; Newton himself never ages, and images of past and present coalesce.

This was followed with *Bad Timing* (1980). It is set in Vienna where Alex (Art Garfunkel), a cool, rationalist, chain-smoking psychologist becomes obsessed with the hedonistic fellow American, Milena. Roeg had initially wanted Sissy Spacek to play the part, but she was acting in *Coal Miner's Daughter*,[36] and Theresa Russell was cast. The view of the relationship is bleak, as they are unable to take it beyond the sexual intensity it has. Milena says to Alex, 'I wish you would understand me

---

.[36] Directed by Michael Apted, 1980.

less, and love me more.' The two are totally unsuited to each other, but are unable to separate or detach themselves.

Roeg was deeply upset by the public comments of someone at the distribution company: 'I guess the only time I have been really hurt was when he said, "This is a sick film, made by sick people for sick people." I thought of the people who had worked on it, and the actors, and all that they had put into it . . . it hurt a lot.'[37] The actual structure of the film is based on flashbacks revealed through a police investigation into Milena's suicide. Throughout the film Milena is in the operating theatre, her life in the balance. The detective, Inspector Netusil (which means someone who knows everything save one small detail), is played by Harvey Keitel, cast only a week before filming. He is a sort of *alter ego* for Alex and through his enquiries the whole relationship between Alex and Milena is slowly revealed.

Eureka (1982) had a $10 million budget but it was a financial disaster, in large part due to its abortive release.[38] The film's central character, Jack McCann (Gene Hackman), is someone who fulfils his obsession—absolute wealth—early in life. He is seen in the terrible struggle of searching for gold through the snowy wastelands of Canada. He finds it in a wonderfully visual scene where water, dense with gold dust, literally liquid gold, pours all over him. He has thereby achieved his ultimate ambition, and in a way his life is now over, yet he must live on trying to find some meaning, some new purpose. It is that quest which the film is really about. But to Roeg, this poses a further problem; 'Achievement and separation happen at the same time; one is put on another plane. The actor who wins his Oscar is separated from those who are still struggling for it.'[39] All McCann can do is retire to a Caribbean island as a tormented billionaire; he has an alcoholic wife (Jane Lapotaire), and a daughter (Theresa Russell) with a playboy husband, whose business associates desperately want to discover the source of McCann's wealth. An essential problem of the film, however, was that the story-line was just not strong enough to hold the emotions and themes together.

Insignificance (1985) was an adaptation from a stage play by Terry

---

[37] *NME*, 14 May 1983.

[38] After a brief showing, it was withdrawn in Britain by the distributors, UIP, at the request of its backers, MGM. It was also withdrawn from the video market. Although there have been no formal explanations for this action, there seem to be several possible reasons. First, due to its relatively large budget and its lukewarm theatrical reception, the backers may have wanted to cut their losses, rather than incur the additional expenses of a large number of prints and of advertising, which often costs as much as the film itself. Secondly, the film was caught up in the buy-out of MGM by UA. It is not unusual for the incoming management to distance themselves from projects previously commissioned. Certainly the new studio boss, Freddie Fields, disliked the film. It was eventually handed to the 'Classics' department, which meant that it would only be released in art-house cinemas with little or no publicity.

[39] *NME*, 14 May 1983.

Johnson. However, Roeg's intentions were very different from those of the original play. Once again the film returns to those broad questions about the clash between the concealed identity and the image: '*Insignificance* is about not knowing anything about anyone or anything. How mistaken we are about everyone's identity.'[40] It has more comic moments than his other films, but is not significantly more optimistic. It is about four 'fictional' characters who just happen to resemble Marilyn Monroe (Theresa Russell), Albert Einstein (Michael Emil), Senator McCarthy (Tony Curtis), and the baseball player Joe DiMaggio (Gary Busey). They are archetypes, myths, or icons but the film reverses some of the accepted notions of them. They are given memory and character through flashbacks which reveal them as beset by fear, or guilt, or both.

Castaway (1987) was based on Lucy Irvine's best-selling book of her experiences with Gerald Kingsland when she replied to his advertisement for 'a wife' to spend a year with him on a tropical island. In the story of this one-year challenge, throwing two very different people together (played by Amanda Donohue and Oliver Reed), Roeg provides a metaphor for twenty years of married life. Initially happy in London, when they get to their island in search of their paradise, they drift apart, emotionally and sexually, and only just survive the dangers of the ordeal. But in the end they learn to forgive each other. It offers some very perceptive insights into the dilemmas felt by so many couples. It seems, however, that Roeg's heart was not in the film. Amanda Donohue commented, 'He never explained one facet of my part to me. I personally feel he lost a lot of direction while we were making that story. He was very cryptic. I felt like I was working in the dark.'[41]

Track 29 (1988) was scripted by Dennis Potter[42] who shares, in his writing style, Roeg's dislocated technique. Once again, this well acted film tackles the theme of identity given to us or forced on us by other people. It has been described as 'a post-modernist American update of the Oedipus myth filtered through a flux of shifting fantasy, reality, and memory'.[43] Dr Henry Henry (Christopher Lloyd) plays a pompous doctor in a geriatric hospital. In secret he has a passion for having his bottom smacked by Nurse Stein (Sandra Bernhard), while at home we see him engrossed, hour after hour, with his toy trains. His younger wife, the heavy drinking Linda (Theresa Russell) is desperate for a baby

---

[40] Interview with the authors.

[41] *Time Out*, 22–9 Mar. 1989.

[42] It is an Americanization of one of his sixties television plays. Potter has also scripted the highly acclaimed BBC serials *Pennies from Heaven* (1978), and *The Singing Detective* (1986), as well as *Gorky Park* (1983, dir. Michael Apted).

[43] Graham Fuller in *American Film*, Apr. 1987.

Amanda Donahoe and Oliver Reed suffering from a lack of food and water in *Castaway*, 1987. © Pathé Releasing Ltd. Photograph: BFI.

to give her life in tedious suburban America some meaning. Her long lost son, the demonic Martin (Gary Oldman), the result of her rape as a teenager years before, who had been snatched into adoption, appears on the scene. He turns out to be a product of her deluded imagination—a son who will come and murder his step-father and make love to her. The final ending shows her seemingly escaping from her madness and world of fantasy and also from her husband. But the feeling one is left with is ambiguous—isn't she just putting on a white suit and walking out of the house?

Tennessee Williams's *Sweet Bird of Youth* (1989), made for the American NBC, marked the first time Roeg had directed a television film. Elizabeth Taylor, the film's lead, commented that Roeg 'has a wonderfully macabre sense of humour. He has a jagged blackness in his soul . . . and that dark element is central to fully understanding Tennessee . . . I'm surprised they haven't merged before.'[44] Certainly the play focuses on Roeg's long-term preoccupations with sexuality, youth, and identity. Although the film was slightly disappointing, it did mark an interesting development for Roeg, in that he was turning his hand to filming classic theatre.

Roeg had made *The Witches* (1990), an adaptation of Roald Dahl's best-selling children's story, in 1988, but as with others of his films, its release was delayed.[45] The story is a typically wicked Dahl yarn. An old Norwegian woman tells her grandson, Luke, tales of witches disguised as innocent old ladies, who have a sinister plan to turn children into mice. When the boy's parents are killed in a car crash, he and his grandmother decide to recuperate in an English seaside resort. There they find themselves in the midst of a conference apparently organized by the Royal Society for the Prevention of Cruelty to Children, but in fact, a gathering of the witches, hatching their wicked plans. Luke is turned into a mouse, and sets out on the daunting task of giving the witches their just come-uppance. Roeg was 'fascinated with the idea of the book. It's a wonderful fairy story that also deals with all our basic emotions—jealousy, greed, love, and hate. We are born with these emotions as human beings and children have them exactly the same as adults. Although as we grow older, we tend to see through a glass a bit more darkly.'[46] The cast includes Mai Zetterling, who was lured back into acting after a fifteen-year absence, 'mostly because of my respect for

[44] *Los Angeles Times*, 1 Oct. 1989.

[45] The film was financed by Lorrimer, but during the production, the studio was bought by Warner Brothers. As with *Eureka*, which suffered during the take-over of MGM by United Artists, the delayed release of *The Witches* could have been due to changes of policy brought about by the new owners.

[46] Press Pack for *The Witches*.

Nick'.[47] Although Roeg gathered a familiar crew around him—the writer Allan Scott,[48] director of photography Harvey Harrison,[49] production designer Andrew Sanders,[50] and editor Tony Lawson[51]—the film marks an interesting development for him. It is the first time since *Walkabout* that he has tackled a theme which appeals to all ages, which, with a cast including Rowan Atkinson, seems to be aimed at entertaining a wider audience than with his previous films, and, aided by the late Jim Henson, it is his first foray into puppetry.

Throughout his career, Roeg has steadfastly refused to offer crass alternatives just to satisfy an audience's thirst for an easy solution to a problem. This is itself closely connected to his general refusal to 'rationalize' or over-analyse. Ironically his greatest problem, however, is that he tends to view the situations and dilemmas in his films from afar. The audience knows what motivates the characters, but those motivations are often not fully intelligible or adequate for the actions involved. Of course, people do often act from inadequate or irrational motives, but if one is to empathize with the people in the film, one has to comprehend the motivation, inadequate though it is, as stemming from their inner, essential nature; the character must be portrayed with sufficient richness and depth that one can perceive that that person, being who he is, and what he is, could do no other. Roeg seems to believe that human beings have no inner nature, or identity, or at least that it is unfathomable. The result is that the audience is often left unmoved; we may perceive the misfortune that is happening to a character, but we do not see it as a tragedy because we do not apprehend any inner inevitability. Roeg himself said (with specific reference to *Bad Timing*), 'We are just observers. We can be sympathetic for them (the characters in the film) if we see a reflection of our understanding of human character.'[52] This is important, because it does mean that the audience can be deeply moved when they have faced similar predicaments to those of the people in the film—then, in a sense, the drama takes on a more personal meaning.

Roeg's greatest significance has arguably been in his contribution to film grammar, to breaking down some of the more fundamental and restrictive conventions of cinema. This is particularly evident in the ways in which he has abandoned linear narrative without ceasing to explore the 'human condition', and in the way that he has given his films multiple levels of meaning and symbolism. Unsurprisingly, his ideas

[47] Press Pack.
[48] Scripted *Don't Look Now*, *Castaway*, and Roeg's current project *Cold Harbour*.
[49] Photographed *Castaway* and Roeg's section in *Aria*.
[50] Designer of *Castaway*.
[51] Editor of *Bad Timing*, *Eureka*, *Insignificance*, *Castaway*, *Aria*, *Track 29*.
[52] NME, 14 May 1983.

have not gone uncondemned. In replying to the criticism (specifically of *The Man Who Fell to Earth*) that his films had too many 'ideas', he replied, 'If you go and look at a Bosch painting, you don't say "Oh, he has put too many devils in it, too many angels." That's what it is. You look at it and enjoy it, or you walk down the gallery and see something else.'[53]

[53] *Screen International*, 27 Mar. 1976.

NICOLAS ROEG

How did you get involved in the film industry?

I would never say that film-making is actually my vocation. In fact, I never totally believe those people who do. We just fall into things—odd shifts in life just take you on different paths. I didn't come in wanting to direct. I didn't even know what the structure of the industry was. I had no idea who did what—what everyone's role was. But that was not unusual then. The media wasn't part of one's life in those days. It didn't have, as it has today, a certain social caché. People got jobs in film studios because they lived in Borehamwood or whatever—milkman one day, working in a film studio the next, and something else the next. What first really hooked me into thinking that this was a job that I would like to become deeply involved in was as a young man sitting at an editola at 'Lingua Synchrome', where they dubbed French films into English. Running the films backwards and forwards to get the words right, I realized that film is a time machine. Film is nothing to do with the theatre at all—it has much more potential. The audience can be rushed out of their seats just to have a look at someone's watch, and then back again. I remember when *Last Year in Marienbad*[1] was shown in England for the first time. It had a screening at the Cameo Polytechnic in Regent Street. It was an exciting time. But I remember people saying 'Sasha Pitoeff walked upstairs, but when he walked downstairs he was in a dinner jacket. This man doesn't know how to make films.' But within eighteen months commercials would be showing Mum putting a pie in the oven, and the next shot would be the cooked pie being put down on the table. Before all this it would have shown the family go out, then a subtitle 'four hours later' and so on. Resnais really excited me.

Was he a big influence?

Influence me . . . ? That point was achieved by osmosis. We are creatures of our time—the sum total of our past, and my influences were anything and everything. But he was a confirmation that there are no experts, that there is no one way to do things.

---

[1] 1961, directed by Alan Resnais.

368 · NICOLAS ROEG

It was a long time before you directed your first film. Did you feel creatively stifled?

No, No. When I started, the idea of 'burning ambition' that came in with the 1960s—the idea that Napoleon was a general at 25, rather than the Victorian-Edwardian idea that you weren't a man until you were 40—had not arrived. I enjoyed what I was doing—working as a junior assistant on the camera. I liked watching the scenes develop and being part of something.

Did you experiment with making small films, perhaps on Super 8 mm., on your own?

No, I couldn't afford to. But I did a lot of stills photography, thought a lot, and later started to write some scripts. I then got into directing second unit and waited for a break, which came when Donald Cammell did *Performance* with me.

Since then you've concentrated on directing films. What is it that you like, or find satisfying about this?

I think that I find it very difficult to give the reasons why I make films. When I was working with Harold Pinter on something, he asked me what I would have been if I had been born in the nineteenth century, and of course he would still have been Mr Pinter the scrivener. But what would I have been? I don't know. I wouldn't have gone into the army, or the church; I'm not a scholar, or a painter, or of the theatre . . . Perhaps I would have emigrated to the colonies—I like adventure. It's difficult to say. I'd probably have found something to express myself through—I might have been a gambler on the Mississippi, playing poker and trying to read other people's characters. You just have a job, most people don't analyse why they do it . . . I haven't got hobbies—my job is my life . . . it's a way of life . . . I can't do anything else.

You have said that you actually find the process of making a film a melancholy thing, but what aspects of film do you actually enjoy?

I think any form of reflection in life is melancholic—any form of introspection is *necessarily* melancholic. I also notice with people who see psychiatrists that the period with their analyst is a melancholy time. But it's not melancholic in a sad way . . . I like what I do.

So you would say that all your films are, in some degree, introspective?

With film, certainly the way I approach it, one has to delve into one's life to put 'truth' onto the screen. One delves into one's emotions and tries to translate that to the story one wants to tell. All our imagination is bound by experience. And when all that is ultimately portrayed in the characters of the film, it becomes a melancholic affair.

So is it the drive to explore these things inside you that pushes you into making films?

I suppose so . . .

Don't you find the exploration and expression of your innermost feelings some form of release as well?

It is a release—for a while it's out of one's system. But then after a period of time one realizes that one never did completely resolve the situation or problem, and that there is still another facet to look at. I think that directors basically make the same film all the time—the same things trouble, interest, or involve them again and again.

So would you describe your films as, in a sense, 'autobiographical'?

I don't know if they're 'autobiographical'—that is a nailing word, but they certainly deal with aspects of the human condition that I'm either nervous of recognizing in myself or am sympathetic with. There are things that I can recognize in the films which are part of my character. But straight autobiography is a curious phenomenon—it leaves all sorts of things out.

Is that why so many of your films are adaptations from books, plays, or, in the case of Castaway, *from an actual biographical account?*

Between autobiography and biography, I think that great biography is probably closer to the truth, and even then one only scratches the surface. However, whether one works from a book, an original screenplay, or just a thought of one's own, one must be touched by something in that story, and that thing has nothing to do with the plot. So, for example with *Castaway*, I liked the idea of two actors naked on the empty stage of a desert island with no place to hide from each other—a generically attractive place but in fact really poisonous and dangerous. The film that one finally makes must have some connection with one's own emotions while remaining truthful to the characters and emotions of the story itself. *Don't Look Now* was my adaptation of a Daphne du Maurier story, *The Witches* of a Roald Dahl book, *Castaway* of a true story, and so on. The films are based on the originals in the same way that Van Gogh's *Sunflowers* are based on sunflowers. One can't say, 'That's not like that.' When I did *Don't Look Now*, the producer was very frightened of Daphne du Maurier seeing it—'She'll freak out,' he said. But she did see it and wrote me a wonderful letter—'Dear Mr Roeg, I've seen *your* film of *my* book and you captured in it the emotions of John and Laura.' In the development of the film one always uses one's own experiences or feelings. Moreover, when one is doing an adaptation, changes to the actual events of the plot often bring one closer to the real emotional truth of the situation. The story is just the shell which the characters inhabit. I suppose the great French thriller *Rififi*[2] is the film with the really classic flat plot—it is as tight as can be. It is almost a documentary of a novel. But it didn't deal with the emotional life of the burglars inside that story. I don't like dealing with that kind of thing. I like to find the plot and then find out what makes the people involved in it tick.

But you also tackle some very broad, almost philosophical issues in your films as well as having this psychological element.

The films are about the human condition; an attempt to confront ourselves, to try to understand something of the state of man.

We are all so lonely really . . . Human beings are so lonely . . . Words aren't enough are they? The only thing one can try to do in any work is to attempt to make a contact—even if never a complete understanding at least some form of mutual knowledge—with an audience so that one feels less alone. The films struggle to say, 'Hello? Is there anybody out there who feels the way I do?', and I hope that they also allow the audience to feel less alone because people can also understand through the characters on the screen that other people have that same predicament. So when I receive letters, as I did particularly after *Bad Timing*, which respond to my films in this way, it makes me feel good.

Very satisfying, I imagine . . . ?

Well, satisfying is too conceited a word—it is confirming in some way. It's a nice thing but it's not a big thing to me at all.

Sometimes you must find that people are just not responding.

One always thinks 'Hey, they'll love it' [laughs] . . . And because of that, its an amazing thing afterwards when you find that they didn't . . . If all you are trying to do is communicate but you get this reaction, it really makes you feel lonely. But you shouldn't feel bitter. It also makes one realize that people don't always see it in the way I see it. With *Bad Timing* I thought that the film would be universal in its appeal—from doctor to don—without social or racial barriers. But I soon realized that I was in fact taking too hard a line. People would respond—'Oh, my God, I don't want to have this dilemma brought into my life again'—and this was especially the case for those seeing the film with their partners. Maybe it was my mistake, but you cannot go around thinking, 'I know the audience like this sort of thing so I'll put in a car chase.' That's superficial. One must be as honest as one can about the internals of a

---

[2] 1955, directed by Jules Dassin. It is a classic heist thriller, with a group of thieves getting together for a daring robbery and falling out afterwards.

Nicolas Roeg and Art Garfunkel on location of *Bad Timing*, 1980. Courtesy of the Rank Organisation plc. Photograph: BFI.
Art Garfunkel and Theresa Russell in *Bad Timing*, 1980. Courtesy of the Rank Organisation plc. Photograph: BFI.

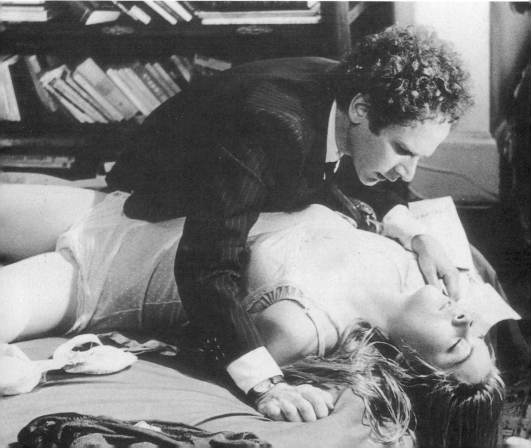

piece. I make films as truthfully as I can. I remember when I was a child, going to see *Lives of a Bengal Lancer*,[3] and thinking 'C. Aubrey Smith'—he was the colonel—'Why wasn't he as important as the young Lieutenant?' These stories were *real* to me.

Do you in any sense see yourself resolving any of these dilemmas of the human condition for the audience?

I present them. There are no answers. I have six sons and three are fully grown. I've been waiting for the question and now, the other night the 5-year-old asked the same thing as the others had done. 'Why am I here, Daddy?' and I reply, 'Because Mummy and Daddy love you, that's why. Now don't you worry yourself about that.' But really it's because of one wild night . . . There is no answer! But as you shut the bedroom door, you see his little eyes looking up at the ceiling saying, 'What kind of answer is that?'

This refusal to give easy answers also reveals itself to some extent in your plots?

The idea that a story has a beginning, a middle, and an end is wrong. A real story continues on either side. The idea of Charlie Chaplin and Paulette Goddard walking off into the sunset [laughs]—that wasn't really the end. There are few stories that manage to end without that marvellous euphemism, 'They lived happily ever after.' But 'they' don't. Life isn't that way. It goes on until death. The beginning is birth. The middle is the events in between.

I think that film has only just reached us. Even now it is not universally thought of as an art form separate from its business side. It takes years and years for this to happen. This must have been true of painting. It probably began as a form of business, scratching on cave walls—'Hey, I like that. I'll give you a chicken if you do that in my cave.' We are scratching around at the surface of the language of film.

Would you say that your films are all tragic in some way?

I'd never make a film without hope.

Tragic with a glimmer of hope? Or perhaps pessimistic?

Well, it depends on what you mean by tragic. I think that my films are funny, in the way that life is 'funny'. Tragedy for me is something which in the end has no hope, like *King Lear*. But I like to be able to save it at the last minute with hope. So right at the end of *Don't Look Now*, Julie Christie stands on the funeral barge of her husband, and she smiles! The audience knows that her life will not be ruined by what has happened. Death is not tragic.

[3] Directed by Henry Hathaway (1934); with Gary Cooper, Franchot Tone, and C. Aubrey Smith.

What about the tragedy of a dying child?

That is a tragedy to the parents. What is really dreadful there is if one says to the child, 'It's a shame that you'll never be able to do such and such.' Today medicine has meant that we expect so much more. In the Middle Ages people were quite happy to die at 30. We would all be in terrible shit if death was 'tragic'—because we are all going to go that way. When Julie Christie smiles she is saying, 'I couldn't have had it forever. My husband and child were wonderful. Life is glorious.' Any other reaction would have been self-pity but that was life triumphant. Let me give you another example. At the end of *Walkabout*, Jenny Agutter stands by the kitchen window cutting up liver. It is the only scene where she is totally at ease. She is happily married—in contrast with her parents at the beginning of the film. She looks back to when she and the aborigine were swimming in the waterhole. She is not being melancholic about what had happened. She was thinking, 'That was a wonderful thing. It has made me who I am. I didn't see it at the time but I can see it now. I look back and can make use of it.' It's the nearest thing to a non-melancholic nostalgia that I can thnk of . . . Between nostalgia and melancholia I myself lean towards melancholia.

Which of your films is your favourite?

That's difficult because they're all related to each other—all facets of, not exactly the same problem, but they're all linked. No one problem, or film, stands alone. All knowledge, all things, are connected in life . . . I don't know . . . it would probably be either . . . I don't know. It's not time to look back. I want to get to the end without looking back . . . *Walkabout* was possibly the most enjoyable actually to make because it was so isolated; so much physical effort; so detached from the technical paraphernalia; so less full of any sophistry in the process of making it; and it was nice to work with the young girl and young boy . . . he was my son. I don't know, where did we start? I've got such a grasshopper brain.

FILMOGRAPHY

BRIEF BIOGRAPHICAL DETAILS

Born in 1928 in London. Educated at Mercers School. In 1947 he began work at Marylebone Studio, dubbing French films and making tea. In 1950 he was hired by MGM's Borehamwood Studio. He worked up the ranks and by 1960 was doing second unit work for *Lawrence of Arabia*. From 1961–8 he was Director of Photography on numerous films, including Truffaut's *Fahrenheit 451* (1966), and Schlesinger's *Far from the Madding Crowd* (1967).

FEATURE FILMS DIRECTED BY NICOLAS ROEG

1970   *Performance*. 102 mins. Goodtimes Enterprises/Warner. SCRIPT: Donald Cammell. PRODUCER: Sanford Lieberson. CO-DIRECTORS: Donald Cammell and Nicolas Roeg. PHOTOGRAPHY: Nicolas Roeg. PRODUCTION DESIGNER: John Clark. EDITOR: Anthony Gibbs. MUSIC: Jack Nitzsche. CAST: James Fox, Mick Jagger, Anita Pallenberg, Michele Breton.

1971   *Walkabout*. 100 mins. Max L. Raab-Si Litvinoff Films Pty Ltd (20th Century Fox). SCRIPT: Edward Bond (based on the novel by James Vance Marshall). PRODUCER: Si Litvinoff. PHOTOGRAPHY: Nicolas Roeg. PRODUCTION DESIGNER: Brian Eatwell. EDITORS: Anthony Gibbs, Alan Patillo. MUSIC: John Barry. CAST: Jenny Agutter, Lucien John, David Gumpilil.

  Glastonbury Fayre. A Crossal Pictures/Goodtimes Enterprises Production. PRODUCERS: Si Litvinoff, David Speechley, David Puttnam, Sanford Lieberson, Nicolas Roeg. PHOTOGRAPHY: Nicolas Roeg, Peter Sinclair, Mike Molloy, Tony Richmond, Frank Simon, Les Parrot, Terry Gold, Eric von Harem Honay, Misha Norland, Anthony Stern. EDITORS: Peter Neal and Elizabeth Kozmia.

1973   *Don't Look Now*. 110 mins. Cassey Productions (London)/Eldorado Films (Rome). SCRIPT: Allan Scott, Chris Bryant (based on the story by Daphne du Maurier). PRODUCER: Peter Katz. PHOTOGRAPHY: Anthony Richmond. PRODUCTION DESIGNER: Giovanni Soccol. EDITOR: Graeme Clifford. MUSIC: Pino Donaggio. CAST: Julie Christie, Donald Sutherland, Hilary Mason.

1976   *The Man Who Fell to Earth*. 138 mins. British Lion. SCRIPT: Paul Mayersberg (based on the novel by Walter Tevis). PRODUCERS:

Michael Deeley and Barry Spikings. PHOTOGRAPHY: Anthony Richmond. PRODUCTION DESIGNER: Brian Eatwell. EDITOR: Graeme Clifford. MUSIC: Various excerpts from classical, pop, and modern music. CAST: David Bowie, Rip Torn, Candy Clark, Buck Henry.

1980 *Bad Timing*. 123 mins. Rank Organization. SCRIPT: Yale Udoff. PRODUCER: Jeremy Thomas. PHOTOGRAPHY: Anthony Richmond. PRODUCTION DESIGNER: David Brockhurst. EDITOR: Tony Lawson. MUSIC: Various extracts. CAST: Art Garfunkel, Theresa Russell, Harvey Keitel, Denholm Elliott.

1982 *Eureka*. 129 mins. Recorded Picture Company (London)/J.F. Productions (Los Angeles)/A Sunley feature for MGM/UA. SCRIPT: Paul Mayersberg (based on the book *Who Killed Sir Harry Oakes?* by Marshall Houts). PRODUCER: Jeremy Thomas. PHOTOGRAPHY: Alex Thomson. PRODUCTION DESIGNER: Michael Seymour. EDITOR: Tony Lawson. MUSIC: Stanley Myers. CAST: Gene Hackman, Theresa Russell, Rutger Hauer, Jane Lapotaire, Mickey Rourke.

1985 *Insignificance*. 105 mins. Zenith/Recorded Picture Company/Island/Palace. SCRIPT: Terry Johnson (based on his play). PRODUCER: Jeremy Thomas. PHOTOGRAPHY: Peter Hannan. PRODUCTION DESIGNER: David Brockhurst. EDITOR: Tony Lawson. MUSIC: Stanley Myers, Hans Zimmer. CAST: Gary Busey, Tony Curtis, Michael Emil, Theresa Russell, Will Sampson.

1987 *Castaway*. 120 mins. Canon Screen Entertainment/United British Artists. SCRIPT: Allan Scott (based on the autobiographical book by Lucy Irvine). PRODUCER: Rick McCallum. ASSISTANT DIRECTOR: Waldo Roeg. PHOTOGRAPHY: Harvey Harrison. PRODUCTION DESIGNER: Andrew Sanders. EDITOR: Tony Lawson. MUSIC: Stanley Myers. CAST: Oliver Reed, Amanada Donohue.

Aria. 98 mins. (directed one of ten segments). RVP Productions. PRODUCER: Don Boyd. ASSOCIATE PRODUCERS: Richard Bell, Luc Roeg. PHOTOGRAPHY: Harvey Harrison. MUSIC: Verdi's *A Masked Ball*. CAST: Theresa Russell, Stephanie Lane, Ron Hyatt.

1988 *Track 29*. 91 mins. Handmade Films. SCRIPT: Dennis Potter. PRODUCER: Rick McCallum. EXECUTIVE PRODUCERS: George Harrison, Dennis O'Brien. PHOTOGRAPHY: Alex Thomson. PRODUCTION DESIGNER: David Brockhurst. EDITOR: Tony Lawson. MUSIC: Stanley Myers. CAST: Theresa Russell, Gary Oldman, Colleen Camp, Sandra Bernhard, Seymour Cassel, Christopher Lloyd.

1989 *Sweet Bird of Youth*. NBC. SCRIPT: Gavin Lambert (from the play by Tennessee Williams). PRODUCER: Fred Whitehead. PHOTOGRAPHY: Francis Kenny. PRODUCTION DESIGN: Penny Hadfield.

EDITOR: Pamela Malouf-Cundy. MUSIC: Ralph Burns. CAST: Elizabeth Taylor, Mark Harmon, Valerie Perrine, Kevin Geer, Seymour Cassel, Rip Torn.

1990   *The Witches*. 92 mins. A Jim Henson Production for Lorrimer. SCRIPT: Allan Scott (from the novel by Roald Dahl). PRODUCER: Mark Shivas. EXECUTIVE PRODUCER: Jim Henson. PHOTO-GRAPHY: Harvey Harrison. PRODUCTION DESIGNER: Andrew Sanders. EDITOR: Tony Lawson. CAST: Angelica Huston, Mai Zetterling, Bill Paterson, Rowan Atkinson, Brenda Blethyn.

OTHER FILM WORK

1950   Second Assistant Camera on *The Miniver Story* (dir. H. C. Potter).

1955   Photographed: *Police Dog* (TV); *Ghost Squad* (TV).

1956   Assistant on *Bhowani Junction* (dir. George Cukor).

1958   Camera operator on: *The Great Van Robbery*; *A Woman Possessed*; *Passport to Shame*; *Moment of Indiscretion*; *The Man Inside* (dir. John Gilling).

1959   Camera operator on: *Tarzan's Greatest Adventure* (dir. John Guillermin); *Jazz Boat* (dir. Ken Hughes); *The Child and the Killer*.

1960   Camera operator on: *The Trials of Oscar Wilde* (dir. Ken Hughes); *The Sundowners* (dir. Fred Zinnemann); *Dr Blood's Coffin* (dir. Sidney Furie).

    One of three directors of 2nd Unit Photography on *Lawrence of Arabia* (dir. David Lean).

1961   Director of Photography on *Information Received* (dir. Robert Lynn).

    Co-writer of *Prize of Arms* with the director, Cliff Owen.

1962   Director of Photography on *Doctor Crippen* (dir. Robert Lynn).

1963   Co-writer of *Death Drums along the River* with Lawrence Huntingdon.

    Director of Photography on: *The Caretaker* (dir. Clive Donner, sc. Harold Pinter); *Nothing but the Best* (dir. Clive Donner); *The Masque of the Red Death* (dir. Roger Corman); *Just for You* (dir. Douglas Hickox).

1964   Director of Photography on: *The System* (dir. Michael Winner); *Every Day's a Holiday* (dir. James Hill).

1965   Second unit director and additional photography on: *Judith* (dir. Daniel Mann); Director of Photography on *Victim Five* (dir. Robert Lynn).

1966   Director of Photography on: *Fahrenheit 451* (dir. François Truffaut); *A Funny Thing Happened on the Way to the Forum* (dir. Richard Lester).

1967   Director of Photography on: *Far from the Madding Crowd* (dir. John Schlesinger); *Petulia* (dir. Richard Lester).

           Additional photography on *Casino Royale* (dirs. John Huston and others).

RECENT COMMERCIALS INCLUDE

Empathy Shampoo, Esso, Moonlight Chocolates, Coca Cola, AIDS (Tombstone) and a commercial for BT shares which was not released because of a fear that it would over-stimulate the market.

ABANDONED FILMS INCLUDE

Early 1970s:

Deadly Honeymoon (sc. W. D. Richter) — cancelled five days before shooting.

Late 1970s:

Flash Gordon (producer Dino de Laurentiis). Script and artwork were completed and the singer Debbie Harry cast. The film was abandoned after disagreements with the producer.

Nabokov's *Despair* — eventually filmed by Fassbinder.

Julie — eventually made by Zinnemann.

Hammett — eventually made by Wim Wenders.

The Strange Voyage of Donald Crowhurst, a story of a lone yachtsman who gave out false information about his position, and died after aimlessly wandering the oceans.

BIBLIOGRAPHY

Daily Cinema, 14 Aug. 1968. Details of career.

Guardian, 3 Apr. 1971. Article by Timothy Wilson.

Films Illustrated, Oct. 1971. Roeg talks about *Walkabout*.

Films and Filming, Jan. 1972. 'Identity', Roeg in an interview with Gordon Gow, including a filmography.

Kinder Marsha and Beverle Huston, *Close Up: A Critical Perspective on Film* (New York, 1972). Essay on *Performance*.

Focus on Film, No. 13, 1973. Biofilmography.

Sight and Sound, Winter 1973/4. The director discusses the making of *Don't Look Now* and describes how individual sequences evolved, with reference to his previous films. Interview by Tom Milne and Penelope Huston.

Cinema Papers, Apr. 1974. Study of his work.

The Velvet Light Trap, Autumn 1974. 'The Search for Self in the Films of Roeg'.

Jump Cut, Sept.–Oct. 1974. Study of the style and themes of his feature films.

Sight and Sound, Autumn 1975. 'The story so far . . . *The Man Who Fell to Earth*: a commentary by the screen-writer', by Paul Mayersberg.

Time Out, 12–18 Mar. 1976. Interview about his films, in particular *The Man Who Fell to Earth*.

Guardian, 22 Mar. 1976. Article by Derek Malcolm.

Time Out, 26 Mar.–1 Apr. 1976. Letter referring to the article in issue 12–18 Mar. 1976 about the artistic genius behind the film *Performance*.

Screen International, 27 Mar. 1976. Interview in which Roeg discusses his work and *The Man Who Fell to Earth*.

Interview, Mar. 1976. Location interview on the set of *The Man Who Fell to Earth* especially about Roeg's attitude to Jagger and Bowie.

Films Illustrated, Apr. 1976. Brief biofilmography.

Film Criticism, Summer 1976. Career notes, influences on his work, survey of his films and unrealized projects.

New York Times, 22 Aug. 1976. Article by Mel Gussow on the cuts made to *The Man Who Fell to Earth*.

Spectator, 9 Oct. 1976. Article by Charles Champlin.

Sight and Sound, Spring 1977. Discussion of Roeg's films.

NFT Booklet, July 1977. 'British Cinema. Part 2: Nicolas Roeg'.

Quarterly Review of Film Studies, Summer 1978. Article on 'insiders' and 'outsiders' in Roeg's films.

Neil Feineman, *Nicolas Roeg* (Twayne Publishers, 1978).

GBCT News, Apr. 1979. Account of an attempt to interview Roeg.

BFI Special Collection, 1980, on *Bad Timing*.

Show Biz, 21 Mar. 1980. James Cameron-Wilson profiles the cinematographer turned director, who discusses film-making in general, and *Bad Timing* in particular.

American Film, Jan.–Feb. 1980. Interview by Harlan Kennedy, 'The Illusions of Nicolas Roeg', in which Roeg discusses the visual style of his films, with particular reference to *Bad Timing*.

Time Out, 4–10 Apr. 1980. Interview about the themes of his films, particularly *Bad Timing*.

The Times, 10 Apr. 1980. Article by Glenys Roberts.

Films Illustrated, July 1980. A look at Roeg's work, with comments from people who have worked with him.

Films, Sept. 1981. Interview on *Bad Timing*, and some of Roeg's thoughts on Freud and actors. Includes filmography.

Peter Cowie, *International Film Guide 1981* (Tantivy Press, 1981). Essay on Roeg.

Guardian, 1 July 1982. Article by Christopher Keats.

Screen International, 24–31 July 1982. Roeg talks about *Eureka*.

City Limits, 11–17 Mar. 1983. Article on a film about Roeg's work.

Film Comment, Mar.–Apr. 1983. Interview on *Eureka*.

City Limits, 29 Apr.–5 May 1983. Roeg talks about his career and *Eureka*.

The Times, 2 May 1983. Article by Clare Colvin.

Time Out, 6–12 May 1983. Interview on *Eureka*.

New Musical Express, 14 May 1983. Article by Richard Cook.

Monthly Film Bulletin (50), May 1983. Interview on *Eureka*.

Video Viewer, May 1983. Assessment of his career as photographer and director.

Photoplay (UK), June 1983. Article on Roeg's career and *Eureka*.

Scotsman, 4 Aug. 1983. Article by Ian Bell.

James Fox, *Comeback: An Actor's Direction* (Hodder and Stoughton, 1983). A chapter on his role in *Performance*.

Sight and Sound, Winter 1984/5. Roeg recalls working with Truffaut on *Fahrenheit 451*.

Stills, June/July 1985. Interview with Roeg on *Insignificance*.

Films and Filming, July 1985. Interview with Roeg on his work and *Insignificance*. Filmography.

City Limits, 26 July–1 Aug. 1985. Roeg and Theresa Russell talk about *Insignificance*.

Monthly Film Bulletin (52), Aug. 1985. Interview with Roeg and Terry Johnson about *Insignificance* and the problems of adaptation.

The Times, 3 Aug. 1985. Article by Glenys Roberts.

Time Out, 8–14 Aug. 1985. Article on Roeg's career and films.

Listener, 15 Aug. 1985. Assessment of Roeg's talents through his film *Don't Look Now*.

Radio Times, 17–23 Aug. 1985. Career article.

Cinema Papers, Sept. 1985. Roeg talks about his career and films.

Broadcast, 24 Jan. 1986. Article on Roeg's recent commercials work for

Robertshaw Needham's Moonlight chocolates.

Sunday Times, 15 Feb. 1987. Article by Minty Clinch.

Independent, 18 Feb. 1987. Article by Paul Nathanson.

New Musical Express, 28 Feb. 1987. Article by Don Watson.

Listener, 26 May 1988. Interview about *Track 29*.

Screen International, 18 June 1988. Location report on *The Witches*.

Monthly Film Bulletin (55), July 1988. Article comparing and contrasting the working styles of Roeg and Dennis Potter.

Time Out, 3–10 Aug. 1988. Criticisms of Roeg's later films.

Hollywood Reporter, 21 Sept. 1988. Comments from Roeg on his planned project, *Chicago Loop*.

Time Out, 22–9 Mar. 1989. Amanda Donohue comments on working for Roeg on *Castaway*.

10 JOHN SCHLESINGER

John Schlesinger on location of *Madame Sousatzka*, 1988. © Curzon Film Distributors Limited.

ESSAY ON

JOHN SCHLESINGER

IN the early seventies Schlesinger was one of Britain's pre-eminent directors. With a string of powerful films behind him, he was respected internationally. Throughout the previous decade he had been seen as a pioneer. His films, which were associated with the 'New Wave', were marked by realism, sincerity, and compassion, reflecting his acute and sympathetic understanding of human behaviour. The Oscar-winning *Midnight Cowboy* (1969) had secured a world-wide reputation, and in the following year he had been awarded the CBE for an outstanding contribution to British cinema. Since then he has frequently directed films in America where he has found it much easier to keep working, but his work has seen some significant changes.

Schlesinger is strongly emotional, 'I react to things intuitively, not intellectually,'[1] and his character is a fascinating mixture of contrasts. He seems confident but full of self-doubt, affable yet capable of being blunt even crude, he can be cynical but is also a romantic. As he admits, 'I'm a very great mixture of things. I have an enormous love of something that's incredibly vulgar, as well as a love of things that are beautiful.'[2] Such feelings have, of course, been the formative influences on Schlesinger's attitude to film.

Born in 1926, Schlesinger made his first film, aged 20, just before he went off to do military service in the Far East. *Horror* is a short melodrama about two escaped convicts, one of whom was played by Schlesinger himself. In the army he served in Combined Services Entertainments, where he was able to carry on his early interests in magic and acting by producing variety shows for the troops. Subsequently, at Oxford University he made two more short, 16 mm. films: *Black Legend* (1948), about a seventeenth-century hanging,[3] and *The Starfish* (1950), a modern fairy tale set in a Cornish sea village. Both had a very small release, and neither did particularly well, although *Black Legend* had some favourable reviews.[4] But Schlesinger's main interest at this time was acting rather than film making. On leaving

---

[1] Interview with the authors.
[2] *Films and Filming*, Nov. 1969.
[3] Starring a University contemporary, Robert Hardy.
[4] Dilys Powell acclaimed it as 'brains not money', and later joked that it was his best film.

Oxford he landed a series of small acting parts, often playing Germans because of his German sounding surname. But he admits that 'I wasn't a very good actor. I wouldn't have cast myself in my own film.'[5] While acting in one small part,[6] he spent his spare time making a fifteen-minute film about a typical Sunday afternoon in Hyde Park (1956). 'It was a fairly candid but lyrical look at people; how they behave, and their reactions to things and to each other.' The film was shown on the BBC, and subsequently Schlesinger found himself working there as a documentary director on the *Tonight*[7] and *Monitor*[8] programmes from 1957 to 1961. As with many feature directors, it was here in television that Schlesinger learned his craft. 'I was regularly thrown out on my own to do a little film, perhaps with one day shooting . . . The speed at which I was obliged to work taught me a sort of basic film grammar, like weekly repertory theatre does for an actor. Also, because first impressions were invariably the last ones, I got used to making quick decisions and was able to sum up an atmosphere in my own way.'[9] It was to be an invaluable skill which he extensively utilized in his feature films.

Schlesinger made twenty-six films for the BBC, all reflecting his curiosity about people. He was sacked from *Tonight* for being too much of a perfectionist, insisting on dubbing his own films, a job not usually done by the director. But he also did not get on with the producer, Donald Baverstock, nor with the assistant producer, Alistair Milne, and admits that after twelve films, 'We parted company and I was relieved.'[10] He felt more at home when he moved on to working for *Monitor* under the paternal guidance of Huw Wheldon[11] – 'my first really marvellous producer'.[12] He then joined *The Four Just Men* series as a second unit director, ending up working with the great Italian director, Vittorio de Sica, whom he admired enormously. The following year he was asked to make *Terminus* (1961),[13] a documentary about a day in the life of Waterloo Station. It focused on life's misfits; the lost boy, the rummaging tramp, the bewildered old ladies. As Schlesinger said, 'We wanted to juxtapose the different kinds of people to be found in a railway station. Under one roof we found all of the misery, happiness, loneliness,

---

[5] Interview with the authors.
[6] In Peter Hall's production of *Mourning Becomes Electra*.
[7] BBC news magazine programme.
[8] BBC arts programme. When Schlesinger left, he was replaced by Ken Russell.
[9] Interview with the authors.
[10] Interview with the authors.
[11] Series editor, and an important figure in bringing 'art' to the attention of the popular audience.
[12] Interview with the authors.
[13] Commissioned by British Transport Films in 1961, it won first prize at the Venice Film Festival, and a British Academy Award.

Formal days: Schlesinger (partly hidden) with a well-dressed camera crew shooting one of his films for the BBC's *Monitor* programme in the late 1950s. The cameraman, Walter Lassally, also worked on many of Anderson's documentaries. Copyright © BBC. Photograph: BFI.

bewilderment and lost people to be found anywhere in the world.'[14] This was the break he needed, and in the following year, 1962, he was asked to direct his first feature film, *A Kind of Loving*.[15]

Schlesinger's passion for film-making derives in part from a desire to communicate emotionally with an audience. 'I'm not interested in being esoteric. I want to make something understandable which communicates human experiences that one can identify with. If there is any common thread in my work, it is that I'm basically very interested in people and their relationship to one another, in complex human beings who never quite get their lives together. All my films have been about people who are pushed to the edge of some sort of experience and who are facing some kind of emotional crossroads.'[16] In particular, he wants to inspire a strong sense of compassion in the audience. It is a common theme of his films that characters unwilling to face up to reality, are forced to confront the truth of their situations. Compromise is often the painful solution for them. 'They're about coping with half measures, or compromising, or what's behind one's fantasies.'[17]

Schlesinger is also attracted by the potential for spectacle in cinema. He wants to entertain in his films, to engage the emotional attention of the audience, 'to disturb, frighten, amuse, or make them think'.[18] This basic desire has encouraged him to attempt thrillers, a comedy, and has even drawn him to work in opera,[19] which he has described as the greatest fantasy of all because it is so entirely theatrical and stylised. With the emphasis on entertaining, he has little interest in being politically, socially, or aesthetically didactic:

I agree with Goldwyn[20]—if you want to send a message send it Western Union. I just like the magic of the world of entertainment. I like the surprise of cinema; that's why I was a conjurer when I was young. I like the way that those old cinemas were built with fanciful architecture and lots of rouched curtains; one went there to enter a magical world, and that's always held an excitement for me.[21]

The mood and approach of Schlesinger's films combine these two potentially divergent interests; his fascination in exploring the complexities of the human spirit, and his desire to entertain. 'There's a mixture of something that's quite sharp, where the darker side of one's nature

---

[14] *American Cinematographer*, June 1975.
[15] He was actually invited to direct by the producer, Joseph Janni.
[16] Interview with the authors.
[17] Interview with the authors.
[18] Interview with the authors.
[19] He has directed three operas: *Tales of Hoffmann* (1980), *Der Rosenkavalier* (1984), and *A Masked Ball* (1989).
[20] A celebrated remark by Hollywood's Samuel Goldwyn (1882–1974).
[21] Interview with the authors.

comes out, and at the same time there is a sort of romanticism—bordering sometimes on the sentimental.'[22]

Sharp realism can be seen in Schlesinger's incisive observations about human nature, particularly the obsessions and weaknesses which torment many of his characters. His treatment of this varies from films such as *The Day of the Locust* (1974), where he concentrates on how callous and bitter people can become, to *Sunday, Bloody Sunday* (1971), where despite the traumas, the mood is one of quiet acceptance. To Schlesinger the actors' characterization and performance is his primary concern. He spends much time working on their specific mannerisms and idiosyncrasies. As Dustin Hoffman observed, 'Schlesinger's the master of subtext. A goddamn butterball. Nobody can touch him when it comes to subtleties and exploration of emotions.'[23] This is partly done through using extensive improvisation with the actors to discuss their characters and motivations, a method first used on *Midnight Cowboy*, and on every film since. 'It was a very worthwhile method of working. We were able to consider what areas were missing from the script, and reshape it accordingly.'[24] Frequently Schlesinger highlights the predicaments which his characters face by using representative motifs: the telephone answering service in *Sunday, Bloody Sunday* epitomizes the way the characters are out of step with each other; the incessant radio and television in *Midnight Cowboy* reflect the need individuals have to stave off loneliness.

His creation of the societies which his characters inhabit is painstaking; there is his stark portrayal of a stifling northern community in *A Kind of Loving* or *Billy Liar* (1963), the rough, callous underbelly of New York in *Midnight Cowboy*, the biting depiction of Hollywood society in *The Day of the Locust*, and the vacuous world of the rich in *Darling* (1965). He draws on his documentary background with an observant attention to details. In *Midnight Cowboy* there is a man lying helplessly outside Tiffany's while passers-by ignore him; or there is the bus which brings the ill-fated Joe Buck to New York and passes a crumbling roadside shack with 'Jesus saves' anachronistically painted on the roof. He is concerned with what is happening in the background of each shot. As his cameraman on *Honky Tonk Freeway* (1981), John Bailey noted, 'Schlesinger's used to creating depth in his pictures; usually as far back as you can see in the frame, there's something happening. That's the way he wants it and rightly so. It was a real challenge for me to work in depth the way he wanted.'[25]

[22] Interview with the authors.
[23] *American Cinematographer*, June 1975.
[24] *Films and Filming*, Nov. 1969.
[25] Schaefer and Salvato, *Masters of Light* (University of California Press, 1984).

There is also the strongly romantic side to Schlesinger's films. It is there in the very subject matter of many of them; in his adaptation of Hardy's *Far from the Madding Crowd* (1967), the romances between American GIs and English women in Britain during the last war in *Yanks* (1978), or the allure of Hollywood in *The Day of the Locust*. Then there are the films which are purely aimed to entertain: his fast-paced action thrillers, *Marathon Man* (1976) and *The Believers* (1987); and his comedy, *Honky Tonk Freeway*. In these three films particularly, Schlesinger's ability to suspend the disbelief of the audience and to entertain them is supported by his high standard of craftsmanship, particularly his editing and lighting.

His romanticism is emphasized by many of his characters, who are larger than life, although extremely believable because of the sensitivity with which they are portrayed. In *Midnight Cowboy*, the central characters' tragic struggle for survival results in an operatic quality, combining romanticism, bordering on melodrama, with sharp and often painful realities. Many of these characters have a romanticized view of themselves, either living fantasy lives in their minds, like Billy in *Billy Liar*, or distorting reality, like Diana in *Darling*, who has an idealized view of her selfish actions.

Over Schlesinger's career, the balance between his realism and his romanticism has shifted decisively towards the latter. In itself, such a transformation need not change the films' potential power. Despite *Yanks*' nostalgia, it is a moving and believable story largely due to the complexity of the characters. What has marked this change in Schlesinger's films is the weak characterizations which have tended to accompany it. There is considerable difference between the depth of the characters in *Sunday, Bloody Sunday* and the shadows that inhabit *The Believers*. Clearer still is the contrast between his two espionage films, *The Falcon and the Snowman* (1985) and *An Englishman Abroad* (1983). Timothy Hutton's character in the former is superficial when compared with Alan Bates's. As a result, the audience merely observe the human dilemmas rather than experience them emotionally. This change is emphasized by the fact that many of his earlier films had been pioneering works in the social issues that they tackled: *A Kind of Loving* looked at the problems of pregnancy trapping an ill-suited couple into marriage in a cloying working class community; *Midnight Cowboy*, with brutal honesty, dealt with vagrancy; and *Sunday, Bloody Sunday* gave a compassionate portrayal of homosexuality.

The most important explanation for this change is Schlesinger's determination to keep working, which is so strong that it dominates his motivations in choosing feature films. He recognizes that this implies undertaking projects which he is less passionate about than others, but regards the rewards of just being able to work as more important. 'I don't think it's possible to have the same level of passion for everything without a terrible period of waiting. I don't want to get rusty waiting for that kind of inspiration. I couldn't bear the idea of not working.'[26] Schlesinger is a workaholic. He has directed thirteen features (the first six films made virtually back to back in nine years), as well as numerous television documentary films, half a dozen plays, three operas, a musical, and countless commercials. But despite this activity, there have been bitterly frustrating periods when he just could not get his feature projects off the ground. Throughout much of the 1970s Schlesinger spent time fruitlessly trying to undertake *Hadrian VII*[27] and *A Handful of Dust*,[28] and what projects were realized saw lengthy pauses in between. After *Sunday, Bloody Sunday* in 1971, it took three years to finance his next film, *The Day of the Locust*. During this time the only film that he made was a short piece on the marathon at the Munich Olympics for the film *Vision of Eight*. He also directed his only stage musical, the disastrous *I and Albert*. Not surprisingly, he was delighted to accept the offer to make his first thriller, *Marathon Man*, immediately after *The Day of the Locust*. There was a similar lean period after the failure of *Honky Tonk Freeway* in 1981. It took four years to get *The Falcon and the Snowman* started, and the only work that Schlesinger did in the meantime was the highly acclaimed *An Englishman Abroad*, and *Separate Tables* for television. Again, he followed this period with another thriller, *The Believers*. In fact, both these thrillers seem to have been attempts by Schlesinger to restore his bankability with studios that had previously backed films which had not done well commercially. *Marathon Man* was made for Paramount, who had backed *The Day of the Locust*, and *The Believers* was for Orion, backers of *The Falcon and the Snowman*. Schlesinger has an extremely strong need to remain active. He has no family of his own, and as his brother says, 'I think John is a lonely person';[29] in Schlesinger's own words, 'Film-making is my life.'[30]

A Kind of Loving proved to be an auspicious entrance into feature films, and marked the beginning of Schlesinger's association with Joseph Janni,[31] the producer of almost all his English films. Alan Bates plays the

[26] Interview with the authors.

[27] A film version of Peter Luke's play about a reclusive eccentric who fantasizes that he has become the first English pope. Schlesinger had even persuaded Dustin Hoffman to play the title role.

[28] Based on the novel by Evelyn Waugh. A film version was made in 1988 by Charles Sturridge.

[29] Roger Schlesinger in *Sunday Times Magazine*, 17 Mar. 1985.

[30] *Screen International*, 1 Nov. 1980.

[31] The Italian producer of many of the British 'Neo-Realist' films of the 1960s including Ken Loach's *Poor Cow*.

young, working class man, fed up with his dull office job, who is trapped into a loveless marriage when he gets his girl-friend (June Ritchie) pregnant. Although the subject matter was typical of British 'Neo-Realist' films at the time, many of which looked at northern, working class life, it was less polemical than most of them. In *Saturday Night and Sunday Morning*[32] Arthur Seaton immediately establishes the tone when he says, 'Don't let the bastards get you down.' But Schlesinger chose to treat the whole subject with sensitivity, focusing equally on the dilemmas of the two protagonists. In fact, *A Kind of Loving* has a greater sense of authenticity than *Saturday Night and Sunday Morning* in that Bates's character, unlike Seaton, is trapped by his chosen method of escape—his vacuous girl-friend. Ultimately there are no easy solutions, and Bates, like a healthy mind within a paralysed body, is forced into a life of bitterness and frustration. The considerable power of the film rests on the sophistication of the characters and the subtlety of the script. Its realism is emphasized through the introduction into the background of the film of issues revolving around class tensions: aspirations for material self-betterment, the sense of claustrophobia in a small community, and the impact of change hitting the industrial north. The film was very successful, grossing three times its cost in Britain alone.[33]

Billy Liar, made the following year, was adapted by Keith Waterhouse and Willis Hall from their original stage play.[34] Billy (Tom Courtenay) is totally out of step with society. Schlesinger saw him as 'a dissatisfied, irritating and imaginative boy, without much talent or true ambition. He lacks the effort to concentrate on anything he really wants to do . . . There seems to me to be so much of all of us in this. It is for me a highly personal theme.'[35] Billy's means of escape is to live a life of fantasy; he may be a humble funeral clerk, but in his mind's eye he is really a rebel, a military leader, or a president. But this is not actually a solution to his problems; merely a means of avoiding them. His girl-friend (Julie Christie[36]), tries to persuade him to leave for London and a new life, but at the last minute he decides to stay at home, consciously preferring the sanctuary of his fantasy world, which is also ultimately his prison. As in *A Kind of Loving* Schlesinger again dealt with the frustrations of northern working class society, although the humour of the film ensures that the mood is more optimistic.

Darling (1965), was the first original script that Schlesinger had

---

[32] 1960, directed by Karel Reisz.
[33] Its budget was £165,000, and it took about £450,000 at the box office.
[34] The play was originally directed by Lindsay Anderson, but he was passed over as director of the feature film.
[35] *Films and Filming*, May 1963.
[36] Brought in after shooting had already begun to replace Topsy Jane after she fell ill.

Alan Bates, the star of many of Schlesinger's films, in *A Kind of Loving*, 1962. He was to appear in virtually the same shot some twenty years later in the closing sequence of *An Englishman Abroad*, 1983. © Weintraub Screen Entertainment. Photograph: BFI.

worked with.[37] Again he cast Julie Christie—this time to play a social
climber, Diana Scott, always dissatisfied with what she has. She passes
from the stable influence of her relationship with a gentle, intelligent,
and respected radio interviewer (Dirk Bogarde) to a sophisticated but
shallow advertising executive, and finally to the apparent peak of success, an Italian prince. To Schlesinger, 'Darling was really about choice,
about a girl who had the possibility of always thinking that there was
something better around the next corner, but was never capable of
settling for anything.'[38] As in Billy Liar the themes of illusion and reality
lie at the heart of this film, and in fact it had much in common with
Fellini's La Dolce Vita, made several years earlier.[39] Darling hit a
responsive chord in Britain, and was a great financial success in
America. In retrospect Schlesinger feels that it has dated badly, and
wishes that he had made the original idea for the film: a true story of a
girl eventually driven to suicide by a syndicate of men who formed a
limited company and set her up as their mistress.

After dealing persistently with British 'Social-Realist' themes of the
1960s, Schlesinger, like Tony Richardson before him,[40] was keen to
make something that the audience and critics would not expect, and
settled on an adaptation of Hardy's Far from the Madding Crowd
(1967). 'At a time when everyone was running around making social
comment films about life as it is now, I thought, "What a relief to get
away and look slowly at people who lived a hundred years ago".'[41]
Although the film was more romantic than anything he had done before,
its melodrama was tamed by the sensitivity of his direction and the
subtle performances of Alan Bates, Peter Finch, and Julie Christie. It was
particularly successful in capturing the mood of the period, and rarely
has the English countryside been so beautifully photographed as here, by
Nicolas Roeg. Schlesinger clearly delighted in exploring the background
characters, and as with Tom Jones there is much humour in their daily
lives and attitudes. But the film was greatly underestimated at the time.
Although it again used the old team of Schlesinger, Janni, Raphael,
Christie, and Bates, part of the problem with the film was its striking
resemblance to Darling in theme, though set in nineteenth-century
countryside. In retrospect, Schlesinger thinks that it was 'too slavish to
Hardy. We didn't take enough liberty with the film because we were too
worried about taking liberties with a classic. In this way it was my least

[37] Written by Frederic Raphael, who later wrote the screenplay for Far from the Madding
Crowd.
[38] Alexander Walker, Hollywood, England (Harrap, 1974).·
[39] La Dolce Vita, directed by Federico Fellini, 1960.
[40] Tony Richardson made an adaptation of Henry Fielding's Tom Jones (1963) after a series
of 'Neo-Realist' films: Look Back in Anger (1958), A Taste of Honey (1961), and The
Loneliness of the Long Distance Runner (1963).   [41] Interview, No. 7, 1970.

Schlesinger directs the shearers' feast scene from *Far From the Madding Crowd*, 1967. At the head of the table is Julie Christie (this was her second film with Schlesinger), at the foot, Alan Bates. Behind the camera (looking straight ahead) is the director of photography, Nicolas Roeg. © Weintraub Screen Entertainment. Photograph: BFI.

personal film.'⁴² The film flopped in America. 'Everybody told me that after the failure of *Far from the Madding Crowd*, it wouldn't be wise to choose as my next project such a risky film as *Midnight Cowboy*, I was surrounded by Job's comforters.'⁴³

Midnight Cowboy (1969) marked the beginning of Schlesinger's relationship with America. 'I had no intention of going to America at all until someone gave me *Midnight Cowboy* to read, which could only be made there.⁴⁴ But since then I've found it easier to get an American film financed than an English one. I've had better opportunities there.'⁴⁵ Schlesinger has become an Atlantic commuter, making most of his films in America. He feels that working there has broadened his horizons and put him in touch with the international audience. 'I resent people who criticize me for working in America. For years, one of my contemporaries, Lindsay Anderson, sat in London castigating us all for liking it in America, and then he finally went and made *The Whales of August*, and then *Glory! Glory!*'⁴⁶ But it proved to be a long haul to get *Midnight Cowboy* off the ground. In fact, it was only after Schlesinger, the producer, Jerome Hellman, and writer, Waldo Salt, agreed to defer a large part of their salaries for a major percentage of the profits that United Artists were prepared to fund the $2.8 million budget. This proved to be a blessing in disguise as the phenomenal and unexpected financial and critical success of the film, which grossed over $45 million, gained them each a fortune and an academy award.⁴⁷ Joe Buck (Jon Voight), naïve to the ways of New York, comes into town imagining himself to be a tough cowboy certain to make his fortune as a stud. Within a few days he is penniless and forced to live on the streets, where he meets Ratso (Dustin Hoffman), a tubercular Bronx conman who dreams of 'making it' in Florida. Ultimately it is their friendship which makes their lives bearable. As Schlesinger said, 'the film is about loneliness and delusion and the emergence of some sort of human dignity from degradation.'⁴⁸ It is a powerful and original film, and its success enabled Schlesinger to pry out of the same backers, United Artists, a slim budget of $2 million for his next project *Sunday, Bloody Sunday*, which he has described as 'probably the best thing I've ever done.'⁴⁹

⁴² Interview with the authors.
⁴³ *Films and Filming*, Nov. 1969.
⁴⁴ It was also his first feature film without Janni as producer. Janni only wanted to take on the project on the condition that Schlesinger would transfer it to a British setting, which he refused to do.
⁴⁵ Interview with the authors.
⁴⁶ Interview with the authors.
⁴⁷ It won the awards for Best Director, Best Screenplay, and Best Film; in fact it was the first 'X' rated film to win the award for Best Film.
⁴⁸ *Films and Filming*, Nov. 1969.
⁴⁹ *Literature/Film Quarterly*, Spring 1978.

Sunday, Bloody Sunday (1971) was the second of Schlesinger's films to be taken from an original script. He gave an outline to the script-writer, Penelope Gilliatt, and together they worked for a few years developing the script through four drafts. Not feeling confident enough to write scripts himself, but still wanting a large degree of involvement, this is the way that Schlesinger likes to work with the writer. The film received great praise particularly from the New York critics.[50] The acclaim in fact helped prevent an early withdrawal of the film by its doubtful backers. It is the compassionate and unsentimental story of Daniel Hirst (Peter Finch), a sophisticated, Jewish, homosexual doctor, and Alex Grenville (Glenda Jackson), an employment counsellor, who are both involved in a love affair with the boyish and selfish psychedelic artist, Bob Elkin (Murray Head), and of how they cope, with calm resignation, when he leaves them. It is, as Pauline Kael wrote, 'a plea on behalf of human frailty—a movie that asks for sympathy for the non heroes of life who make the best deal they can'.[51] Schlesinger describes it as 'a piece of slightly uncommercial chamber music'.[52] It is amongst his most passionate films, with the most sophisticated of all his charac-terizations and a sensitivity which marks it as one of the most outstand-ing films of that decade.[53]

The following year Schlesinger contributed to *Vision of Eight*, a study by eight international directors of events at the Munich Olympics. Schlesinger's section, entitled 'The Longest', was about the marathon and followed the fortunes of Britain's entry, Ronald Hill, from training in the hills of northern England to the actual event. The piece reflected his interest in lonely and obsessed characters, intercutting coverage of the massacre of the Israeli athletes by Palestinian terrorists with Hill's frustration that his race had been postponed for one day and his determination not to be distracted. Schlesinger also showed the despera-tion of the race itself, intercutting the award ceremony with shots of struggling runners yet to finish. It was an interesting and compelling piece, and a welcome chance for experimentation.

It took three years to find a backer for his next project *The Day of the Locust* (1974).[54] Based on Nathaniel West's novel, the film looks at pre-war Hollywood. 'It is really an allegory about values at a particular time . . . Nathaniel West saw Hollywood as a lotus land where people were

[50] Particularly Pauline Kael; Penelope Gilliatt, the script writer of *Sunday, Bloody Sunday*, was her partner on the film column in the *New Yorker* at that time.

[51] Paulene Kael, *Deeper into Movies* (Little, Brown, 1973).

[52] *American Film*, Dec. 1979.

[53] Billy Williams's understated camera style was particularly appropriate to the mood of the film.

[54] As with *Midnight Cowboy*, Schlesinger was working with Salt and Hellman, as writer and producer respectively.

beckoned by the easy life,—instant fame and an abundance of everything. When their illusions and hopes turned out to be something else,
then I think they became violent and frustrated.'[55] *The Day of the
Locust* is a tremendously powerful and polished film. Its glossy visual
style, brilliantly captured through Conrad Hall's photography, and the
romantic soundtrack are effective in heightening the contrast between
the allure of Hollywood and the sordid reality.[56] Unlike his other
films where characters cope, here the pressures on them are too great.
Schlesinger wryly compares Hollywood to a marshmallow: 'Too late
you find there is something soft and sticky at its center, and you can get
your feet stuck if you're not careful.'[57] But the film did poorly at the box
office, and critical opinion was polarized, partly because of its length,
and partly due to its mood of prevailing doom. It is however one of
Schlesinger's favourite films.

The following year Schlesinger turned his hand to something very
different. The thriller, *Marathon Man*, based on William Goldman's
best-selling novel, was offered to him by the head of Paramount, who
had backed *The Day of the Locust*. 'I read the book and thought, "This
is a film which I'm not going to have a lot of trouble getting financed".'[58]
So he did it, although he came in for much criticism for tackling such a
lightweight albeit exciting subject. Schlesinger refers to it now as 'My
Jewish thriller';[59] it is a story of the Jewish hero's hunt for Nazis in New
York, using the strong cast of Dustin Hoffman, Laurence Olivier, and
Roy Scheider. 'The real problem for me in doing a melodrama of this
kind is that you are dealing essentially with plot rather than with characters, which appeal to me so much. Every time you try to develop the
characters in a scene a little further it holds up the plot, and so you just
go ahead and do it primarily with plot in mind.'[60] Nevertheless, it was an
experience that he thoroughly enjoyed. After going to a regular screening of the film in New York he said that 'It was thrilling. This was the
first time I'd made a genre picture of that kind, and to see the audience
reacting exactly as they should have done was terrific.'[61]

Yanks (1978) was Schlesinger's first film made in Britain and produced by Janni since *Sunday, Bloody Sunday*, seven years earlier. With an
original script by Colin Welland[62] he returned to northern England, to
focus with much warmth and affection on the American GIs billeted

[55] *Screen International*, 9 July 1977.
[56] Hall claims that this style was in fact chosen in order to foster a greater audience appeal,
by trying to sweeten the film's bitterness.
[57] *Screen International*, 9 July 1977.
[58] *American Film*, Dec. 1979.
[59] Interview with the authors. Paulene Kael referred to it as a 'Jewish revenge fantasy'.
[60] *Literature/Film Quarterly*, Spring 1977.
[61] *American Film*, Dec. 1979.
[62] Later script-writer of *Chariots of Fire*.

there during World War II and their relationship with the local community. Schlesinger saw the film as reflecting a great personal dilemma. 'One of the reasons I wanted to make the film so badly was that I'd been looking for a subject which expressed my own dichotomy. I am divided—I'm an English director who loves working in England, but who also loves working in the States where I've been given a lot of opportunities.'[63] Despite the theme and setting, no British finance could be raised for the project—the £3 million budget was American and German money. Although entertaining and sensitive, the film was not a great success. Schlesinger is right in regarding it as being overly sentimental.

After making *Yanks*, Schlesinger commented, 'I'd like to do something that was just very funny, and I never have. I love to laugh, and I adore comedy. I always seem to want to give people something to think about rather than something just to say "Wasn't that fun?"'[64] He got his opportunity three years later in 1981, when he was offered an original script, *Honky Tonk Freeway*. 'I read the script twice in an afternoon, rang them up and said "I'll do it"'[65] . . . 'I usually make snap decisions like that. It doesn't take me long to reject or jump at a script.'[66] But he could not have envisioned the disaster that it turned out to be, in fact becoming the greatest commercial failure in film history.[67] Edward Clinton's original screenplay was a 'road' movie about a number of people making their way across the country towards a small resort town in Florida. At the same time, the leading inhabitants of this town have to face being bypassed by a new freeway. The cinematographer, John Bailey, praises Schlesinger for his handling of the tortuous script: 'He's a very good storyteller . . . there's a very intricate weaving of the characters, especially as they all come together . . . Schlesinger knows how to do that and all his films have that same kind of quality.'[68] The film also had a particularly strong visual style. As Bailey noted,

Although Schlesinger is a very actor orientated director, he is also incredibly visually orientated. He has very specific ideas about how he wants a shot. . . . Performance of course was paramount but it had to be in concert with composition, camera movement and lighting. A sense of detail about all those elements and being able to juggle them is what makes a film-maker an artist.[69]

[63] *American Film*, Dec. 1979.
[64] *Literature/Film Quarterly*, Spring 1978.
[65] Interview with the authors.
[66] *Screen International*, 1 Nov. 1980.
[67] Patrick Robertson, *Guiness Book of Movie Facts and Feats* (Guiness Books, 1988). In terms of budget/box office ratio. It cost $23 million and earned $0.5 million in North American rentals, although this was compounded by the fact that it was withdrawn from distribution at a very early stage.
[68] Schaefer and Salvato, *Masters of Light*.
[69] Ibid.

Schlesinger knew that the film's unusual theme and structure would be a risk, but as he said, 'I like the idea of yet another deep end to plunge into.'[70] But there were numerous problems built into the production; a British company, EMI, satirizing America, yet marketing it at an American audience. Added to this was EMI's insistence on funding a large budget film in order to break into the American market with Hollywood style product. Costs escalated to some $23 million, at which level the prospect of commercial success was slim indeed. After the film had been edited in England it was previewed to a sample American audience. The results were not favourable and so it was re-edited. The consequence, according to the producer Don Boyd, was that 'we diluted our original instinct, not because of some objective opinion we could argue with, but because of those damn computerised preview cards.'[71] Certainly they might have done better to keep to their original instincts, because the final film did suffer from a script whose humour lacked consistency. Part of the problem with any director attempting a new genre is that he often does not succeed the first time.[72] Hitchcock may never have mastered comedy, but he did improve after his first disastrous attempt.[73] *Honky Tonk Freeway*'s budget ensured that Schlesinger would not get a chance to do likewise. The death knell of this comedy was ironically its timing. After the recent failure of the $40 million *Heaven's Gate*, critics were gunning for extravagant projects.[74] Bailey was surprised at its reception: 'I think it's one of those films that five years from now somebody will look back on it and say, "Why did that film take such a beating; it's kind of interesting".'[75] Schlesinger displays a characteristic mood of defiance—'I hadn't seen it since checking a print in the Labs before it opened, and, watching it again on British television this Easter, I stand by every frame.'[76]

Schlesinger's next venture, *An Englishman Abroad* (1983), was his first work for the BBC since his documentary days. With a magnificent script by Alan Bennett—'The best written script I've ever had'[77]—it was based on an actual chance meeting in Moscow between the actress Coral Brown and defected spy Guy Burgess. Shot in Dundee, the film had a

[70] *American Film*, Dec. 1979.

[71] Alexander Walker, *National Heroes* (Harrap, 1985).

[72] Although Schlesinger had pulled it off with his first thriller, *Marathon Man*.

[73] The failure of his first straight comedy, *Mr and Mrs Smith* (1941), did not stop him later using black comedy, most notably in *The Trouble with Harry* (1955).

[74] Probably the most harmful was in *Variety*, 19 Aug. 1981—'The overriding question about EMI's *Honky Tonk Freeway* is why anyone should want to spend over $25 million on a film as devoid of any basic humorous appeal . . . Its long-term commercial appeal appears to be almost nil.'

[75] Schaefer and Salvato, *Masters of Light*.

[76] Interview with the authors.

[77] Interview with the authors.

William Devane on a water-skiing elephant in the commercially disastrous *Honky Tonk Freeway*, 1981. Devane also starred in Schlesinger's previous two films: *Yanks*, 1978, and *Marathon Man*, 1976. © Weintraub Screen Entertainment. Photograph: BFI.

great feeling of authenticity. It represented something of a rebirth for Schlesinger, deservedly achieving phenomenal critical praise, winning nine BAFTA awards, and even bringing him back to working with Alan Bates. In fact, the closing sequence in *An Englishman Abroad*, with Bates walking nonchalantly across a bridge in his new suit from London to the accompaniment of Gilbert and Sullivan's 'He remains an Englishman', bears a striking visual resemblance to a scene in *A Kind of Loving*.

In the same year, he consolidated this success with another television film, *Separate Tables*, based on Terence Rattigan's play. Again he worked with a familiar cast, including Alan Bates and Julie Christie. It is a film of two plays, unified by their setting in a genteel seaside hotel in Bournemouth in the mid-1950s, with the same actors playing different characters in each. The first part, entitled *Table by the Window*, tells of a faded fashion model (Julie Christie) fearful at the thought of growing old alone, who arrives at the hotel in the hope of winning back her ex-husband (Alan Bates), now a washed-out alcoholic and former Labour politician, who lives there and is having an affair with the hotel manager (Claire Bloom). The second part, *Table Number 7* is set eighteen months later. Here, Christie plays a demeaning spinster who is staying at the hotel with her overpowering mother. She takes a liking to the pompous Major Pollock (Bates), who is in fact a fraud and minor sex offender. Her mother wants him ostracized and sides are taken amongst the hotel guests. As with *An Englishman Abroad*, this film is marked by great subtlety and sophistication, and although the narrative seems slight, it deals with profound themes such as ageing, unfulfilled promise, and most importantly for Schlesinger, the reality beneath the veneer of civilized society.

The success of these films bolstered Schlesinger's confidence and undoubtedly helped him in the struggle to finance his next project, *The Falcon and the Snowman* (1985), which he had been trying to get off the ground for four years. Like *An Englishman Abroad* it deals with espionage, the true story of two young Americans from prosperous middle class homes, who come to sell secrets to the Russians. There were many interesting themes for Schlesinger. There was the contrast in motivation of the two central characters. For Timothy Hutton's character espionage is a means of protest when he discovers proof of the CIA's involvement in Australian politics, whereas for Sean Penn's it is merely a way of supporting his drug addiction. There is also the larger issue of loyalty to a country which is far from perfect, and the psychological effect disloyalty has on Hutton. Although it was a valuable and entertaining film, it suffered from a script, and from characterizations, which lacked sophistication. Schlesinger recognizes this: 'It was too ambitious. The problem was that we didn't have time within the context of the movie to deal sufficiently with Hutton's youth: to set up the complexity of his motives for doing what he did, and to establish that it was not a revenge whim.'[78] He had a particularly difficult time during the four-month shoot in Mexico City. The combination of the death of his father, his mother being taken seriously ill, and the terrible frustrations due to Sean Penn's preoccupations with 'method' acting, led him near to a nervous breakdown.

The Believers (1987) was Schlesinger's second thriller; unlike *Marathon Man*, it had a dismal critical reception. It is a flat, cliché-ridden story of sinister occultism which threatens the lives of Martin Sheen, a psychologist specializing in the occult, his son, and his girl-friend. The film does have a powerful visual mood, more so than his previous works, and Schlesinger seems to have been primarily attracted by this potential. 'What's interesting about suspense is the way you use the camera to tell the story. You're really thinking in cinematic terms.'[79]

Madame Sousatzka (1988) was only Schlesinger's second return to Britain to make a feature film since *Sunday, Bloody Sunday*, and, as with *Yanks*, he was unable to raise any British finance for it. Based on the book by Bernice Rubens, it is about the triangular relationships between an eccentric, possessive, and demanding piano teacher, her gifted young pupil, and his mother, who is reluctant to lose him. The film was made for £3.5 million, a slim budget by Schlesinger's standards. It marked a refreshing change of pace compared with *The Believers* and *The Falcon and the Snowman*. 'I felt very anxious after two dark films to get back to making something that was calmer, more intimate and optimistic.'[80] It was his most personal project for some years, incorporating his love of music, complex central characters, and a hint of social comment (such as the destructive impact of gentrification on the community, and commercial pressures on musicians). Although the film suffered from obtrusive secondary characters and unnecessary subplots which diluted the tension of the main drama, it clearly reflected the devotion, enthusiasm, and sophistication he had put into it.

Making a variety of films has always been important to Schlesinger. 'I'm not particularly interested in repeating myself. I want to try my hand at a lot of things, because I think it's very important to take risks.'[81] And he has done this, making *Midnight Cowboy* and *Sunday, Bloody Sunday* after the failure of *Far from the Madding Crowd*, *Honky Tonk Freeway* after *Yanks*, and *Madame Sousatzka* after *The Believers*.

[78] Interview with the authors.
[79] *Dialogue on Film*, Nov. 1987.
[80] At a preview of *Madame Sousatzka* at the University of Southern California, Los Angeles, 13 Oct. 1988.
[81] *Screen International*, 1 Nov. 1980.

Martin Sheen (centre) in Schlesinger's second thriller, *The Believers,* 1987. Orion Pictures Corporation © 1987.

While some of these risks have paid off in great films such as *Midnight Cowboy* and *Sunday, Bloody Sunday*, most have not. Schlesinger finds it very difficult to gauge what will be popular. 'I wouldn't ever be too sure about what's commercial.'[82] But he views himself 'essentially as a practical man. I've never regarded myself as an artist out on a limb.'[83] As such he recognizes a need to accept compromise in order to get the films he wants to make off the ground. 'You have to play games to get people to back your fantasy. Occasionally you have to make a film for "them".'[84] So, like other directors before him, he has, on occasions been a 'director for hire' hoping to restore his financial credibility. Such apparently 'safe' films have included his thrillers, *Marathon Man* and *The Believers*. But ironically the failure of *The Believers* shows how commercially unpredictable these types of film can actually be. Under production pressures throughout his career, his driving spirit, previously hurled at the underlying issues of his films, now increasingly appears to be manifested in a compulsion to keep working, and in overcoming specific production problems. He has less passion for his films, although not for the process of film-making itself. Schlesinger acknowledges this with characteristic frankness: 'People change,' he says.

[82] *Films in Review*, Oct. 1981.
[83] Interview with the authors.
[84] Interview with the authors.

JOHN SCHLESINGER

Before you went into feature films you spent several years making documentaries for the BBC's Tonight *and* Monitor *programmes. What kind of things were you doing?*

I did a variety of things. For *Tonight* I started experimenting with setting poems to films; such as 'The Charge of the Light Brigade' set against the rush hour in the City. That was the first thing that I think caught people's eyes as a slick piece of counterpoint. But it worked! Later I was asked to do something for Armistice Day. I felt that this had been previously treated in a rather hackneyed fashion. I don't think it's enough to see the impressive but cold ceremony at the Cenotaph. Our film was set in the Imperial War Museum, and seen through the eyes of a child, to whom war was just a series of exciting models, and we then contrasted the models with the real thing.

What do you think are the most important skills you learned from this experience?

I've always found that using one's powers of observation are very important. In documentary films, because you don't usually have much time, first impressions are invariably the last. You see a director is really an observer, at least I am, and for me observing is not only my profession but my hobby. I love watching people, it's like sitting with a sketchbook. You store away incidents in your head, never quite knowing when you're going to use them. It was like that in *Terminus*, when I used shots of the little boy crying because he was lost. I'd seen such a thing happen, just by chance, as one does all the time.

So you set it up?

Yes, first of all we tested a few little dramatic school kids who were all hard boiled and couldn't care less if their mothers left them. And then I suddenly found a relative of mine, a boy of 6 or 8. His mother brought him along to the railway station. She told me he was terribly possessive and I said, 'Well, what would happen if you left him for a bit?' She said he would probably cry, and after she left, the boy started to get tearful rather quickly. So I thought, 'Oh, splendid.' I suppose one is so intent upon getting the effect one wants that one ceases to be a humanist at that moment.

What did you see as the advantage of working in television?

Well, television is such a hungry mouth to feed that you're on a kind of treadmill, making films continuously. You were onto your next before the previous one was shown. I made about a film a month for *Monitor*. So, unless something disastrous went wrong, you were afforded a valuable opportunity to learn from your mistakes; a luxury not shared in the film industry. There, because of the expense, the old adage 'You're only as good as your last picture' is absolutely true. If you want to get your next project off the ground, it's extremely difficult not to be concerned about how successful a picture is commercially. So you're looking over your shoulder much more in film than television. It is so uncertain. *Sunday, Bloody Sunday* and *The Day of the Locust* were both financially unsuccessful, but with *Midnight Cowboy* and *Marathon Man* I found myself bankable again. All film-makers find themselves in the same boat: you dive in at the deep end every single time—providing you can get near the water.

You made An Englishman Abroad *for the BBC. Would you like to make other films for television?*

Well, I haven't often made films for television because the short shelf life makes it very frustrating. You work just as hard, and with just as much care, to make a film for television. *An Englishman Abroad* was one of my most successful films, but it got three showings, and that was it. I also think that television has changed, forever, the audience's perception of watching. Their patience and attention span is much less than it ever was, because television requires no effort, you can change channels so easily. It's just not the same experience as paying money to go to the cinema and sitting down with other people in a darkened room, and really experiencing something. There, you think twice before getting up to get popcorn—or maybe the present audience doesn't.

At best what reaction do you think you could hope for from an audience?

I think that you want to give the audience a special experience. You want to affect them in some way; whether to alarm them, or make them think about something. I think that's what one hopes films are doing, and therefore you want the audience's maximum attention. I've seen a number of my films at special occasions such as festivals. There you can tell when the film is playing well, when you've got the attention of the audience and they're with it. This happened when we played *Madame Sousatzka* in Toronto to an audience of two thousand, before it opened publicly. It was a tremendous moment. I think that the audience's reactions are terribly important. I'm not someone who says, 'Well, I don't give a damn what anyone else writes or says.' I do, otherwise why are we in this business? But I also want to have some fun out of making films. It's a great source of pleasure to me. And therefore I don't think one's got to get too 'arty-farty' about it. I'm not a particularly political animal. I resent being criticized for doing films which I thought would be fun to do. I'm not committed to saying, 'I've got to make movies communicating *that* particular kind of message.' I have never been a great idealist. I love working too much for that.

Does that imply being prepared to do anything?

No, but I love work. For me it's terribly important, particularly as I get older, to continue to work, as long as it's not shoddy work. I'm not really happy unless I've got a project on the go, and am thinking about the creation of something. And I like to do different things. I like the fact that I can make a small picture here, a big picture there, come back and go to Salzburg to do an opera. It's a very stimulating life. I can't relax, I enjoy the challenges too much; that's my life. I'm not tired, I'm as enthusiastic about working as I ever was.

Do you think that a result of wanting to work constantly is that it is not possible to keep up the same level of passion? Your earlier films seem much more personal than some of your recent ones.

I think it's easy to say that earlier films are filled with more identifiable passion; although I do think that some of them are better than the later ones. You see I don't think it's possible to expend the same level of passion on everything without working much less and waiting for that personal passion to come to you.

After all, those kinds of projects are usually very difficult to get off the ground, and I don't want to get rusty waiting to get the finance through. You see, as one gets older and wants to carry on working, I think one loses patience with the system, and I recognize that there are certain films which I have much more passion for than others. It's very difficult to say, 'Well I'll only do personal films, and I won't touch anything else.' There are people who have done that, and perhaps they are the top filmmakers in the world; but I don't consider myself part of that group. I'm not a particularly political animal, and I don't want to make 'important' films all the time. But because it is so difficult to get films off the ground financially, you've got to have several oranges in the air. So I develop more than one script at a time.

What kind of projects are you working on at the moment?

I'm working with David Leavitt on an AIDS project, which I think will be very difficult to get off the ground. I'm working on *Oscar and Lucinda*, the Peter Carey book, and we have just had a breakthrough on the script, so maybe that has a future, although again it's a very difficult subject to get off the ground. I'm also working on a film about a psychopath, which has suddenly come to me and which I like a lot, called *Pacific Heights*.[1] It's about a couple who invest their savings in a property which they convert into apartments. Into one of them moves a psychopath played by Michael Keaton who wrecks their lives in an unscrupulous and manipulative manner. It's about tenants' rights; the ironies of real estate; the problems of owning anything. There is a line from *Sunday, Bloody Sunday* which I like—'Possessions, possessions, it's getting to be a disease.'

How do you maintain your enthusiasm for such a variety of projects?

I think enthusiasm evolves very often. I don't really think that one gets madly enthusiastic about a project until you know that you're actually going to be able to make it. For example, with *The Day of the Locust*, which initially was going to be a Warner Brothers film; when they were complaining about one of the early drafts of the script, and weren't in the least bit interested in spending enough money to make it properly, I tried to withdraw from the project. The writer, Waldo Salt, who was a dear friend, said, 'You're not playing fair by me. Come back and do the work we used to do together on *Midnight Cowboy*.' I knew that to get a page of script out of him was like pulling teeth, and that I'd have to nurse him along, but once I started to work, and saw things coming together I got hooked. But I certainly don't believe in going around with one script and saying, 'This is the thing that I most want to do in the world, and I'll lay down for it.'

Do you think that you've had to make many compromises working for the Hollywood studios?

No, because I've never really been a 'Studio' man. If you look back at the films that I've made, quite a lot of them have been put into 'turnaround'. For instance, *Yanks*, which I was passionate about, was initially a Paramount picture. I remember persuading the head of the studio to give the property back to us, because it was clear that he was never going to make it. The only studio pictures that I've ever really made were *Marathon Man* and *The Believers*.

But in order to make the films you particularly want to make, you seem to be prepared to make a more commercial project? Isn't that a compromise?

---

[1] Went into production in January 1990.

I think that's true; I'll do one for me and one for them. But I don't regard it as a compromise. After all, everything is a compromise. If you're doing a low budget picture and there isn't really enough time and money to do it the way you really want to, then that's a compromise too. But I think that what it does mean is that studio films are less personal.

Which do you think is your most personal film?

Sunday, Bloody Sunday. That was a film which I made at the end of a relationship where I said, 'I want to make a film about that.' I knew all those people in that film extremely well.

In many of your films you seem to focus on characters out of step with their environments. Is this also a personal theme?

I can empathize with people who've had difficulties in their lives. Many times in my life I've been assailed by self-doubt and lack of confidence. At public school I wasn't good at those things expected of a middle class boy. I always felt hopeless, particularly over physical things. I was a bit ungainly. I couldn't ride a bicycle until I was 11; I didn't get a commission in the army; I felt conscious about being persecuted at school in the thirties for being Jewish. There were many times, and it still happens, when I suffer from a sense of frustration at being made to feel inadequate or unequal to the job, therefore when success comes it's all the more enjoyable. It also means that I'm not very good at dealing with failure.

In many of your films you've been very interested in looking at the games people play: the ways that they fail to face up to reality, or try to fool themselves by living in a fantasy world. It's there in many of your films: Billy Liar, Darling, Midnight Cowboy, The Day of the Locust . . .

Well, I've always been interested in digging below the glitter of something to look at the dirt underneath. Occasionally there's gold, but usually it's dirt. One of the appalling things about this age is the way we kid ourselves, the way so often we simply don't tell the truth. We're so skilful at keeping up appearances, while behind them all sorts of interesting and curious things happen. No film of mine is going to have a totally happy ending. There is a lurking bit of optimism, but I am essentially a rather pessimistic person, although I'm not without hope, and a lot of things amuse me. Sometimes I have a rather cynical way of looking at things, but eventually there is something in me which says that we have got to cope with whatever is our lot, otherwise why go on? And most people do, and that's one of the incredible miracles of mankind. Human beings, through providence (and I don't know if I'm really talking about God or one's beliefs), have the most extraordinary inner resources that come to bear at times of crisis. Most people don't

Schlesinger checks a shot on location of *Billy Liar*, 1963. © Weintraub Screen Entertainment. Photograph: BFI.

give up because they are so resilient to the appalling things that happen to them and are always happening to them.

Was this the dominant idea in Midnight Cowboy?

Yes, what attracted me to *Midnight Cowboy* was that it was a story about a fantasy of what New York represented to this boy from out of town and what in fact the reality was.

You seem to have worked closely with writers such as Salt and Raphael. How deeply involved are you in the script-writing of a film?

I do an enormous amount with the writer. Most scripts go through a period of development and change, and I work a lot with the script-writer laying out the initial film, or if they want a certain amount of freedom, I work with them on the later drafts. I can't actually sit down and write good dialogue, but I always know when it's not right, or when a 'beat' is missing. So much about human relationships deals with what isn't said, with the subtext, that you've got to make sure that it's there.

I have always loved this period of preparation, when the film is still a fantasy in your head. Then you have none of the pressures of shooting, which can be agony as you become something between a wet nurse and a psychiatrist. Trying always to keep your first intentions clear in your head, you must remain flexible. One should be a kind of father to an enormous family that can be allowed to be creatively helpful. The cinema *is* such a collaborative medium, feeding off so many other people's talents. Film directors are terrible vultures.

You pride yourself on a high quality of acting in your films. How is this achieved?

First of all, you've got to cast well. I generally ask the actor to do two scenes and an improvisation. But casting is all about balance—balancing the personalities, so that as well as finding a good actor for the part, you have to make sure they balance with whoever they will be acting with. Very often I'll use a totally unknown actor in a big role because it surprises the audience. If a character is going through a certain experience for the first time, as happens a lot in my films, I like the idea of a fresh image. We knew that with *Midnight Cowboy* we needed a new face for the cowboy and not Warren Beatty, good actor though he is.

But the reason I get good performances from actors is because I care desperately about what they're doing. I find it very irritating when people write about my films and say, 'We all know he's good with actors, but . . .'—who else is good with them? Not that many! As a director you must 'make space' for the actors. You have to make them feel that the pressure is not on *them*, that they've got room to create something freely, room to make contributions. And one of the interesting things about any great actor is that they're just as hungry for direction as anyone else. Olivier was one of my idols and I grabbed at the chance to work with him in *Marathon Man*. I remember directing him . . . when I thought he was over the top I used to say, 'Larry, I wonder if we could make this scene a little more "intimate"', and he'd say, 'You mean cut off the ham-fat, dear boy.'

Do you consider that you rehearse much?

I try to allow two weeks for rehearsals, although after make-up and costume tests it's normally about ten days. I use this time to work quietly with the actors, to deal exclusively with them and their problems; to work with the subtext of the scenes. This is the period when you make discoveries . . . to improvise if you need to, to deal with events not in the script, but which are critical in understanding the motivation of a character in a particular scene which is in the movie. This is something I've done since *Midnight Cowboy*, and it's critical because the art of directing the actors is to let them *be* their character, not *act* it, as they do in the theatre. I don't like too much pre-rehearsal on set, there is something very intimidating about the atmosphere created by all the technicians standing around watching. I prefer to rehearse in dressing rooms, or in odd corners away from the general activity. But when you get on the set you deal with totally different, technical problems. Most people find these boring, and I have even less tolerance for them. But I am concerned with the visual side of things, especially finding visual images to tell the story and develop the characters. You are thinking all along how to tell the story in a cinematic way. I love the effect of changes of light on the actors—lights going on and off; an actor walking into a dark cupboard and out again, and so on.

In the past you've been very pessimistic about the viability of a film industry in Britain. Do you still hold this view?

The basic problem is not a lack of talent, because there is an enormous pool of it here, but in the lack of opportunities to make films. Even the people who have been making the small, parochial kind of movies are now saying, 'We've got to make bigger pictures, with bigger horizons in order to survive in an international market.' Which is right, because the industry just can't survive on low budget pictures. And it is true that there have been a few films recently that have successfully crossed the Atlantic, but you can hardly say that *Dangerous Liaisons* is a British film.

I think recently 'British film' has been given more of a high profile than the reality it deserves. People are so desperate to claim that there is some life in the poor old coffin; the lid is being prized open a little from the inside. I don't think that basically the British as a nation are that interested in 'The Cinema', especially when compared with the US or France. Of course there are pockets of resistance—the National Film Theatre—but it's interesting that the regional film theatres didn't work. Why should Manchester support an orchestra and various theatre companies, but not the National Film Theatre? We get a film industry we really deserve and want.

In retrospect how do you feel about your films and the film-making process?

When I look at my films, which is seldom now, I often say 'I wish I could re-do that,' but then they are of a certain time and there is no point in going back. The film which is nearest to being absolutely right is *An Englishman Abroad*; the best script I ever had, and one of the greatest pleasures to make.

I feel that if one goes through the agony of making a film, you want it to say something in the end. I hope my films have provided people with something to engage their attention and to remember. If they come away feeling they've learned a little more about other people, that's fine. If they remember enough to say, 'Oh, I liked that,' or 'That moved me,' then I think I've fulfilled what I hoped I might.

As to the film-making process, I *am* tired of having to struggle to get people to believe in my fantasies, and often discouraged, but I still have enormous enthusiasm—thank God.

FILMOGRAPHY

BRIEF BIOGRAPHICAL DETAILS

Born 1926, educated at Uppingham School. National Service in the Royal Engineers and Combined Services Entertainment. Studied English Literature at Balliol College, Oxford (1947–50). BBC director 1957–61 for the *Tonight* and *Monitor* programmes.

FEATURE FILMS DIRECTED BY JOHN SCHLESINGER

1946   *Horror*. Shot on 16 mm.

1948   *Black Legend*. Shot on 16 mm. 60 mins. A Mount Pleasant Production. DIRECTORS, PRODUCERS, WRITERS: Alan Cooke and John Schlesinger. PHOTOGRAPHY: John Schlesinger. CAST: Robert Hardy, Charles Lepper, Hilary Schlesinger, Roger Schlesinger, Wendy Schlesinger, Susan Schlesinger, John Schlesinger, Alan Cooke.

1950   *The Starfish*. Shot on 16 mm. 75 mins. A Mount Pleasant Production. DIRECTORS, PRODUCERS, WRITERS: Alan Cooke and John Schlesinger. PHOTOGRAPHY: John Schlesinger. EDITORS: Malcolm Cooke and Richard Marden. CAST: Kenneth Griffith, Nigel Finzi, Susan Schlesinger, Ursula Vaughan-Williams.

1956   *Sunday in the Park*. Shot on 16 mm. 15 mins. DIRECTORS, PRODUCERS, WRITERS: Basil Appleby and John Schlesinger. PHOTOGRAPHY: John Schlesinger.

(Note: For all Schlesinger's work for *Tonight* the producer was Donald Baverstock, and the assistant producer, Alistair Milne.)

1957   *Petticoat Lane*. Tonight (BBC). EDITOR: Jack Gold.

Song of the Valley. Tonight (BBC). (Based on a song recorded by Dorothy Squires.)

Rush Hour. Tonight (BBC).

Armistice Day. Tonight (BBC).

1958   *Day Trip to Boulogne*. Tonight (BBC).

(Note: All Schlesinger's work for *Monitor* was produced by Huw Wheldon.)

1958   *The Circus*. Monitor (BBC). PHOTOGRAPHY: John Turner. EDITOR: Allan Tyrer.

Cannes Film Festival. Monitor (BBC). PHOTOGRAPHY: Ken Higgins. COMMENTARY: Robert Robinson. EDITOR: Allan Tyrer.

414 · JOHN SCHLESINGER

1959   *Benjamin Britten at Aldeburgh.* Monitor (BBC). PHOTOGRAPHY: Ken Higgins. EDITOR: Allan Tyrer.

Brighton Pier. Monitor (BBC).

Brussells Exhibition. Monitor (BBC). EDITOR: Allan Tyrer. SCRIPT: Robert Robinson.

Foreign Artists in Paris. Monitor (BBC). SCRIPT: Robert Robinson.

Hi-Fi Fo Fum. Monitor (BBC). SCRIPT AND COMMENTARY: Robert Robinson.

Simenon in Switzerland. Monitor (BBC).

The Innocent Eye: A Study of the Child's Imagination. Monitor (BBC). SCRIPT: Michell Raper. PRODUCER: Peter Newington. PHOTOGRAPHY: John McGlashan. EDITOR: Allan Tyrer. COMMENTARY: Michell Raper.

1960   *The Italian Opera.* Monitor (BBC). PHOTOGRAPHY: Charles de Jaeger. EDITOR: Allan Tyrer.

Danger Man (as Second Unit Director). ITV. PRODUCER: Ralph Smart.

The Four Just Men. ITV. (Second Unit Director).

1961   *The Class.* Monitor (BBC). PRODUCER: Huw Wheldon. PHOTOGRAPHY: Tony Leggo. EDITOR: Allan Tyrer.

A Study of Three Painters. Monitor (BBC).

The Making of The Guns of Navarone. Columbia.

The Valiant Years. American Broadcasting Company. (Interviewed Montgomery, Mountbatten, Lord Slim, and Alan Brooke for the series.)

Terminus. 30 mins. British Transport Films/A British Lion Presentation. WRITER AND DIRECTOR: John Schlesinger. PRODUCER: Edgar Anstey. PHOTOGRAPHY: Ken Higgins. EDITOR: Hugh Raggett. MUSIC: Ron Grainer.

1962   *A Kind of Loving.* 112 mins. Vic Films/Waterhall Productions/Anglo Amalgamated. SCRIPT: Willis Hall, Keith Waterhouse (from Stan Barstow's novel of the same name). PRODUCER: Joseph Janni. PHOTOGRAPHY: Denys Coop. PRODUCTION DESIGNER: Ray Simm. EDITOR: Roger Cherrill. MUSIC: Ron Grainer. CAST: Alan Bates, June Ritchie, Thora Hird, Bert Palmer, Gwen Nelson.

1963   *Billy Liar.* 98 mins. Vic Films/Waterhall Productions. SCRIPT: Keith Waterhouse, Willis Hall (from Keith Waterhouse's novel and original stage play). PRODUCER: Joseph Janni. PHOTOGRAPHY: Denys Coop. PRODUCTION DESIGNER: Ray Simm. EDITOR: Roger Cherill. MUSIC: Richard Rodney Bennett. CAST: Tom Courtenay, Julie Christie, Wilfred Pickles, Mona Washbourne, Leonard Rossiter, Ethel Griffies.

1965   *Darling.* 127 mins. Vic Films/Appia Films/Anglo Amalgamated.

SCRIPT: Frederic Raphael (from a story by Frederic Raphael, John Schlesinger, Joseph Janni). PRODUCER: Joseph Janni. PHOTOGRAPHY: Ken Higgins. PRODUCTION DESIGNER: Ray Simm. EDITOR: James Clark. MUSIC: John Dankworth. CAST: Julie Christie, Dirk Bogarde, Laurence Harvey, Roland Curram, Alex Scott.

1967 *Far from the Madding Crowd*. 175 mins. Panavision 70 mm. Vic Films/Appia Films/EMI. SCRIPT: Frederic Raphael (from Thomas Hardy's novel of the same name). PRODUCER: Joseph Janni. PHOTOGRAPHY: Nicolas Roeg. PRODUCTION DESIGNER: Richard Macdonald. EDITOR: Malcolm Cooke. MUSIC: Richard Rodney Bennett. CAST: Julie Christie, Terence Stamp, Peter Finch, Alan Bates.

1969 *Midnight Cowboy*. 113 mins. A Jerome Hellman/John Schlesinger Production. SCRIPT: Waldo Salt (based on James Leo Herlihy's novel of the same name). PRODUCER: Jerome Hellman. PHOTOGRAPHY: Adam Holender. PRODUCTION DESIGNER: John Robert Lloyd. EDITOR: Hugh A. Robertson, Jr. MUSIC: John Barry. CAST: Dustin Hoffman, Jon Voight, Brenda Vaccaro, Sylvia Miles, John McGiver.

1971 *Sunday, Bloody Sunday*. 110 mins. A Vectia Film/Vic Film/United Artists. SCRIPT: Penelope Gilliatt. PRODUCER: Joseph Janni. PHOTOGRAPHY: Billy Williams. PRODUCTION DESIGNER: Luciana Arrighi. EDITOR: Richard Marden. MUSIC: Ron Geesin. CAST: Glenda Jackson, Peter Finch, Murray Head, Peggy Ashcroft.

1972 *Vision of Eight*. 110 mins. ('The Longest'—Marathon section). MGM/EMI. SCRIPT: John Schlesinger. PRODUCER: Stan Margulies. PHOTOGRAPHY: Arthur Wooster, Drummond Challis. EDITOR: Jim Clark.

1974 *The Day of the Locust*. 144 mins. Paramount Pictures in association with Ronald Sheldo. SCRIPT: Waldo Salt (from Nathanael West's novel of the same name). PRODUCER: Jerome Hellman. PHOTOGRAPHY: Conrad Hall. PRODUCTION DESIGNER: Richard Mac-Donald. EDITOR: Jim Clark. MUSIC: John Barry. CAST: Donald Sutherland, Karen Black, Burgess Meredith, Geraldine Page.

1976 *Marathon Man*. 126 mins. Paramount. SCRIPT: William Goldman (based on his novel). PRODUCERS: Robert Evans, Sidney Beckerman. PHOTOGRAPHY: Conrad Hall. PRODUCTION DESIGNER: Richard MacDonald. EDITOR: Jim Clark. MUSIC: Michael Small. CAST: Dustin Hoffman, Laurence Olivier, Roy Scheider, William Devane, Marthe Keller.

1978 *Yanks*. 141 mins. United Artists/Joe Janni—Lester Persky/CIP. SCRIPT: Colin Welland, Walter Bernstein. PRODUCER: Joseph Janni. PHOTOGRAPHY: Dick Bush. PRODUCTION DESIGNER: Brian Morris. EDITOR: Jim Clark. MUSIC: Richard Rodney

Bennett. CAST: Richard Gere, Lisa Eichhorn, Vanessa Redgrave, William Devane, Rachel Roberts.

1981   *Honky Tonk Freeway.* 107 mins. Kendon Films/EMI/HTF Co. (Boyd and Koch Jr.) SCRIPT: Edward Clinton. PRODUCERS: Don Boyd and Howard W. Koch Jr. PHOTOGRAPHY: John Bailey. PRODUCTION DESIGNER: Edwin O'Donavan. EDITOR: Jim Clark. MUSIC: George Martin, Elmer Bernstein. CAST: Beau Bridges, William Devane, Beverly D'Angelo, Teri Garr, Hume Cronyn, Jessica Tandy, Howard Hesseman, Geraldine Page.

1983   *An Englishman Abroad.* 65 mins. BBC TV. SCRIPT: Alan Bennett. PRODUCER: Innes Lloyd. PHOTOGRAPHY: Nat Crosby. PRODUCTION DESIGNER: Stuart Walker. MUSIC: George Fenton. CAST: Alan Bates, Coral Browne, Charles Gray.

1983   *Separate Tables.* 125 mins. HTV/With Ely and Edie Landau for HBO. DIRECTORS: Schlesinger and Ken Price. SCRIPT: Terence Rattigan. PRODUCERS: Edie and Ely Landau. PRODUCTION DESIGNER: Julia Trevelyan-Oman. CAST: Alan Bates, Julie Christie, Claire Bloom, Irene Worth.

1985   *The Falcon and the Snowman.* 123 mins. Hemdale Film Production/Orion Pictures. SCRIPT: Steven Zaillan (from Robert Lindsey's novel of the same name). PRODUCERS: Gabriel Katzka and John Schlesinger. PHOTOGRAPHY: Allen Daviau. PRODUCTION DESIGNER: James Bissell. EDITOR: Richard Marden. MUSIC: Pat Metheny, Lyle Mays. CAST: Timothy Hutton, Sean Penn, David Suchet.

1987   *The Believers.* 113 mins. Orion Pictures. SCRIPT: Mark Frost (from Nicholas Conde's book *The Religious*). PRODUCERS: John Schlesinger, Michael Childers, Beverly Camhe. PHOTOGRAPHY: Robby Muller. PRODUCTION DESIGNER: Simon Holland. EDITOR: Peter Honess. MUSIC: J. Peter Robinson. CAST: Martin Sheen, Helen Shaver, Robert Loggia.

1988   *Madame Sousatzka.* 122 mins. Cineplex Odeon/Curzon. SCRIPT: Ruth Prawer Jhabvala (from Bernice Rubens's novel of the same name). PRODUCER: Robin Dalton. PHOTOGRAPHY: Nat Crosby. PRODUCTION DESIGNER: Luciana Arrighi. EDITOR: Peter Honess. CAST: Shirley MacLaine, Peggy Ashcroft, Navin Chowdhry.

Schlesinger has also directed numerous commercials including ones for Polo Mints, Black Magic, Fray Bentos, Enos, Kraft, Kellogg's, Danish Bacon, Stork Margarine, L&M Cigarettes, Nescafé, After Eight, Manikin.

PLAYS DIRECTED

1964   *No, Why!,* by John Whiting. Royal Shakespeare Company at the Aldwych.

1965   *Timon of Athens*, by William Shakespeare. Royal Shakespeare Company at Stratford.

1966   *Days in the Trees*, by Marguerite Duras. Royal Shakespeare Company at the Aldwych.

1975   *Heartbreak House*, by George Bernard Shaw. National Theatre at the Old Vic.

1977   *Julius Caesar*, by William Shakespeare. Royal Shakespeare Company.

    True West, by Sam Shepard. National Theatre.

MUSICAL DIRECTED

1972   *I and Albert* (from a book by Jay Allen). Piccadilly Theatre, London. MUSIC: Charles Strausse. LYRICS: Lee Adams. PRODUCTION DESIGNER: Luciana Arrighi.

OPERAS DIRECTED

1980   *Tales of Hoffmann*, by Offenbach. Royal Opera House, Covent Garden.

1984   *Der Rosenkavalier*, by Richard Strauss. Royal Opera House, Covent Garden.

1989   *A Masked Ball*, by Verdi. Salzburg Festival.

FILM PERFORMANCES

1952   *Single-Handed*. Directors: John and Roy Boulting.

1955   *Oh, Rosalinda!* Director: Michael Powell.

1956   *Battle of the River Plate*. Director: Michael Powell.

    Brothers in Law. Directors: John and Roy Boulting.

1957   *Seven Thunders*. Director: Hugo Fregonese.

TELEVISION PERFORMANCES INCLUDE

1956   *Robin Hood*. Director: Lindsay Anderson.

    The Buccaneers.

1957   *Ivanhoe* (with Roger Moore).

Has also appeared in *Coronation Street* and in a performance of *As You Like It* (dir. Rudolf Cartier), date unknown.

STAGE PERFORMANCES

1950   *The Alchemist*, by Ben Johnson. Oxford Players.

    Julius Caesar, by William Shakespeare. Colchester Repertory Company.

King Lear, by William Shakespeare. Oxford Players.

The Lady's Not for Burning, by Christopher Fry. Colchester Repertory Company.

1951   *The Devil's Disciple*, by George Bernard Shaw. British Commonwealth Theatre Company.

Twelfth Night, by William Shakespeare. British Commonwealth Theatre Company.

1952   *Twelfth Night*, by William Shakespeare. Nottingham Repertory Company.

On tour with the Wilson Barratt Company: *Beyond the Cage*, by Neil Gunn; *By Candlelight*, by Seigfried Geyer; *The Constant Wife*, by Somerset Maugham; *The Housemaster*, by Ian Hay; *The Importance of Being Earnest*, by Oscar Wilde; *Message from Margaret*, by James Parish; *The Sacred Flame*, by Somerset Maugham; *Who Goes There*, by John Dyson.

1955   *Mourning Becomes Electra*, by Eugene O'Neill. DIRECTOR: Peter Hall.

BIBLIOGRAPHY

The Times, 7 Feb. 1949. On *Black Legend* being shown at the House of Commons.

Sight and Sound, Feb. 1950, 41–2. Mount Pleasant Productions, report on the production of *Starfish*.

Daily Express, 31 Mar. 1962. 'With the Censor's Approval'. Michael Wale on controversy surrounding *A Kind of Loving*.

Films and Filming, June 1962. Profile of Schlesinger after completing his first feature film.

Behind the Scenes: Theatre and Film Interviews from the Transatlantic Review, 1962. Schlesinger interviewed by Robert Rubens.

Films and Filming, May 1963. Interview with Schlesinger about *Billy Liar* and the current state of film-making in Britain: the lack of original material for films, and the lack of inventiveness compared with European directors. He argues for the usefulness of cheap short films to take more risks with in terms of subjects and film-making techniques.

Films and Filming, July 1963. 'How to get into films by the people who got in themselves'.

Film Heritage, Autumn 1965. John Schlesinger at the 6th Montreal Film Festival, by Gretchen Weinberg.

Daily Mail, 14 Sept. 1965. 'What we set out to do', by Freddy Raphael, script-writer of *Darling*.

Bob Thomas (ed.), *Directors in Action* (Indianapolis, 1968). Interview with Schlesinger.

After Dark (Boston), No. 5, 1969. Schlesinger talks to Neal Weaver about making films in the US, particularly *Midnight Cowboy*, the problems of casting, crew, and exhibition.

Films and Filming, Nov. 1969. 'A Buck for Joe'. Excellent article and interview reviewing Schlesinger's work and themes in detail.

Film Comment, Winter 1969. 'John Schlesinger, Social Realist', by Gene Phillips.

London Evening Standard, 26 Mar. 1970. On *Sunday, Bloody Sunday*. Ian Bannen falls ill after six weeks of shooting and is replaced by Peter Finch.

Today's Cinema, 14 Apr. 1970. On Schlesinger's working methods on *Sunday, Bloody Sunday*.

Guardian, 21 Apr. 1970. 'Schhh . . . you know who', by Mark Shivas.

Screen (Britain), Summer 1970. Very interesting interview by David Spiers covering Schlesinger's career to date.

Interview (USA), No. 7, 1970. Interview on *Midnight Cowboy*.

Action (USA), July–Aug. 1970. 'John Schlesinger, Award Winner'. Interesting article by William Hall.

Film World, June–July 1971. 'John Schlesinger Takes a Look at Hindi Cinema'. Interview.

Nova, Sept. 1971. John Coleman interviews Schlesinger about *Sunday, Bloody Sunday*.

America, 16 Oct. 1971. 'The Personal Vision of John Schlesinger', by Gene Phillips.

Daily Express, 25 Oct. 1971. Only copy of *Black Legend* lost in the post.

Cinema 71, Nov. 1971. Interview on *Sunday, Bloody Sunday*.

Today's Cinema, 11 Dec. 1971. Report on a John Player Lecture at the National Film Theatre given by Schlesinger.

London Evening Standard, 23 June 1972. Article by Sydney Edwards in which Schlesinger talks about his only West End musical, *I and Albert*.

Screen, Summer 1972. Career notes.

Daily Mail, 29 Sept. 1972. Schlesinger's response to the murder of Israeli athletes while filming *Vision of Eight*.

American Cinematographer, Nov. 1972. Interview about his filming of the marathon at the 20th Olympic Games in Munich for *Vision of Eight*.

Film Library Quarterly (New York), Winter 1972. 'Belles, Sirens, Sisters', by Lillian Gerard.

International Film Guide (1973). Profile by Peter Cowie.

Gene Phillips, *The Movie Makers: Artists in an Industry* (Nelson-Hall, 1973).

Hollywood Reporter, 31 Aug. 1973. Comments by Schlesinger about the present financial situation of the British film industry, and his future projects.

George Perry, *The Great British Picture Show: From the '90's to the '70's* (Hart-Davis, MacGibbon, 1974).

Alexander Walker, *Hollywood, England: The British Film Industry in the '60's* (Harrap, 1974).

London Evening Standard, 3 Jan. 1974. Article by Alexander Walker about Schlesinger working in the US and *The Day of the Locust*.

Photoplay Film Monthly, Apr. 1974. Interview about the making of *The Day of the Locust*.

New York Times Magazine, 2 June 1974. Interview with Schlesinger about *The Day of the Locust*.

Interview, July 1974. Schlesinger discusses the difficulties encountered in casting *The Day of the Locust*.

Action (Los Angeles), Sept.–Oct. 1974. 'What the directors are saying'.

The Times, 25 Feb. 1975. John Higgins on Schlesinger's first production for the National Theatre, *Heartbreak House*.

Film Comment, May–June 1975. Interview about his career, his attitude towards actors, and the film business.

American Cinematographer, June 1975. Interview on *The Day of the Locust*.

Penthouse (New York), June 1975. 'John Schlesinger', by Ken Kelley.

Film Makers Newsletter, July 1975. Interview on *The Day of the Locust*.

Films Illustrated, Aug. 1975. Interview on *The Day of the Locust*.

Eric Sherman, *Directing the Films: Film Directors on their Art* (Little, Brown, & Co., 1976). Chapter on Schlesinger.

London Evening Standard, 7 May 1976. Sydney Edwards on Schlesinger's life, house, and loves in Hollywood, and his current project.

The Thousand Eyes Magazine, Oct. 1976. A detailed study of the themes and stylistics of Schlesinger's films.

London Evening Standard, 3 Nov. 1976. Article by Sydney Edwards and Michael Owen on *Marathon Man* and other current projects, as well as Schlesinger's comments on the differences in attitudes he sees between the US and Britain.

The Times, 19 Mar. 1977. On Schlesinger directing *Julius Caesar* for the National Theatre, and a comprehensive account of his pre-feature film career.

Literature/Film Quarterly, Spring 1977. 'Exile in Hollywood: John Schlesinger', by Gene Phillips. Article and interview with Schlesinger, about *Marathon Man*, *Midnight Cowboy*, and *Sunday, Bloody Sunday*, and working in America.

NFT Booklet, June 1977. 'British Cinema Season: Part 1. John Schlesinger'.

Screen International, 9 July 1977. Career article, with comments by Schlesinger on his films.

Nancy Brooker, *John Schlesinger: A Guide to Reference and Resources* (G.K. Hall & Co., 1978).

Literature/Film Quarterly, Spring 1978.

Screen International, 8 July 1978. Schlesinger discusses his difficulties in getting *Yanks* off the ground, and reflects on his earlier films.

Films Illustrated, Aug. 1978. Location interview on *Yanks*.

Times Saturday Review, 16 Sept. 1978. 'John Schlesinger: Divided Loyalties', by Glenys Roberts, with comments about *Yanks*.

Focus on Film, Nov. 1978. 'On *Yanks* and other films'. Interview by Gene Phillips focusing on Schlesinger's early career in television, on working in American studios, on violence in *Marathon Man*, his reason for not directing *Coming Home*, and his current film, *Yanks*.

Films and Filming, Sept. 1979. Interview in which he discusses the treatment of the subject matter of his films (which over the years have been progressive for their times), his views on the state of the British film industry, and the problem of making British films acceptable to the American market.

After Dark (Boston), Nov. 1979. 'John Schlesinger: "I can't keep from diving into the deep end"', by Patrick Pacheco.

What's on in London, 9 Nov. 1979, by James Cameron-Wilson. Schlesinger talks about *Yanks*, his sentimentality, and his desire to make more optimistic films.

American Film, Dec. 1979. Interview in which Schlesinger discusses his work with actors, directing children and love scenes, his favourite producers, the making of *Yanks*, and his next project.

Guardian, 13 June 1980. On Schlesinger's debut as opera director, with *Tales of Hoffmann* at Covent Garden.

Screen International, 14–21 June 1980. Note that Schlesinger has been awarded the David di Donatello film prize, as best director for *Yanks*.

Screen International, 1 Nov. 1980.

The Times, 18 Dec. 1980. John Higgins interviews Schlesinger about his first foray into opera with a production of *Tales of Hoffmann*.

Gene Phillips, *John Schlesinger* (Twayne Publishers, 1981).

Film Makers Monthly, May 1981. Interview on the making of *Honky Tonk Freeway*.

Films in Review, Oct. 1981. Interview on *Honky Tonk Freeway*.

The Times, 14 Nov. 1983. 'Shared Fascination with English Irony'. Interview with Schlesinger by Brian Appleyard about *An Englishman Abroad*.

Scotsman, 14 Jan. 1984. Alison Kennedy on location of *An Englishman Abroad*. Interviews with the author, Alan Bennett, director, John Schlesinger, and set designer, Stewart Walker.

Schaefer and Salvato, *Masters of Light* (University of California Press, 1984).

Sunday Times Magazine, 17 Mar. 1985. 'Brotherly Love and Affection'. John Schlesinger and his younger brother talk to Danny Danziger.

City Limits, 12–18 Apr. 1985. Interview on *The Falcon and the Snowman*, and Schlesinger's views on the British film industry and British Film Year.

Stills, Apr. 1985. Interview on *The Falcon and the Snowman*, and comments on the differences between working in the US and Britain.

Television Today, 9 May 1985. Schlesinger to give the 1985 Edinburgh Television Festival MacTaggart Memorial Lecture.

Cinema Papers, July 1985. Interview on *The Falcon and the Snowman*.

Broadcast, 23 Aug. 1985. Report on Schlesinger's MacTaggart lecture.

Screen International, 24–31 Aug. 1985. Career profile, on the occasion of his serving on the 1985 Venice Festival jury.

Hollywood Reporter, 15 Nov. 1985. Note that *Honky Tonk Freeway* was then considered to be the biggest box office flop of all time, in terms of budget/box office ratio.

Dialogue on Film, Nov. 1987.

American Film, Nov. 1987. Career interview.

Hollywood Reporter, 14 Oct. 1988. Schlesinger talks about *Madame Sousatzka*.

LIST OF DISTRIBUTORS

16mm prints of most of the films covered in this book are available for hire from the following distributors:

Film and Video Library
British Film Institute
21 Stephen Street
London W1P 1PL

Tel. (071) 255 1444
Fax. (071) 436 7950

Curzon Film Distributors
38 Curzon Street
London W1Y 8EY

Tel. (071) 629 8961
Fax. (071) 499 2018

Concord Video and Film Council
201 Felixstowe Road
Ipswich
Suffolk IP3 9BJ

Tel. (0473) 715754

Filmbank Distributors Ltd
Grayton House
498–504 Fulham Road
London SW6 5NH

Tel. (071) 386 5411
Fax. (071) 381 2405

Glenbuck Films
Glenbuck Road
Surbiton
Surrey KT6 6BT

Tel. (081) 399 5266
Fax. (081) 399 6651

INDEX

All proper names of people, films, and organizations have been indexed save those simply listed in the bibliographies and filmographies. For general subject headings see under the names of the directors for the respective chapters, or under 'British Film' for the introduction to the book.